RNA Viruses

Practical Approach Series

Related **Practical Approach** Series Titles

Animal Cell Culture 3e*
Differential Display
Protein Localization by Fluorescence Microscopy
Protein Phosphorylation 2e
PCR3: PCR In Situ Hybridization
RNA—Protein Interactions
Virus culture
DNA viruses
Gene Probes 1
Gene Probes 2
Gene Targeting 2e
Immunodiagnostics
DNA Microarray Technology
Post-Translational Modification
Protein Expression
Transcription Factors 2e
Gel Electrophoresis of Proteins 3e
In Situ Hybridization 2e
HIV Volume I
HIV Volume II
Non-isotopic Methods in Molecular Biology
DNA Cloning 1: Core Techniques
Basic Cell Culture
Protein Blotting
RNA Processing I
RNA Processing II

* indicates a forthcoming title

Please see the **Practical Approach** series website at
http://www.oup.co.uk/pas
for full contents lists of all Practical Approach titles.

RNA Viruses

A Practical Approach

Edited by
Alan J. Cann
Department of Microbiology and Immunology
University of Leicester

OXFORD
UNIVERSITY PRESS

OXFORD
UNIVERSITY PRESS

Great Clarendon Street, Oxford OX2 6DP

Oxford University Press is a department of the University of Oxford.
It furthers the University's objective of excellence in research, scholarship, and education by publishing worldwide in

Oxford New York

Athens Auckland Bangkok Bogotá Buenos Aires Calcutta Cape Town Chennai Dar es Salaam Delhi Florence Hong Kong Istanbul Karachi Kuala Lumpur Madrid Melbourne Mexico City Mumbai Nairobi Paris São Paulo Singapore Taipei Tokyo Toronto Warsaw

with associated companies in Berlin Ibadan

Published in the United States by Oxford University Press Inc., New York

First published 2000

British Library Cataloguing in Publication Data
Data available

Library of Congress Cataloguing in Publication Data

RNA viruses : a practical approach / edited by Alan J. Cann.
(The practical approach series ; 226)
ISBN 0 19 963717 2 (Hbk) ISBN 0 19 963716 4 (Pbk)
1. RNA viruses–Research–Methodology. I. Cann, Alan. II. Series.
QR395.R583 2000 579.2′5–dc21 99-057577

1 3 5 7 9 10 8 6 4 2

Typeset in Swift by Footnote Graphics, Warminster, Wilts
Printed in Great Britain on acid-free paper
by The Bath Press, Avon

Preface

The properties of viruses are distinct from those of living organisms, which makes the study of virology different from other areas of biology. Many specialized techniques have been developed to study viruses, and yet these have almost invariably found their way into mainstream biology. It is impossible to define the techniques of virology in a single volume, but this book and its companions set out to illustrate the major experimental methods currently employed by virologists. In particular, it is hoped that by grouping methods used by those who study RNA viruses and those who study DNA viruses—the same groupings frequently observed at scientific meetings—the maximum utility has been gained.

Inevitably there will be those who feel that this or that should have been included or left out. It is impossible to include everything within the format of this series! Nevertheless, this volume is wide-ranging in scope, from emerging technology such as reverse genetics and retrovirus vectors, to money-saving tips—how to make your own silica particles for high-efficiency RNA extraction (Chapter 6) and liposomes for cell transfection (Chapter 9)!

Chapter 1 covers the fundamentals of investigating RNA virus genome structure at a molecular level. Chapters 2 and 3 describe techniques for the mutagenesis of RNA genomes and analysis of transcription. Chapter 4 deals with RNA virus-encoded proteinases, an important aspect of the control of RNA virus gene expression. Chapter 5 considers retrovirus oncogenesis and Chapter 6 the nalysis of RNA virus quasispecies. Chapter 7 describes systems for investigation of *in vitro* replication of positive-stranded viruses, and Chapter 8 the packaging of RNA virus genomes. In addition to the technical aspects of reverse genetics and retrovirus vectors, both of the final chapters also consider ethical aspects of these new technologies.

My thanks go to David Hames (series editor) for his guidance in shaping this and the accompanying volumes, and to the staff of Oxford University Press. Most importantly, thanks must go to the contributors who were prepared to share their combined expertise with the wider research community.

Leicester University A. J. Cann
January 2000

Contents

Protocol list

Abbreviations

5-FU	5-fluorouracil
ALV	avian leukosis virus
AMV	avian myoblastoma virus
APS	ammonium persulfate
Ara C	cytosine-β-D-arabinofuranoside
ATCC	American Tissue Culture Collection
ATP	adenosine triphosphate
BAC	bacterial artificial chromosome
BHK	baby hamster kidney (cells)
BIV	bovine immunodeficiency virus
BLV	bovine leukaemia virus
BMV	Brome mosaic virus
bp	base pair
BSA	bovine serum albumin
BUN	Bunyamwera virus
CAEV	caprine arthritis-encephalitis virus
CAT	chloramphenicol acetyltransferase
CD	cluster of differentiation
cDNA	complementary DNA
CEF	chick-embryo fibroblasts
c.p.e.	cytopathic effect
c.p.m.	counts per minute
CSPD	disodium 3-(4-methoxyspiro{1,2-dioxetane-3,2′-(5′-chloro)tricyclo[3.3.1.1^{3,7}]decan}-4-yl)phenyl phosphate
CTE	constitutive transport element
DDAB	dimethyldioctadecyl ammonium bromide
ddNTP	dideoxynucleotide
DEAE	diethylaminoethyl
DEPC	diethylpyrocarbonate
DI	defective interfering
DIG	digoxigenin
DMEM	Dulbecco's modified Eagle's medium
DMSO	dimethyl sulfoxide

dNTP	nucleotide triphosphate
DOPE	dioleoyl-L-α-phosphatidyl ethanolamine
DTT	dithiothreitol
ECACC	European Collection of Cell Cultures
EDTA	ethylenediaminetetraacetic acid
EIAV	equine infectious anaemia virus
EMSA	electrophoretic mobility shift assay
EUCIB	The European Interspecific Backcross
FACS	fluorescence-activated cell sorting
FCS	fetal calf serum
FeLV	feline leukaemia virus
FITC	fluorescein isothiocyanate
FIV	feline immunodeficiency virus
GFP	green fluorescent protein
GMEM	Glasgow modified Eagle's medium
GST	glutathione S-transferase
HBS	Hepes-buffered saline
hCMV	human cytomegalovirus
HCV	hepatitis C virus
Hepes	*N*-2-hydroxyethylpiperazine-*N*′-2-ethanesulfonic acid
HGMP	Human Genome Mapping Project
HIV-1	human immunodeficiency virus-type 1
HMA	heteroduplex mapping analysis
HTA	heteroduplex tracking analysis
HTLV-1	human T-cell leukaemia virus-type 1
IBB	immunofluorescence blocking buffer
ICR	internal control RNA
IF	initiation factors
IL	interleukin
IPTG	isopropyl β-D-thiogalactopyranoside
IRES	internal ribosomal entry site
kb	kilobase
kbp	kilobase-pair
kDa	kilodalton
LB	Luria (or Lennox) broth
LPA	length polymorphism analysis
LTR	long terminal repeat
Luc	luciferase
MBP	maltose binding protein
MDCK	Madin–Darby canine kidney (cells)
MDE	mutation detection enhancement
MHV	mouse hepatitis virus
MLV	murine leukaemia virus
MMTV	mouse mammary tumour virus
m.o.i.	multiplicity of infection
MoMLV	Moloney murine leukaemia virus
MOPS	3-[*N*-morpholino]propanesulfonic acid

MPMV	Mason–Pfizer monkey virus
M_r	molecular weight
N	nucleocapsid
NA	neuraminidase
nef	*ne*gative *f*actor
NIH	National Institutes of Health
NLS	nuclear localization signal
NP	nucleoprotein
NP-40	Nonidet P-40
NS	non-structural
nt	nucleotide
NTP	nucleotide triphosphate
ORF	open reading frame
p.f.u	plaque-forming unit
p.i.	post-infection
P	phosphoprotein
PAGE	polyacrylamide gel electrophoresis
PBS	phosphate-buffered saline
PCR	polymerase chain reaction
PCV	packed cell volume
PEG	polyethylene glycol
Pipes (buffer)	1,4-piperazinediethanesulfonic acid
PVDF	polyvinylidene fluoride
QHTA	quantitative heteroduplex tracking analyses
RCR	replication-competent retrovirus
RdRp	RNA-dependent RNA polymerase
RE	restriction enzyme
rev	*re*gulator of *vi*rus expression
REV	reticuloendotheliosis virus
RFLP	restriction fragment length polymorphism
RLB	Rose lysis buffer
RNP	resulting nucleoprotein complex (also ribonucleoprotein)
rNTP	any ribonucleotide
RRE	responsive RNA element
RT	reverse transcription
RT–PCR	reverse transcription–polymerase chain reaction
RWB	Rose wash buffer
SAP	shrimp alkaline phosphatase
SCID	severe combined immunodeficiency disease
SDS	sodium dodecyl sulfate
SIN	self-inactivating
SIV	self-inactivating vector
SNL	SDS–Na acetate–LiCl
SNV	spleen necrosis virus
SSC	NaCl–sodium citrate
SSCP	single-stranded conformation polymorphism
t.s.	temperature sensitive

TAR	*tat*-responsive RNA element
tat	*t*rans-*a*ctivator of virus *t*ranscription
TBE	Tris-borate–EDTA
TBS	Tris-buffered saline
TCA	trichloroacetic acid
TE	Tris–EDTA
TEMED	*N,N,N′,N′*-tetramethyl-1,2-diaminoethane
TKM	Tris–KCl–$MgCl_2$
TLC	thin-layer chromatography
TMB	3′,3′,5′,5′-tetramethylbenzadine
TMV	tobacco mosaic virus
TNE	Tris–NaCl–EDTA
TPB	tryptose phosphate broth
TPCK	L-(1-tosylamido-2-phenyl)ethylchloromethyl ketone
TSEA	Tris–sodium acetate–EDTA–acetic acid
UNG	uracil-*N*-glycosylase
UV	ultraviolet
vif	*v*irion *i*nfectivity *f*actor
vpr	*virus* *p*rotein *r*egulatory
vpu	*virus* *p*rotein *u*nknown
VSV	vesicular stomatitis virus
VSV-G	vesicular stomatitis virus-G glycoprotein
YAC	yeast artificial chromosome

Chapter 1

Investigation of RNA virus genome structure

A.J. Easton, A.C. Marriott, and C.R. Pringle
Department of Biological Sciences, University of Warwick, Coventy CV4 7AL, U.K.

1 Introduction—the nature of the virus genome

The viruses infecting microorganisms, plants, invertebrates, and vertebrates are diverse in terms of structure and genetic organization. There are six basic genetic strategies that can be recognized according to the type of nucleic acid sequestered in the extracellular particle:

- single-stranded DNA viruses
- double-stranded DNA viruses
- single-stranded, positive-sense RNA viruses
- single-stranded, negative-sense RNA viruses
- double-stranded RNA viruses, and
- reverse transcribing viruses that may have either RNA or DNA in the virion.

Taxonomically, some 1500 distinct species of viruses (among some 3600 named viruses) are recognized at the present time and these are classified into 189 genera (1). Of these genera, 23 have still to be assigned to a family, the next higher taxon. The remaining 166 genera are grouped into 55 families: 5 families contain viruses with single-stranded DNA genomes; 17 families contain viruses with double-stranded DNA genomes; 17 families contain viruses with single-stranded, positive-sense RNA genomes; 7 families contain viruses with single-stranded, negative-sense RNA genomes; 6 families contain viruses with double-stranded RNA genomes; and 3 families contain viruses with a reverse transcription step in their multiplication cycle: i.e. the RNA viruses exhibit greater diversity than the DNA viruses. The methods described in this chapter reflect our interest in negative-strand RNA viruses, but they are applicable to RNA viruses in general with minor modifications. Completely comprehensive treatment is, however, beyond the scope of a single chapter, and therefore the retroviruses have been excluded from consideration. Focusing on the negative-strand RNA viruses has some merit in that major human pathogens are present in all seven families, whereas less than a third of the families in the other five genomic categories are associated with human disease.

Table 1 Characteristics of the genomes of the 30 families of RNA viruses

Family	Genome type	Genome size (kb or kbp)	Special features	Host
Leviviridae	+ve ssRNA	3.5–4.3	Equimolar base composition	Bacteria
Picornaviridae	+ve ssRNA	7.0–8.5	5′ VPg; 3′ polyA (5′ polyC)	Vertebrates Invertebrates
Sequiviridae	+ve ssRNA	9.0–12.0	5′ VPg; (3′ polyA)	Plants
Comoviridae	bipartite +ve ssRNA [in separate particle]	RNA1 5.9–8.4 RNA2 3.5–7.2	5′ VPg; 3′ polyA 5′ VPg; 3′ polyA	Plants
Potyviridae	(a) mono- (b) bi-partite +ve ssRNA	(a) 8.5–10.0 (b) RNA1 7.9 RNA2 4.6	5′ VPg; 3′ polyA [?]; 3′ polyA	Plants
Caliciviridae	+ve ssRNA	7.4–7.7	5′ VPg; 3′ polyA	Vertebrates
Astroviridae	+ve ssRNA	6.8–7.9	[?]; 3′ polyA	Vertebrates
Nodaviridae	bipartite +ve ssRNA	RNA1 3.1 RNA2 1.4	5′ cap; no 3′ polyA 5′ cap; no 3′ polyA	Invertebrates Vertebrates
Tetraviridae	mono- or bi-partite +ve ssRNA	(a) 5.5 (b) RNA1 5.5 RNA2 2.5	[?] [?] no 3′ polyA [?] no 3′ polyA	Invertebrates
Tombusviridae	+ve ssRNA	4.0–4.7	5′ cap; no 3′ polyA	Plants
Coronaviridae	+ve ssRNA	20–30	5′ cap; no 3′ polyA	Vertebrates
Arteriviridae	+ve ssRNA	13	5′ cap; 3′ polyA	Vertebrates
Flaviviridae	+ve ssRNA	9.5–12.5	5′ cap; (3′ polyA)	Invertebrates Vertebrates
Togaviridae	+ve ssRNA	9.7–11.8	5′ cap; 3′ polyA	Invertebrates Vertebrates
Bromoviridae	+ve ssRNA	RNA1 2.9–3.6 RNA2 2.6–3.0 RNA3 2.0–2.2 RNA4 0.9–1.0	5′ cap; 3′ cap 5′ cap; 3′ cap 5′ cap; 3′ cap 5′ cap; 3′ cap	Plants
Barnaviridae	+ve ssRNA	4.4	n.d.	Fungi
Closteroviridae	+ve ssRNA	15.5–20.0	[?]; no 3′ polyA	Plants
Bornaviridae	–ve ssRNA	8.9	[?]; no 3′ polyA	Vertebrates
Rhabdoviridae	–ve ssRNA	11.0–15.0	No 5′ cap; no 3′ poly A	Plants Invertebrates Vertebrates
Paramyxoviridae	–ve ssRNA	15.2–15.9	No 5′ cap; no 3′ poly A	Vertebrates
Filoviridae	–ve ssRNA	19.1	No 5′ cap; no 3′ poly A	Vertebrates
Arenaviridae	Bipartite –ve* ssRNA	L-RNA 7.2 S-RNA 3.4	Circular RNPs; complementary termini; ribosomal RNA	Vertebrates
Bunyaviridae	Tripartite –ve* ssRNA	L-RNA 6.4–8.9 M-RNA 3.2–5.0 S-RNA 1.0–2.9	Non-covalently closed circular RNAs and RNPs	Plants Invertebrates Vertebrates
Orthomyxoviridae	Multipartite –ve ssRNA	10.0–13.6; 6, 7, or 8 0.9–2.4 linear RNAs	partially complementary termini	Invertebrates Vertebrates

Table 1 *Continued*

Family	Genome type	Genome size (kb or kbp)	Special features	Host
Cystoviridae	Tripartite –ve ssRNA	L-RNA 6.4 M-RNA 4.1 S-RNA 3.0	× 3 linear dsRNAs	Bacteria
Reoviridae	Multipartite dsRNA	18.1–23.7; 10, 11, or 12 RNAs	5′ caps; no 3′ poly A	Plants Invertebrates Vertebrates
Birnaviridae	Bipartite dsRNA	RNA-A 3.1 RNA-B 2.8	5′ VPg; no 3′ poly A	Invertebrates Vertebrates
Totiviridae	dsRNA	4.7–7.0	No 5′ cap; no 3′ poly A	Fungi Protozoa
Partitiviridae	Bipartite dsRNA	RNA1 1.4–3.0 RNA2 1.4–3.0	× 2 linear dsRNAs	Fungi Plants
Hypoviridae	dsRNA	10–13	(5′ cap); 3′ poly A	Fungi

n.d., no data.
(...), variable, some species only
*Viruses belonging to the *Arenaviridae* and *Bunyaviridae* have some genetic information encoded in the +ve sense strand and are effective 'ambisense' viruses.
Data compiled from Murphy *et al.* (2).

1.1 Properties of the genomes of RNA viruses

The currently recognized families of RNA viruses are listed in *Tables 1* and *2*. Viruses that have not yet been assigned to a family are excluded: details of these viruses can be found in Murphy *et al.* (2). *Table 1* lists the sizes and nature of the genomes. Encoding genetic information as RNA is unique to viruses (vertebrate, invertebrate, plant, fungal, and bacterial). Comparatively few bacterial viruses have RNA genomes, whereas more than 90% of known plant viruses have RNA genomes. The genomes of the majority of RNA-containing viruses are monopartite, but segmentation of the genome is not uncommon. The single characteristic of animal and plant RNA viruses that has relevance for processing genomic RNA, other than the nature of the tissue of origin, is the circumstance that the individual segments of multipartite viruses are assembled in separate particles in the case of plant viruses and in the same particle in the case of animal viruses.

1.2 Properties of the virions of RNA viruses

The concentration and purification of viruses is a necessary prerequisite to the analysis of genomic RNA. Methods for the concentration and purification of viruses make use of the physical rather than the chemical properties of the virus concerned. The methodology described below is generally applicable, subject to modification according to the relevant physical properties of the virus by extrapolation from the examples given. *Table 2* lists some of the physical properties of the virions of these viruses that are relevant to their concentration and purification.

Table 2 Some physical properties of the virions of the 30 families of RNA viruses

Family	Virion M_r ($\times 10^6$)	Structural type	Size (d × l) (nm)	S_{20w}	Sucrose bouyant density (g/cm³)	CsCl bouyant density (g/cm³)	Composition	Sensitivity
Leviviridae	3.6–4.2	Icosahedral (T = 3)	26	80–84		1.46	No lipid No carb. RNA 30%	Detergents
Picornaviridae	8.0–9.0	Icosahedral (T = 1)	30	140–165		1.33–1.45	No lipid No carb. RNA	(low pH) (heat) Stabilized by divalent cations
Sequiviridae	n.d.	Isometric	30	150–190		(high)	No lipid No carb. RNA 40%	Protease
Comoviridae	T = 3.2–3.8 M = 4.6–5.8 B = 6.0–6.2	Three icosahedral particles (T = 1)	28–30 28–30 28–30	49–63 86–128 113–134		128–130 141–148 144–153	No lipid, (carb.) RNA1 in B RNA2 in M None in T	
Potyviridae	n.d.	Flexuous filaments (helical)	11–15 × 250–900	150–160		1.29–1.30	No lipid No carb. 5% RNA	
Caliciviridae	15	Icosahedral (T = 3)	30–38	170–187		1.33–1.40	No lipid No carb. RNA	Low pH Heat Freeze/thaw (trypsin)
Astroviridae	8	Isometric (5/6 pt. star)	28–30	160		1.36–1.39	No lipid No carb. RNA	
Nodaviridae	8	Icosahedral (T = 3)	30	135–140		1.30–1.34	No lipid No carb. RNA 16%	(1% SDS)

Tetraviridae	16	Icosahedral	40	194–210		1.29–1.30	No lipid No carb. RNA 11%	
Tombusviridae	8.2–8.9	Icosahedral (T = 3)	30	118–140		1.34–1.36	No lipid No carb. 17% RNA	High pH EDTA
Coronaviridae	400	Pleomorphic	120–160	300–500	1.15–1.19	1.23–1.24	Lipid Carb. RNA	Heat Organic solvents Detergents
Arteriviridae	n.d.	Spherical [isometric nucleocapsid]	60	200–230	1.13–1.17	1.17–1.20	Lipid Carb. RNA	n.d.
Flaviviridae	~60	Spherical [isometric core]	40–60	140–200	1.10–1.23		Lipid ~17% Carb. ~9% RNA	Heat Organic solvents Detergents
Togaviridae	52	Sperical [icosahedral core (T = 4)]	70	280	1.18–1.22		Lipid 30% Carb. 5% RNA	Heat Low pH, Organic solvents Detergents
Bromoviridae	(a) 4.6–6.0 (b) 3.5–6.9	(a) Icosahedral (T = 3) (b) bacilliform	(a) 26–35 (b) 18 × 30–57	63–99		1.35–1.37	No lipid No carb. RNA 14–25%	SDS Rnase
Barnaviridae	7.1	Bacilliform	18–20 × 48–53	n.d.		1.32 ($CsSO_4$)	No lipid No carb. RNA 20%	
Closteroviridae	n.d.	Flexuous filaments (helical)	12 × 1200–2200	96–140		1.30–1.34	No lipid No carb. RNA 5–6%	High salt CsCl EDTA RNase

Table 2 *Continued*

Family	Virion M_r ($\times 10^6$)	Structural type	Size (d × l) (nm)	S_{20w}	Sucrose bouyant density (g/cm^3)	CsCl bouyant density (g/cm^3)	Composition	Sensitivity
Bornaviridae	n.d.	Spherical	90	n.d.	1.22	1.22	Lipid Carb. RNA	Heat Organic solvents Detergents
Rhabdoviridae	300–1000	Bacilliform or Bullet-shaped (helical nucleocapsid)	45–100 × 100–430	550–1045+	1.17–1.19	1.19–1.20	Lipids 15–15% Carb. 3% RNA 1–2%	Heat Organic solvents
Paramyxo-viridae	500+	Pleomorphic (helical nucleocapsid)	150+	1000+	1.18–1.20		Lipids 20–25% Carb. 6% RNA 0.5%	Heat Organic solvents Detergents
Filoviridae	420	U-, 6-, or S-shaped, and circular filaments (helical nucleocapsid)	80 × 800–14 000	1400+	1.14 [pot. tartrate]		Lipid Carb. RNA 1%	Organic Solvents Irradiation
Arenaviridae	n.d.	Pleomorphic	50–300 (mean = 120)	325–500	1.17–1.18	1.19–1.20	Lipid 20% Carb. 8% RNA 2%	HeatOrganic solvents Low pH
Bunyaviridae	300–400	Spherical or pleomorphic	80–120	350–500	1.16–1.18	1.20–1.21	Lipid 20–30% Carb. 2–7% RNA 1–2%	Heat Organic solvents Detergents
Orthomyxo-viridae	250	Spherical or pleomorphic	80–120	700–800	1.19		Lipid 18–37% Carb. 5% RNA	Heat Organic solvents Detergents
Cystoviridae	99	Spherical with icosahedral core	86	405	1.24	1.27	Lipid 20% No carb. RNA 10%	Organic Solvents Detergents

Reoviridae	120	Icosahedral; complex shell + inner core	60–80	630		1.36–1.39	No lipid Carb. (?) RNA 15–20%	(low pH)
Birnaviridae	55	Icosahedral; single shell	60	435		1.33	No lipid Carb. (?) RNA 9–10%	
Totiviridae	12.3	Isometric	30–40	160–190		1.33–1.43	No lipid No carb. RNA	
Partitiviridae	6–9	Isometric	30–40	101–145		1.34–1.39	No lipid No carb. RNA	
Hypoviridae	n.d.	Vesicle with dsRNA + polymerase	50–80	200		1.27–1.30	Lipid Carb. No virus structural protein	Heat Organic solvents Low/high pH

+, More than.
(...), Variable, some species only.
carb., carbohydrate; NC, nucleocapsid; n.d., no data.
Data compiled from Murphy *et al.* (2)

1.3 Paradigm for analysis of RNA virus genome structure

The principles and practical procedures employed in the analysis of RNA virus genome structure are best exemplified by the classic investigation of Choo *et al.* (3), which resulted in the identification of the putative infectious agent responsible for non-A, non-B hepatitis as the RNA virus now known as hepatitis C virus. The molecular characterization of hepatitis C virus is now more advanced than that of any other vertebrate RNA virus, despite the fact that it cannot be propagated *in vitro* and can only be visualized with difficulty in the electron microscope. The key stages in its identification were the demonstration of the transmissibility of the agent in experimental animals. Experiments in susceptible animals identified the agent as a small (< 80 nm) enveloped virus, by virtue of its passage through filters and its organic solvent sensitivity. Molecular characterization of the virus required the production of material of high infectivity, in this case plasma from chimpanzees. A large volume of plasma was subjected to conditions of centrifugation likely to concentrate the smallest known enveloped virus. Nucleic acid extracted from the pellet was identified by nuclease digestion. DNA fragments obtained by reverse transcription with random primers were cloned into an expression vector (lambda gt11). Immunoscreening of large numbers of phage plaques (expressing potential peptide sequences of the unknown virus as fusion proteins) were screened using sera from non-A, non-B hepatitis patients. A single immunoreactive clone was identified and used as a probe to detect overlapping sequences in the DNA fragment library. From this beginning the sequence of the entire genome was built up and diagnostic reagents of increasing sensitivity were developed for both detection of hepatitis C virus antigens and for the genotyping of epidemic strains.

The protocols described in the following sections are less all-embracing than the example cited above. We are assuming that an adequate source of genomic RNA will be available and consequently specialized methods requiring immunoscreening of expression libraries and the products of transcription and translation assays have been excluded.

2 Growth, assay, and purification of RNA viruses

No generalized protocols are presented here because of the diverse biological characteristics of the RNA viruses. There are no common hosts nor permissive cell-culture systems even for closely related viruses. However, virus culture has been considered in another recent volume in this series (4), and some general principles can be stated in lieu of specific protocols.

2.1 Source of virus: *in vivo* versus *in vitro* methods

In vivo propagation of virus may be the best (e.g. growth of influenza A virus in embryonated hens' eggs) or even the only source of virus (e.g. hepatitis B virus from human plasma and hepatitis C virus from chimpanzee plasma). The principle advantage of propagation *in vivo* is that large amounts of virus may be obtained with relative ease. Purification of the virus is facilitated by its relative abundance, and the downstream processing of virus RNA may be simpler.

Propagation in cultured cells offers greater scope for control and the possibility of obtaining starting material of greater genetic homogeneity. However, a limiting factor is the period of adaptation required by some viruses before they grow to appreciable titres in cultured cells. This may not be a matter of great consequence in the primary characterization of genomic RNA, but it has to be considered in the analysis of virus variability and in the study of molecular epidemiology. For example, host cell-mediated selection of antigenic variants is a recognized hazard in the isolation of influenza A viruses, and many primary isolates of other common respiratory viruses, such as respiratory syncytial virus and measles virus, never adapt to growth in cultured cells.

Plant viruses are obtained preferably from systemically infected plants, but often local lesion host plants have to suffice. None the less, a large yield of partially purified virus can be obtained from either source which facilitates subsequent analysis of virus RNA. Guidance on the practical aspects of propagation of plant viruses can be found in reviews elsewhere (5, 6).

2.2 Assay of virus yield

Assay of virus infectivity is a necessary adjunct to protocols for the concentration and purification of viruses. The recovery of infectious virus is a measure of the efficiency of the process and of the integrity of the final product. No generalized protocol can be provided for the assay of virus infectivity. The reader should consult the primary literature on the virus in question, or one of the standard textbooks of virology: e.g. *Field's virology* (7), Topley and Wilson, 9th edition (8), the *Encyclopaedia of virology* (9), or a recent ICTV report (2).

2.3 Harvesting and concentrating the virus

Virus can be concentrated from culture fluids or *in vivo* sources by either chemical or physical means. The chemical methods involve either precipitation with ammonium sulfate or polyethylene glycol, while physical methods involve some form of centrifugation or, rarely, partition chromatography. Enveloped viruses generally require more gentle methods than non-enveloped viruses. Tissue-culture fluids harvested at the time of maximum virus cytopathic effect should be clarified by low-speed centrifugation at 4°C. Then picornaviruses, for example, can be precipitated by the addition of 0.4 g ammonium sulfate per ml of culture fluid and the precipitate concentrated by centrifugation at 2000 *g* for 1–2 h at 4°C. The pellet should be resuspended in 1/10th the original volume of culture fluid. Alternatively, precipitation can be achieved by dissolving 2.2 g of NaCl per 100 ml of culture fluid with constant stirring at 4°C, followed by the addition of 6 g of polyethylene glycol 6000 per 100 ml. Stirring should be continued at 4°C for a minimum of 4 h, and the precipitate concentrated by centrifugation at 2000 *g* for 2 h. The pellet should be resuspended in 1/50th to 1/100th the original volume of culture fluid (or an appropriate buffer). The latter method gives a better recovery of infectivity (close to 100% with care) and a greater degree of concentration. It is also suitable for concentrating enveloped viruses. Parainfluenza viruses, for example, can be concentrated by diluting a 36% (w/v) solution of polyethylene glycol 6000 into clarified culture fluid to give a final concentration of 6%. After gentle agitation at 4 °C for a minimum of 4 h, the precipitate can be collected by centrifugation at 4000 *g* for 10 min and resuspended in a 1/100th volume of culture fluid or buffer.

Viruses can also be concentrated directly by ultracentrifugation if some loss of infectivity can be tolerated. Centrifugation at 50 000 *g* at 4 °C for 90 min is sufficient to sediment enveloped, and 100 000 *g* for 1 h for non-enveloped, viruses in a 5-cm path-length tube. The supernatant fluid is discarded and the pellet resuspended in a small volume of medium or buffer by soaking at 4 °C for 4 h with gentle rotation, or alternatively by soaking overnight without rotation (10).

2.4 Purification of virus

Purification of viruses is achieved predominantly by some form of gradient centrifugation. Other procedures such as partition chromatography, gel filtration, or adsorption–elution methods have been used occasionally, but these are not considered here.

2.4.1 Rate-zonal gradient centrifugation

Enveloped viruses are usually purified by rate-zonal centrifugation through preformed density gradients. The solutes used include buffered sucrose, glycerol, potassium bromide, and potassium tartrate; sucrose being the most common. The gradients can be prepared by stepwise layering of decreasing concentrations of sucrose, or using a mixing device producing a continuous linear gradient (11). The nature of the gradient and the duration of sedimentation can be assessed from the physical properties of the viruses (see *Table 1.2*). The following protocol for the purification of an enveloped virus (vesicular stomatitis virus) and a non-enveloped virus (poliovirus) can be used as points of reference.

Protocol 1

Purification by sucrose gradient centrifugation

Equipment and reagents

- Ultracentrifuge
- Gradient mixing device
- Gradient fraction collector (see ref. 12)
- Syringe and needle
- Ribonuclease-free sucrose
- 20 mM Tris–HCl pH 7.6
- Nonidet P40 (NP-40)

A. For an enveloped virus, e.g. vesicular stomatitis virus

1. Resuspend the pellet, concentrated by polyethylene glycol precipitation from 100 ml of the original culture fluid, in 1 ml of 20 mM Tris–HCl (pH 7.6).

Protocol 1 continued

2. Carefully layer on top of a 25 ml gradient of 15–45 % ribonuclease-free sucrose in 20 mM Tris-HCl (pH 7.6). Centrifuge at 40 000 g for 90 min at 4 °C.
3. Harvest the virion zone into a hypodermic syringe by side-puncturing the tube. If multiple bands are observed, the lower band will be the virion zone and the upper bands may contain truncated (defective interfering) particles.
4. Concentrate the virions by centrifugation at 35000 g for 60 min at 4 °C. Discard the supernatant and, after soaking for a minimum of 2 hours at 4 °C, resuspend the pellet in 0.4 ml 20 mM Tris-HCl (pH 7.6). Store at −70 °C.

B. For a non-enveloped virus, e.g. poliovirus

1. Concentrate a virus-containing culture fluid by either ammonium sulfate or polyethylene glycol precipitation. Add a 1/10th volume of 10% NP-40 to dissociate the virions from contaminating membrane fragments.
2. Prepare a 30 ml linear gradient of 15–45% sucrose in 10 mM Tris-HCl (pH 7.6), then layer 1–6 ml of the concentrated virus carefully on top of the pre-formed gradient. Centrifuge at 80 000 g at 4 °C for 4 h.
3. Harvest the gradients by collecting ~ 0.5 ml sequential fractions from the bottom using a fraction collector of the type described by Minor (12). Locate the peak fraction(s) by infectivity assay (12). Store at −70 °C.

2.4.2 Isopycnic gradient centrifugation

Isopycnic gradient centrifugation can be used with advantage for the purification of non-enveloped viruses. In general, this procedure yields a purer product, and relatively larger volumes of sample can be processed per gradient. Caesium chloride is the preferred medium because of its high density and low viscosity. Separation is achieved on the basis of buoyant density; the gradient may be preformed or self-formed during centrifugation provided the maximum density exceeds the particle density (11). Purification of poliovirus (12) is described in the following protocol.

Protocol 2

Purification by caesium-chloride density gradient centrifugation

Equipment and reagents

- Ultracentrifuge
- Gradient mixing device
- Gradient fraction collector
- Caesium chloride (optical grade)
- 10 mM Tris-HCl pH 7.4

Protocol 2 continued

A. Pre-formed gradient

1. Prepare a 40% (w/v) solution of CsCl by dissolving 0.4 g of CsCl in 6 ml 10 mM Tris–HCl (pH 7.4) and a 5% (w/v) solution by dissolving 0.5 g of CsCl in 9.5 ml of the same buffer.
2. Prepare a 10 ml 5–40% linear CsCl gradient in a 12 ml centrifuge tube. Layer 1 ml of the virus/NP-40 mix (see *Protocol 1B*) on top of the gradient.
3. Centrifuge at 120 000 *g* for 4 h or longer, preferably overnight, at 4 °C.
4. Harvest the gradient by fraction collection and locate the virus-containing fraction(s) by the infectivity assay as described by Minor (12).

B. Self-forming gradient

1. Add solid CsCl to the virus-containing sample to a concentration of 0.46 g per ml.
2. Add NP-40 to give a final concentration of 1% (v/v).
3. Centrifuge for 24 hours at 140 000 *g* at 4 °C.
4. Harvest the gradient by fraction collection and locate the virus by a biological assay.

2.5 Radiolabelling

RNA viruses growing in cell culture can be radiolabelled either by incorporating [^{35}S]methionine into virus proteins or incorporating [^{3}H]uridine into virus RNA. The radionuclide is typically added in growth medium depleted for the labelled compound and the labelling period can be from several hours to several days, depending on the growth characteristics of the virus. For viruses that do not require the host DNA-dependent RNA polymerase for their growth cycle, the radionuclide can be preferentially incorporated into the virus by inhibiting host RNA synthesis, and hence protein synthesis, with actinomycin D. Following labelling, the virus can be purified as described above.

3 RNA extraction

Commercial kits are now available from a number of manufacturers for many of the following operations. For experimental work of limited scale, the advantages of quality control and convenience built into these kits outweigh the relatively higher cost of reliance on kits. As a general rule however, the reliability and sensitivity of commercial kits should always be verified by comparison with conventional protocols under local conditions. In particular, students, technicians, and others in training should be given practical experience of the basic methodology of handling virus RNA before resorting to the use of kits.

3.1 Genome RNA extraction from virions

The best source of uncontaminated genomic RNA is from purified virions, or purified intracellular virus nucleocapsids in the case of poorly released viruses. *Protocol 3A* is suitable for the extraction of RNA from small volumes of virus.

In many cases it is desirable or necessary to prepare virus RNA or mRNA from infected cells. Where possible, this should be done by first fractionating the cell into cytoplasmic and nuclear fractions without lysing the nuclear membrane. The high levels of DNA in nuclei frequently interfere with subsequent steps in RNA analysis, but, although it is possible to remove this using RNase-free DNase, in practice this is difficult to achieve without degradation of RNA. Wherever possible it is preferable to use RNA extracted from the cytoplasm of infected cells. Many methods have been described for this, but we have found that the method described in *Protocol 3* is suitable for a wide range of tissue culture cells and yields reproducibly intact RNA, which can be further analysed by reverse transcription or translation *in vitro*.

Protocol 3

RNA extraction

Safety considerations:

(a) Concentrated positive-sense (infectious) RNA is inherently dangerous.

(b) Phenol is corrosive and can be adsorbed rapidly through skin. Wear protective gloves. If phenol contacts skin, irrigate the area of exposure with polyethylene glycol 300/IMS to neutralize the phenol. Do **not** use water.

Equipment and reagents

- Low-speed refrigerated centrifuge
- Microcentrifuge
- Vortex mixer
- Nylon/rubber policeman or sterile glass beads
- Heat-sterilized pipettes and glassware
- Vacuum extractor
- Isotonic lysis buffer: 150 mM NaCl, 1.5 mM $MgCl_2$, 10 mM Tris–HCl pH 7.8, 0.65% NP-40. Autoclave.
- EDTA stock solution: 0.5 M EDTA pH 8.0
- Phenol equilibration buffer: 150 mM NaCl, 1.0 mM EDTA, 10 mM Tris–HCl pH 7.8. Autoclave.
- Phenol extraction buffer: 7.0 M urea, 350 mM NaCl, 10 mM EDTA, 1% SDS, 10 mM Tris–HCl pH 7.8. Autoclave.
- Buffered phenol/chloroform: mix 25 ml phenol with 25 ml chloroform in a 150 ml glass bottle. Add 50 ml equilibration buffer and shake well. When the phases have separated, remove the aqueous (upper) layer and check the pH before discarding. Repeat until the pH of the aqueous layer is 7.8 and then add a final volume of 50 ml equilibration buffer. Aliquot to avoid contamination and store refrigerated in the dark.
- Ethanol
- 10 M lithium chloride
- Syringe fitted with a 19-gauge needle
- Polyethylene glycol 300/IMS (for use in case phenol accidentally contacts skin, see below)

Protocol 3 continued

A. Small-scale RNA preparations

1. Centrifuge 1 ml of a virus preparation or clinical sample in an Eppendorf tube for 3 min at maximum speed.
2. Remove the supernatant and resuspend the pellet in 500 μl of lysis mixture prepared by mixing equal volumes of the isotonic lysis buffer and the phenol extraction buffer. Add 500 μl of buffered phenol/chloroform, mix by vortexing, and centrifuge in a microcentrifuge for 10 min at maximum speed.
3. Transfer the aqueous layer to a fresh tube containing 1 ml ethanol and 1.3 ml 10 M LiCl. Store at −20 °C for 2 hours or longer.
4. Centrifuge in a microcentrifuge for 20 min and pour off the supernatant. Wash the pellet by resuspending in 500 μl of 70% ethanol and repeat the centrifugation. Repeat a third time and dry the final pellet.

B. Large-scale RNA preparations

The volumes given below are suitable for up to 2×10^8 cells and can be scaled up for larger numbers. The presence of high concentrations of phosphate buffers generates a flocculent interface during the phenol extraction step, and these should be avoided in any washes given to the cells prior to the extraction.

1. Carry out all procedures as quickly as possible at 4 °C, apart from the deproteinization by phenol extraction which should be carried out at room temperature. Carry out RNA extraction in a laminar-flow cabinet using heat-treated glassware to reduce the risk of contamination prior to any reverse transcription stage.
2. Harvest the infected cells when the cytopathic effect is advanced using a nylon scraper or sterile glass beads, then sediment the infected cells by centrifugation in a refrigerated centrifuge (e.g. 250 xg for 10 min at 4 °C).
3. Thoroughly resuspend the cells in 5 ml of isotonic lysis buffer and leave on ice for several minutes. While this is sufficient to lyse many types of tissue culture cells, for others it is necessary to shear the cells to ensure lysis. Shearing can be easily achieved by passing the cell suspension twice through a 19-gauge needle. Remove the intact nuclei by centrifugation at 2500 xg for 3–4 minutes.
4. Mix the supernatant containing the cytoplasm immediately with a pre-made mix of 10 ml buffered phenol/chloroform and 5 ml of phenol extraction buffer to give a final volume of 20 ml (10 ml aqueous + 10 ml organic phases). Shake vigorously for 2 minutes at room temperature and then separate the phases by centrifugation at room temperature for 10 minutes at 6000 xg. Transfer the aqueous (upper) layer to a fresh tube containing 10 ml of buffered phenol/chloroform and repeat the extraction. Avoid transferring debris from the interface of the phases and repeat the extraction for a third time if the aqueous phase is not clear.
5. Precipitate the RNA by adding 2–2.5 volumes of cold (−20 °C) ethanol and maintain at −20 °C overnight or in a solid CO_2/ethanol bath at −70 °C for 30 min.

Protocol 3 continued

6. Pellet the precipitated RNA by centrifugation at 4000 xg for 15 minutes in a refrigerated centrifuge. Pour off the ethanol and wash the pellet by resuspending it in 70% ethanol. Repeat twice and dry the final pellet in a vacuum extractor. Resuspend the pellet in 100 μl of sterile distilled water.

3.2 Genome RNA extraction without virion purification

For those viruses that replicate their genome in the cytoplasm, *Protocol 3* can be used to prepare total cytoplasmic RNA. Where the genomic sequence is known and consensus primers can be devised it is possible to dispense with virus purification. Gritsun and Gould (13) describe a rapid method for the production of infectious full-length transcripts of flaviviruses starting from unpurified suspensions of mouse brain. *Protocol 4* describes their rapid RNA extraction procedure. The polymerase chain reaction (PCR) is used to produce two overlapping cDNA products representing the complete genome of the virus. The 5′-portion of the cDNA is designed to contain an SP6 promoter. Full-length cDNA is produced either by ligating the two overlapping DNA molecules in a unique restriction endonuclease site or by fusion-PCR. The RNA transcribed *in vitro* from these templates proved to be infectious when inoculated intracerebrally into mice. By this means, molecularly cloned virus can be generated in less than 10 days. In principle, the method is applicable to any virus with an infectious RNA genome. In the case of flaviviruses, transfection into mouse brain enhanced the frequency and speed of recovery of cloned virus.

Protocol 4

Genomic RNA extraction from unpurified material

Equipment and reagents

- Catrimox (Iowa Biotechnology Corp. Oakdale, IA, USA).
- 2 M LiCl

Method

1. Precipitate RNA from 100 μl of infected mouse brain or cell suspension by incubating with 1 ml of Catrimox for 40 min at room temperature.
2. Wash the precipitate by resuspending in 2 M LiCl to remove DNA, followed by resuspension in 70% ethanol.
3. Concentrate the precipitate by centrifugation at 13 000 g in a microfuge and resuspend the pellet in 50 μl distilled water.

4 Fractionation of RNA, and analysis by Northern (RNA) blotting

4.1 Fractionation of RNA

Fractionation of RNA is usually carried out by electrophoresis under denaturing conditions, using formaldehyde or glyoxal as the denaturant. The agarose gel concentration should be between 0.8 and 1.5% and the thickness of the gel should be 3.0–5.0 mm. RNA can be stained with ethidium bromide; in the protocol below the RNA is stained before electrophoresis. The stain does not interfere with the transfer step, and can also be used to monitor the success of the transfer of RNA to the membrane.

Protocol 5

Fractionation of RNA

Safety considerations:

(a) Formaldehyde gels should be prepared and run in a fume hood.

(b) Protective gloves should be used when handling transfer media and solutions containing ethidium bromide.

Equipment and reagents

- Electrophoresis apparatus
- Agarose (Ultrapure from Life Technologies)
- 10 × MOPS buffer: 0.2 M MOPS (3-[*N*-morpholino]propanesulfonic acid), 50 mM sodium acetate, 10 mM EDTA. Adjust to pH 7.0 with sodium hydroxide. Sterilize by filtering through a 0.22 μm filter or autoclaving.
- Formaldehyde
- 10 mg/ml ethidium bromide in water. Store in the dark.
- Electrophoresis buffer: 1 × MOPS containing 0.22 M formaldehyde
- Bromophenol Blue marker dye (0.1% w/v)
- Ultraviolet transilluminator

Method

1. Dissolve the agarose by boiling in the appropriate volume of 1 × MOPS buffer.
2. Cool to 45 °C and add formaldehyde to 0.22 M final concentration.
3. Pour into a gel-former and allow to set in a fume hood or a closed box.
4. For each 8 μl RNA, add 7.5 μl formamide, 2.5 μl formaldehyde, 2 μl 10 × MOPS buffer, and 0.1 μl ethidium bromide.
5. Denature by heating to 68 °C for 10–15 min.
6. Cool and load on to the gel. Load Bromophenol Blue marker dye into one or more wells.
7. Electrophorese in a fume hood or closed apparatus at 5–6 V/cm until the blue marker dye has migrated the desired distance.
8. Visualize the RNA bands (e.g. size markers and ribosomal RNAs) by illumination with UV light. The gel is now ready for transfer.

4.2 Northern blotting and detection with labelled probes

The RNA is transferred to a membrane support ('blotting') and hybridized with a labelled probe for the detection of specific virus RNAs. Probes can consist of either DNA or RNA, and may be labelled with a ^{32}P radioisotope, or with a non-radioactive hapten (e.g. biotin or digoxigenin). Non-radioactive probes can be detected chromogenically, or with chemiluminescence as in the protocol below.

Protocol 6

Northern blotting

Equipment and reagents

- Charged nylon membrane (e.g. Amersham Hybond-N+)
- 3MM chromatography paper (Whatman)
- Optional: Turboblotter™ device (Schleicher & Schuell)
- Hybridization oven or shaking water bath
- Hybridization bottle or plastic bag
- Bag sealer
- Labelled polynucleotide probe, denatured
- X-ray film
- 20 × SSPE: 3 M sodium chloride, 0.2 M sodium dihydrogen phosphate, 20 mM EDTA. Adjust to pH 7.4 with sodium hydroxide.
- Autoradiography cassette
- Transfer buffer: 7.5 mM sodium hydroxide
- 100 × Denhardt's reagent: 2% w/v each of Ficoll-400, bovine serum albumin, and polyvinylpyrrolidone
- Hybridization fluid: mix 25 ml formamide, 12.5 ml 20 × SSPE, 2.5 ml 100 × Denhardt's reagent, 5 ml 10% sodium dodecyl sulfate, 0.5 ml 10 mg/ml sheared, denatured salmon sperm DNA. Add water to 50 ml final volume. If using a riboprobe, supplement with 0.25 ml 10 mg/ml Torula yeast RNA (Sigma type VI).
- 0.1% SDS

Additional reagents needed for Part C (available from Boehringer Mannheim (now Roche Diagnostics))

- Blocking solution: 20 mM Tris–HCl pH 8.0, 3 M sodium chloride, 0.3% (v/v) Tween-20, 1% w/v blocking reagent (Boehringer Mannheim)
- Conjugate: 1 in 10 000 dilution of anti-DIG–alkaline phosphatase conjugate in blocking solution
- Washing buffer: 20 mM Tris–HCl pH 8.0, 3 M sodium chloride, 0.3% (v/v) Tween-20
- DIG-labelled probe
- Phosphatase buffer: 0.1 M Tris–HCl pH 9.5, 0.1 M sodium chloride
- CSPD: 25 mM solution of disodium 3-(4-methoxyspiro{1,2-dioxetane-3,2′-(5′-chloro)tricyclo[3.3.1.$1^{3,7}$]decan}-4-yl)phenyl phosphate
- Plastic sheets

Method

A. Capillary transfer

1. Rinse the gel in transfer buffer for 2–10 min.
2. Cut 3MM paper and membrane to the size of the gel. Pre-wet the membrane and four pieces of 3MM paper in transfer buffer.

Protocol 6 continued

3. Lay four pieces of dry 3MM paper, then one piece of wetted 3MM paper, then the membrane on top of a stack of absorbent paper towels.
4. Carefully place the gel on top of the membrane, avoid introducing any air bubbles between the gel and the membrane.
5. Lay a further three pieces of wetted 3MM paper on top of the gel, followed by two 3MM paper wicks which dip into a reservoir of transfer buffer at either side of the gel. The Turboblotter™ device is specifically designed for downward capillary transfers. Place a small weight on top, e.g. a plastic box lid.
6. Allow the transfer to proceed at ambient temperature for 1.5–6 hours.
7. Remove the membrane and rinse in 5 × SSPE for 5 min.
8. Air-dry the membrane, then bake in an oven at 80 °C for 1–2 hours.

B. Hybridization with a ^{32}P-labelled probe

1. Seal the membrane in a hybridization bottle or plastic bag containing warmed hybridization fluid; use 10 ml per 100 cm^2 membrane.
2. Pre-hybridize for at least 30 min at the hybridization temperature.
3. Add denatured probe to 5–20 ng/ml (typically 0.5–5 × 10^6 c.p.m./ml).
4. Hybridize at 50 °C (DNA probe) or 68 °C (RNA probe) for 12–20 hours.
5. Remove the membrane from the bottle or bag and wash it at the hybridization temperature: 50–100 ml per wash, twice for 15 min in 2 × SSPE, 0.1% SDS, then twice for 15 min in 0.1 × SSPE, 0.1% SDS (stringent wash).
6. Dry the membrane or seal it in a plastic bag.
7. Expose the membrane to X-ray film in an autoradiography cassette. For the strongest signal, use intensifying screens at −70 °C.

C. Hybridization with a digoxigenin-labelled probe followed by chemiluminescent detection

1. Pre-hybridize as in Part A, steps 1–2.
2. Add denatured probe to 50–100 ng/ml, then hybridize and wash as in Part A, steps 3–5.
3. Rinse the membrane in washing buffer for 1 min.
4. Block for 30–60 min in blocking solution at ambient temperature.
5. Add 10 ml of the conjugate per 100 cm^2 membrane, seal in a plastic bag, and incubate with shaking for 30 min.
6. Wash twice for 15 min each time in 50–100 ml washing buffer.
7. Wash for 2 min in phosphatase buffer.
8. Make up a fresh dilution of CSPD, 1:100 in phosphatase buffer.
9. Drain excess buffer from the membrane, lay it on a plastic sheet, and apply diluted CSPD: 0.5 ml per 100 cm^2 membrane. Overlay with a second plastic sheet to distribute the CSPD as an even thin film over the membrane. Incubate for 5 min then seal between plastic sheets.

Protocol 6 continued

10. Incubate at 37 °C for 15 min before exposure if a stronger signal is required.

11. Expose to X-ray film in an autoradiography cassette at ambient temperature.

5 Further analysis

For further analysis of RNA genomes, such as the determination of nucleotide sequence or expression of gene products, it is necessary to convert the RNA into DNA by reverse transcription. Once in the form of DNA the material can be treated as for standard DNA using the large repertoire of techniques described in a variety of well-known laboratory manuals, the most comprehensive of which is Sambrook *et al.* (14). Protocols for reverse transcription, amplification by polymerase chain reaction, agarose gel electrophoresis, construction of T-vectors, ligation, production and use of supercompetent cells, transformation, DNA minipreps, DNA maxipreps, and purification of DNA by CsCl density gradient centrifugation, can also be found at the authors' laboratory home page (www.bio.warwick.ac.uk).

The following two protocols illustrate specific techniques frequently used in the analysis of the genetic heterogeneity of RNA viruses. *Protocol 7* provides an example of the conversion of RNA to DNA, whereas in *Protocol 8* RNA is analysed without the intervention of reverse transcription.

5.1 Characterization by restriction-endonuclease digest patterns

Rapid analysis of RNA virus isolates can be achieved by the amplification of virus RNA using the polymerase chain reaction (PCR) followed by restriction enzyme cleavage-site mapping. The following protocol is taken from an analysis of the molecular epidemiology of respiratory syncytial virus (15). The method is applicable to the analysis of virus in clinical specimens or tissue-culture propagated virus.

Protocol 7

Characterization by restriction mapping

Equipment and reagents

- 25 cm^2 Petri dishes
- Sterile glass beads
- Microcentrifuge
- Vortex mixer
- Solution A: 3.5 M urea, 20 mM NaCl, 10 mM Tris–HCl pH 7.8, 5 mM EDTA, 0.75 mM $MgCl_2$, 0.5% SDS, 0.35% NP-40
- Solution B: 0.5 ml phenol/chloroform (1:1) equilibrated with 150 mM NaCl, 10 mM Tris–HCl pH 7.8, 1 mM EDTA
- Reverse transcriptase mix: 0.15 μg primer,[a] 10 units AMV reverse transcriptase (Amersham Pharmacia Biotech), and 1 mM of each dNTP (Pharmacia) in the buffer supplied by the manufacturer

Protocol 7 continued

- Thermal cycler (Hybaid TR2)
- Phenol/chloroform
- Ethanol
- Vacuum dryer
- PCR mix: 0.4 μg of each primer, 250 mM each dNTP, 1.5 mM $MgCl_2$, 2 units of *Taq* polymerase (Promega) in the buffer supplied by the manufacturer, and 2.5 μl of the reverse transcribed cDNA
- *Taq* polymerase (Promega)
- Electrophoresis equipment
- Agarose (Ultrapure from Life Technologies)
- Tris–borate buffer: 89 mM Tris base, 89 mM boric acid, 2 mM EDTA, pH 8.3
- Restriction enzymes
- Ethidium bromide (see *Protocol* 5)
- 1 kb DNA ladder

Method

1. When the cytopathic effect is advanced, harvest virus-infected cell cultures in 25 cm^2 Petri dishes by shaking with sterile glass beads. Store suspensions in aliquots at −70 °C.
2. Sediment the lysed cells by centrifugation in a microcentrifuge for 2 min. Resuspend the pellet in 0.5 ml of solution A. Add 0.5 ml solution B and mix by vortexing for 5 sec, then centrifuge in a microcentrifuge for 10 min. Re-extract the aqueous (upper) layer with phenol/chloroform and then add 1 ml cold ethanol. Hold at −20°C for 2–20 hours to precipitate the nucleic acids, then centrifuge in a microcentrifuge at top speed for 10 min to obtain a pellet. Wash with 0.5 ml 70% ethanol and vacuum-dry.
3. Resuspend the precipitate directly into the reverse transcriptase mix and incubate at 41 °C for 30 min.
4. Set up PCRs in 100 μl volumes. Select appropriate primers for the target sequence. Amplify by 30 cycles (94 °C for 45 sec, 54 °C for 45 sec, and 74 °C for 45 sec) using a Hybaid TR2 thermal cycler (or equivalent).
5. Analyse 10 μl of the PCR product by electrophoresis on 2% agarose gels with Tris–borate buffer.
6. Dilute the remainder of the PCR product with 100 μl distilled water. Extract with phenol/chloroform and precipitate with ethanol as described above.
7. Digest the PCR product with the appropriate restriction enzymes, using the recommended buffers supplied by the manufacturer. Select the restriction enzymes according to the nature of the pre-determined target sequence.
8. Analyse the digested PCR products on 2% agarose gels stained with ethidium bromide (see *Protocol* 5) in comparison with a 1 kb DNA ladder.

[a]**NB** The primer 5′(GGCCCGGGAAGC)TTTTTTTTTTTTTTT3′ primes on polyadenylated mRNA.

5.2 Characterization by ribonuclease protection

One-dimensional RNA fingerprinting by RNase A protection assays can be used to monitor and analyse the genetic heterogeneity of RNA viruses, and even to locate single or multiple base changes in the virus genome (16). The following protocol is taken from an analysis of respiratory syncytial virus genetic heterogeneity by Storch *et al.* (17).

Protocol 8

RNase A mismatch cleavage mapping

Equipment and reagents

- Uniformly labelled full-length RNA probe generated by in vitro transcription from an appropriate DNA template
- PBS
- 75 cm Petri dish
- Lysis buffer: 6 M guanidinium isothiocyanate, 50 mM Tris–HCl pH 7.5, 10 mM EDTA pH 7.5, 0.5% sodium sarkosyl, 0.2 M 2-mercaptoethanol
- CsCl
- 0.1 M EDTA pH 7.5
- Beckman SW50.1 rotor
- Buffer A: 10 mM Tris–HCl pH 7.5, 5 mM EDTA pH 7.5, and 1% SDS
- Chloroform/butanol (4:1)
- 3 M sodium acetate pH 5.5
- Ethanol
- 80% formamide buffer
- RNase digestion solution: 10 mM Tris–HCl pH 7.5, 1 mM EDTA pH 8.0, 200 mM NaCl, 100 ηM LiCl, 10 μg/ml RNase A. (2 μg/ml RNase T1 may be added to enhance digestion of the non-hybridized probe.)
- Solution A: 10% SDS, 2.5 mg proteinase K/ml, 104 μg yeast RNA/ml
- Phenol/chloroform/isoamyl alcohol (25:14:1)
- Electrophoresis equipment
- 4 or 8% polyacrylamide/8 M urea gel
- Autoradiography equipment and reagents

Method

1. Rinse infected cell monolayers with PBS pH 7.6 and then lyse the cells with 3 ml of the lysis buffer per 75 cm Petri dish.
2. Add 1 g CsCl to each 2.5 ml lysate, and layer the lysate on top of a 1.2 ml solution of 5.7 M CsCl in 0.1 M EDTA (pH 7.5).
3. Centrifuge at 35 000 r.p.m. (105 000g) in a Beckman SW50.1 rotor for 16 h at 30 °C.
4. Dissolve the RNA pellet in Buffer A. Extract twice with chloroform/butanol (4:1). Then precipitate the RNA by adding 3 M sodium acetate (pH 5.5) and 2.5 volumes of ethanol. Pellet the RNA, re-extract, and precipitate with ethanol.
5. Set up the RNase protection assay by mixing 1 μl of the radiolabelled probe with 2 μl of unlabelled RNA in 29 μl of 80% formamide buffer (pH 7.4). Denature the RNA by incubating at 85 °C for 10 min then incubate at 45 °C for 16 hours to allow hybridization.
6. Cool to ambient temperature, add 350 μl of RNase digestion solution and digest for 30 min at 20 °C.
7. Add 24 μl of Solution A, and continue the digestion at 37 °C for 30 min to exhaust the RNase.
8. Purify the RNA by two cycles of phenol/chloroform/isoamyl alcohol (25:14:1) extraction and ethanol precipitation.
9. After heat denaturation at 95 °C for 2 min separate the protected fragments by electrophoresis in a 4% or 8% polyacrylamide/8 M urea gel, and visualize by autoradiography.

References

1. Mayo, M. A. and Pringle, C. R. (1998). *J. Gen Virol.*, **79**, 649.
2. Murphy, F. A., Fauquet, C. M., Bishop, D. H. L., Ghabrial, S. A., Jarvis, A. W., Martelli, G. P., Mayo, M. A., and Summers, M. D. (ed.) (1995). *Virus taxonomy, Sixth Report of the International Committee on Taxonomy of Viruses*. Springer-Verlag, Vienna.
3. Choo, Q.-L., Kuo, G., Weiner, A. J., Overby, L. R., Bradley, D. W., and Houghton, M. (1989). *Science*, **244**, 359.
4. Cann, A. J. (ed.) (1999) *Virus culture: a practical approach*. Oxford University Press.
5. Hull, R. (1985). In *Virology: a practical approach* (ed. B. W.J. Mahy), p. 1. IRL Press, Oxford.
6. Rybicki, E. P. and Lennox, S. (1999) In *Virus culture: a practical approach* (ed. A. J. Cann), pp. 239–65. Oxford University Press.
7. Fields, B. N. and Knipe, D. M. (ed.) (1990). *Field's virology* (2nd edn). Raven Press, New York.
8. Mahy, B. W. J. and Collier, L. (ed.) (1998). *Topley and Wilson's microbiology and microbial infections* (9th edn), Vol. 1 Virology. Arnold, London.
9. Webster, R. J. and Granoff, A. (ed.) (1994). *Encyclopaedia of virology*. Academic Press, San Diego, CA.
10. Wunner, W. H. (1985). In *Virology: a practical approach* (ed. B. W. J. Mahy), p. 79. IRL Press, Oxford.
11. Killington, R. A., Stokes, A., and Hierholzer, J. C. (1995). In *Virology methods manual* (ed. B. W. J. Mahy and H. O. Kangro), p. 71. Academic Press, London.
12. Minor, P. D. (1985). In *Virology: a practical approach* (ed. B. W. J. Mahy), p. 25. IRL Press, Oxford.
13. Gritsun, T. S. and Gould, E. A. (1995). *Virology*, **214**, 611.
14. Sambrook, J., Fritsch, E. F., and Maniatis, T. (1989). *Molecular cloning: a laboratory manual* (2nd edn). Cold Spring Laboratory Press, New York.
15. Cane, P. A. and Pringle, C. R. (1992). *J. Virol. Methods*, **40**, 297.
16. Lopez-Galindez, C., Lopez, J. A., Melero, J. A., de la Fuente, L., Martinez, C., Ortin, J., and Perucho, M. (1988). *Proc. Natl. Acad. Sci. (USA)*, **85**, 3522.
17. Storch, G. A., Park, C. S., and Dohner, D. E. (1989). *J. Clin. Invest.*, **83**, 1889.

Chapter 2
Mutagenesis of RNA virus genomes

Wendy S. Barclay* and Jeffrey W. Almond†
*Department of Microbiology, School of Animal and Microbial Sciences, University of Reading, PO Box 228, Reading RG6 2AJ U.K.
†Vice President of Research and Development (France) Pasteur Merieux Connaught, 1541 Avenue Marcel Merieux, 69280 Marcy-L'Etoile, France

1 Introduction

Since the pioneering experiments of Burnett (1) and Cooper (2) in the 1950s and 1960s, the isolation of RNA virus mutants has been a cornerstone of studies aimed at determining the function of virus genomes and their encoded proteins. Not only have the phenotypes of individual mutants been informative about the replication and biology of the wild-type virus, but, collectively, the exploitation of mutants in complementation and recombination assays has provided useful information on the number of genes and the sizes of genomes. Nowadays the generation of mutants is a routine procedure exploited widely in structure-function studies and for the development of designer viruses as vaccines, vectors, and reporter systems for studies on gene expression and genome replication. Basically, there are two approaches to the generation of mutant RNA viruses:

(a) Mutant viruses can be generated and isolated in cell culture *in vivo* by classical methods. This approach requires an appropriate selection procedure to identify mutants of the desired phenotype. Where necessary, the frequency of generating mutants by this route can be enhanced by treatment with various mutagens such as chemicals or irradiation.

(b) Mutants can be generated *in vitro* by site-directed mutagenesis of cDNA copies of the RNA genome and then rescued into live virus using the techniques of reverse genetics. This approach is described in Chapter 9 but will discussed in limited detail in this chapter.

For the meaningful comparison of RNA virus mutants with their parental wild-type viruses, it is important to have methods for isolating populations of virus that are as homogeneous (clonal) as possible. Because RNA genomes are generally highly variable, it is essential to confirm the sequence of the mutant virus genome. This is so that the phenotype can be related to a particular mutation rather than to an inadvertent co-mutation in the genome. In this chapter we will illustrate the general principles of RNA virus mutagenesis using two well-studied RNA virus families, the picornaviruses and the orthomyxoviruses, as examples. Poliovirus and the related rhinovirus are examples of picornaviruses that have single-stranded, positive-sense RNA genomes. In contrast, the orthomyxovirus influenza virus has a genome comprised of negative-sense, single-stranded RNA which is divided into discrete segments. Both of these viruses can be cultivated in cell culture, and clonal populations of virus can be isolated by plaque purification, or by limit dilution. *Protocol 1* describes the procedure for plaque purification of influenza virus and *Protocol 2* describes the limit dilution of a human rhinovirus.

Protocol 1

Plaque purification of influenza virus

Equipment and reagents

- MDCK cells (ECACC or ATCC)
- 6-well, sterile cell-culture plates
- DMEM (Dulbecco's modified essential medium) containing 10% FCS
- Oxoid agar (Whittaker)
- PBS containing 0.2% BSA, penicillin, streptomycin, and added $CaCl_2$ and $MgCl_2$ to 0.01%
- Influenza virus
- Overlay mix: 2 × DMEM, 0.42% BSA, 2 × L-glutamine; 0.3% $NaHCO_3$, 20 mM Hepes, 2 × penicillin/streptomycin (100 IU/ml each)
- 1 mg/ml TPCK trypsin (Worthington)
- Crystal violet stain: 0.1% crystal violet in 20% methanol

Method

1. Seed freshly subcloned MDCK cells into 6-well, cell-culture plates at 5×10^5 cells per 35 mm well.
2. Incubate at 37 °C overnight. Check that the cells are 80% confluent by the next morning.
3. Make a series of 10-fold dilutions of the influenza virus stock in the supplemented PBS diluent.
4. Remove the cell growth medium and wash the cells once with PBS pH 7.3.
5. Inoculate each well with 100 μl of the virus dilution.
6. Leave to adsorb for 1 hour at 37 °C.
7. Meanwhile assemble the overlay medium. Dissolve 2 g Oxoid agar in sterile water by heating in a microwave oven. Place in a 50 °C water bath. Warm the DMEM overlay mix to 37 °C. When ready to overlay, add 7.5 ml 2% Oxoid agar to 17.5 ml of the DMEM overlay mix.
8. Remove the virus inoculum and overlay each well with 2 ml of Oxoid agar in the overlay mix containing trypsin to a final concentration of 2 μg/ml.
9. Incubate for 2–3 days at 37 °C in a 5% CO_2 incubator. The plaques should be clearly visible by eye as opaque areas within the cell monolayer.
10. Using a sterile Pasteur pipette, remove the agar overlaying the virus plaque and expel the plug into 500 μl of the supplemented PBS.

Protocol 1 continued

11. Visualize the remaining cells in the monolayer. Remove the agar overlay with a spatula and overlay the cells with crystal violet stain. Incubate for 30 minutes at room temperature. Wash away the stain and dry the plates.

The ease with which virus plaques can be visualized before staining depends on both virus and cell factors. For example, certain subcloned lines of MDCK cells allow much clearer plaque formation of influenza viruses than others. The Ohio subclone of the HeLa cell line is favoured for detecting poliovirus and rhinovirus plaques. For some cell lines, it may be difficult to visualize the plaques at step 9 and it may be advisable to use a viable stain at this stage. For example, the addition of a second layer of DMEM in agar containing 0.1% Neutral Red, followed by further incubation at 37 °C for 4–6 h should make the plaques highly visible. If this altered protocol is used it is not then necessary to progress to step 11. If the virus under study does not readily form plaques, but virus replication can be detected by cytopathic effect or other methods, then the procedure of limit dilution for obtaining clonal virus populations can be applied (see *Protocol 2*).

Protocol 2

Limit dilution of a rhinovirus

Equipment and reagents

- Sterile, 96-well, cell-culture plates
- DMEM containing 10% FCS, pH 7.2
- Maintenance medium: DMEM containing 2% FCS and 30 mM $MgCl_2$, pH 7.2
- Rhinovirus
- Ohio HeLa cells (ECACC no. 84121901)
- Multichannel pipette

Method

1. Seed freshly subcloned Ohio HeLa cells into 96-well, cell-culture plates at 2×10^4 cells per well (DmEM + 10% FCS).
2. Incubate at 37 °C overnight. Check that the cells are 80% confluent the next morning.
3. Remove the cell growth medium and wash the cells once with PBS.
4. Replace the medium with DMEM plus 2% FCS plus $MgCl_2$. Use 90 μl in each well.
5. Add 10 μl rhinovirus to the first well in each row on the plate.
6. Using a multichannel pipette mix the well contents, but take care to avoid disturbing the cell monolayer. Use a fresh tip to remove 10 μl from the first well and place it in the second well. Carry on across the plate, to make a series of 10-fold dilutions.
7. Add a further 100 μl of the maintenance medium to each well
8. Incubate for 3–4 days at 34 °C in a CO_2 incubator.

Protocol 2 continued

9. Analyse the appearance of the cells with a microscope.
10. Remove and store the medium from the wells containing the highest dilution of virus in which you can see a cytopathic effect as evidence of virus replication.
11. Discard the medium from the plate and replace with 100 μl crystal violet stain in each well.
12. Incubate for 30 minutes at room temperature. Wash away the stain with tap water and dry the plate.

2 Generation of RNA virus mutants

2.1 Selection of mutants from the RNA quasispecies

RNA viruses (other than the retroviruses) replicate their genomes using RNA-dependent RNA polymerases. These enzymes typically lack the proof-reading activity associated with DNA polymerases and are thereby somewhat error prone. The error rate of such enzymes in copying their templates is high, being of the order of 1 mistake per 10 000 nucleotides synthesized. Thus, an RNA virus with a genome size of 10 000 bases (which approximates to the size of the genomes of the picornaviruses and influenza viruses) will generate a single nucleotide mutation in every new genome replicated. The result is that almost every genome is different from the rest of the population at one or more positions, and the RNA sequence of the virus population should be regarded as a 'consensus sequence'. These heterogeneous populations of RNA viruses have been termed 'quasispecies' (3).

If necessary, the diversity of genomes in the quasispecies, particularly of double-site mutations, can be increased by chemical mutagenesis. For example, virus can be grown in cell cultures in the presence of RNA-damaging chemicals such as 5-fluorouracil at 10–50 μg/ml (4). Mutant viruses can then be either selected from the mixture or plaque-purified and screened.

The above principles have been frequently applied for isolating temperature-sensitive (t.s.) mutants. Various methods have been used, the simplest of which is to pick plaques as described above (see *Protocol 1*) and plate them in pairs of small multiwell plates, one of which is incubated at the permissive and the other at the non-permissive temperature (e.g. 34 °C and 40 °C) (5). Candidate t.s. mutants should form plaques at 34 °C but not at 40 °C. Using this method it may be necessary to screen several thousand plaques to obtain a few t.s. mutants. This laborious plaque picking can be short-cut to some extent by staining plaques formed at 34 °C using a viable stain such as Neutral Red, marking their outline on the outside of the plastic dish with a marker pen, incubating overnight at 40 °C, and selecting those plaques that fail to show an increase in size. Such plaques are highly likely to contain t.s. mutants. The disadvantage of this approach, however, is that the further incubation at 40 °C may select for the emergence of t.s.-positive revertants within the plaque, therefore further plaque purification will be necessary.

Occasionally, the particular mutation required may confer a selective growth advantage. In such cases, then progeny of that genome will emerge as a dominant species when the selective pressure is applied. The selective pressures on a virus vary naturally during its replication inside the host, and may also be artificially varied by the experimenter when propagating the virus in the laboratory. The selection applied can be positive or negative. Positive selection utilizes an environment where only the mutant can grow. For example, the ability of influenza virus to form plaques in MDBK (Bovine kidney) cells in the absence of exogenous trypsin requires the possession of a particular neuraminidase gene sequence (6). Negative selection creates an environment where any virus other than the mutant is inhibited in growth. An example of the latter is the selection of monoclonal antibody-resistant viruses, which (usually) have a mutation in the antibody binding site of the target protein. Such mutants have been widely exploited for obtaining information on the number, size, and complexity of antigenic sites on the surface of viruses. Early examples include the definition of four antigenic sites on the influenza haemagglutinin (7) and the identification of a major neutralizing site on the surface of poliovirus (8). The same principles can be used for the selection of drug-resistant variants. These may provide information about the site of action of anti-viral agents.

Provided the virus can be propagated to titres in excess of 10^5 infectious units, chemical mutagenesis is unnecessary before selecting single-point RNA virus mutants such as antibody escape mutants. Sufficient diversity will exist within a high-titre population so that mutants of interest are readily isolated as in *Protocol 3* below.

Protocol 3

The isolation of rhinovirus monoclonal antibody-escape mutants

Equipment and reagents

- Sterile cell-culture plates (35-mm diameter wells)
- Medium for growth, maintenance, and agar overlay of Ohio HeLa cells (see *Protocols 1* and *2*)
- Neutralizing monoclonal antibody—ascites, or hybridoma supernatant, or purified immunoglobulin
- Rhinovirus

Method

1. Infect a 35-mm dish containing approximately 10^6 Ohio HeLa cells with 10^5 p.f.u. of rhinovirus.
2. Incubate for 1 hour at 34 °C.

Protocol 3 continued

3. Overlay the cells with DMEM + 2% FCS + $MgCl_2$ media containing monoclonal antibody at neutralizing concentration.[a]
4. Propagate the virus and cells at 34 °C for 4 days.
5. Plaque-purify viruses that have grown in the presence of antibody as in *Protocol 1*. Include the antibody in the agar overlay at the pre-determined neutralizing concentration.
6. Perform three plaque-to-plaque purifications before analysing the genotype and phenotype of the mutant virus.

[a] This should be pre-determined by a plaque assay of wild-type virus in which increasing concentrations of monoclonal antibody are included in the agar overlay. The amount of antibody used for selection should completely inhibit plaque formation by wild-type virus.

The virus mutants recovered in this way are expected to contain a single amino-acid change corresponding to a single nucleotide change from the wild-type RNA genome sequence. If double mutations are required to escape the selective pressure applied, then it is unlikely that mutants will be isolated unless extremely high starting infectious titres in excess of 10^{10} are used. Alternatively, mutagenesis using 5-fluorouracil (5-FU) can be attempted.

2.2 Recombination of RNA virus genomes

An alternative to direct mutagenesis for the generation of viruses with novel phenotypes is provided by recombination. Recombination occurs at homologous regions of the genomes of related viruses that replicate simultaneously in the same cell. High rates of recombination have been observed for positive-stranded RNA viruses such as picornaviruses and alphaviruses. Recombination occurs much less frequently in negative-stranded RNA genomes.

For positive-stranded RNA viruses, *in vitro* recombination can be achieved experimentally by co-infection of cell cultures with a high multiplicity of two or more different but related viruses. As with mutants, recombinants will represent only a small subpopulation of progeny and must be selected or screened for using temperature, drug resistance, antibody, etc. Once isolated, recombinants can be characterized by sequencing, or by performing a reverse transcription-polymerase chain reaction (RT–PCR) from virus RNA templates using primers specific for two different parent genomes as described below in *Protocol 9*.

2.3 Reassortment of segmented RNA virus genomes

The RNA viruses with segmented genomes can undergo a special form of recombination known as reassortment. Viruses in this class include the orthomyxoviruses, bunyaviruses, reoviruses, and arenaviruses. When cells are infected with two different strains of a related virus, the progeny virus particles can contain mixtures of RNA segments derived from either parental strain. In many instances it is most convenient to study single-gene reassortants in which a single segment of RNA has been exchanged between the parental virus strains. The chances of generating virus with this genotype after mixed infection can be enhanced by UV-irradiation of one of the parental viruses. The largest segments have a greater target size and are more likely to be destroyed in this way. This procedure and the generation of reassortants is described in *Protocol 4*.

Protocol 4

Generation of influenza virus single-gene reassortants

Equipment and reagents

- Germicidal ultraviolet lamp
- Sterile cell-culture plates
- Two different influenza A virus strains
- DMEM, 10% FCS, pH 7.3
- Serum-free DMEM, pH 7.3
- MDCK cells

Method

1. Expose 10^6 p.f.u. of one of the parent viruses to a time course of UV-irradiation by placing in an open Petri dish on ice beneath the germicidal ultraviolet lamp at a distance of approximately 10 cm.
2. Quantify the reduction in titre by plaque assay as in *Protocol 1*.
3. Choose an aliquot of irradiated virus for which the infectious titre has been reduced by 4–5 $\log_{10}$.
4. Co-infect 35 mm dishes of freshly grown MDCK cells with the two virus stocks at an m.o.i. >3. (Base the calculation of infectivity for the UV-irradiated virus on the original virus titre, not that after irradiation.)[a]
5. Incubate overnight at 37 °C in serum-free DMEM containing 1 μg/ml trypsin if necessary, and harvest the cell supernatant containing released virus.
6. Plaque purify the culture supernatants as in *Protocol 1*.

[a]Recovery of certain genotype reassortants may be enhanced by applying appropriate selective pressure at step 4, for example, incubation in the presence of neutralizing monoclonal antibody specific for one of the parental viruses, or at a non-permissive temperature as in step 5 of *Protocol 3*.

The genotype of reassortant viruses can be analysed by separating the RNA segments on polyacrylamide gels containing high urea concentrations and comparing the mobility of each segment with the corresponding segment of each parent virus (see *Figure 1*).

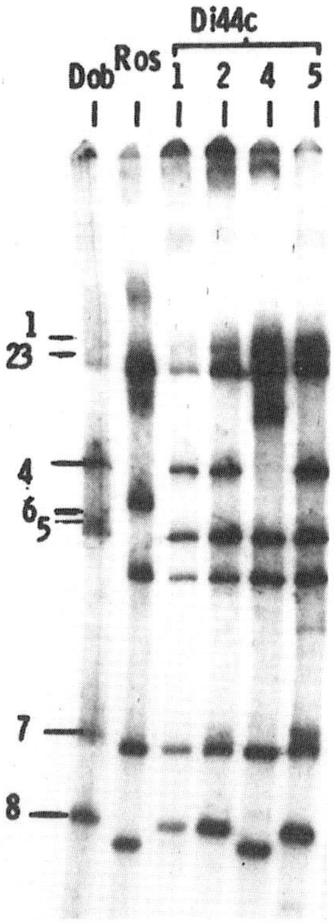

Figure 1 Reassortant influenza viruses (Di44-c 1, 2, 4, and 5) analysed by denaturing RNA gel electrophoresis show RNA segments derived from each parental virus, Rostock and Dobson 4H.

Protocol 5

Denaturing RNA gel electrophoresis for the analysis of reassortant influenza virus genotypes

Equipment and reagents

- Glass plates approximately 5 cm × 5 cm
- Plastic spacers and comb 1.5 mm thick
- Vertical electrophoresis equipment
- DEPC-treated water: add 0.1% (v/v) diethylpyrocarbonate (DEPC) to sterile distilled water and incubate at room temperature overnight. Inactivate DEPC by autoclaving.
- 10% ammonium persulfate and TEMED
- 2.8% polyacrylamide gel containing 6 M urea.
- Formamide loading buffer: 95% formamide, 20 mM EDTA containing 0.05% (w/v) Bromophenol Blue and 0.05% (w/v) xylene cyanol, pH 8.4
- 10 × TBE: 108 g Tris, 55 g boric acid, 40 ml 0.5 M EDTA pH 8.0, in 1 litre DEPC-treated water

Method

1. Assemble two glass plates with three 1.5 mm spacers, one at each side and one across the base.

Protocol 5 continued

2. Dissolve 9.2 g urea in 9.8 ml DEPC-treated water. Heat may be required.
3. Add:

30% acrylamide containing 0.8% bisacrylamide	1.8 ml
10 × TBE	2 ml
10% ammonium persulfate	150 μl
TEMED	25 μl

4. Pour the gel mix between the plates. Insert a comb to make wells at the top of the gel. Allow to polymerize.
5. Mix 150 ng virus RNA purified as in *Protocol 8* with 5 μl formamide loading buffer.
6. Heat the RNA to 95 °C for 1 minute and load into the wells formed by the comb.
7. Remove the base spacer from the gel. Insert the gel plates in a vertical electrophoresis apparatus with 1 × TBE running buffer bathing the top and bottom of the gel.
8. Electrophorese at 150 V for 90 minutes.
9. Remove the gel from the plates and visualize the RNA. The Quicksilver silver stain kit available from Amersham International is suitable for RNA staining. Alternatively, the RNA can be metabolically labelled by propagating virus in the presence of ^{32}P-orthophosphate.

2.4 Generation of defective RNA virus populations by passage in culture

If an essential virus function is provided *in trans* by the addition of exogenous material to the cell-culture medium, RNA virus deletion mutants, which have arisen through natural polymerase error, may be favoured over those that retain the coding capacity for the virus function. For example, by propagating influenza virus in the presence of exogenous bacterial neuraminidase, mutants were selected that contained deletions within the virus neuraminidase gene, and which were now dependent on the addition of the exogenous enzyme (9). These smaller viruses are usually termed *defective* since they have lost an essential virus function and can no longer replicate by themselves. Defective viruses can be also be complemented by co-infection with a wild-type virus that provides the deleted function *in trans*. Propagation of defective viruses in this circumstance is only possible at high m.o.i. to ensure co-infection of individual cells with helper and defective. If the defective competes with the helper for a factor(s) from the host cell which is limiting, then the defective might interfere with the helper virus replication. *Defective interfering* genomes (DIs) retain all the *cis*-acting signals required for replication.

Based on the analysis of the sequence of DI genomes of poliovirus, it has been possible to design poliovirus replicons that contain all the gene products and *cis*-acting functions essential for replication competence, but in which the non-essential sequences have been replaced with those of a reporter gene, such as chloramphenicol acetyl transferase (CAT) (10). Similar replicons have been extremely useful in a range of virus systems for studying translation, replication, and encapsidation. The recovery of such replicons by reverse genetics techniques is described in *Protocol 7*.

2.5 Recovery of RNA virus mutants from infectious cDNA

As discussed elsewhere in this volume, the development of reverse genetics techniques for both positive- and negative-stranded RNA viruses has provided a dramatic increase in the potential for the construction of mutant genomes. Using these techniques it is now possible to create highly novel genome structures such as replicons, chimeras, specifically engineered mutants, and viruses carrying foreign genes.

2.5.1 Site-directed mutagenesis of infectious clones of RNA virus genomes

The *in vitro* techniques that facilitate the mutagenesis of RNA viruses rely on site-directed mutagenesis of cloned cDNA. This is performed using standard recombinant DNA methodology with oligonucleotide-directed mutagenesis of an infectious cDNA clone of the virus genome. Often the mutagenesis is facilitated by PCR, in which one of the primers encodes the desired mutation as a mismatch to the wild-type sequence. If the mutation is close to a naturally occurring restriction enzyme site within the cDNA sequence, then the replacement of the wild-type sequence with the mutation is relatively simple.

If there are no convenient restriction enzyme sites near to the region of the cDNA to be altered, then the PCR fragment containing the mutation can be joined to other PCR products that do extend to a usable restriction enzyme site. The fragments can be joined by overlapping-PCR, or by ligation of PCR products. In all cases of PCR-mediated mutagenesis, the PCR should be performed with proof-reading polymerases and low cycle numbers (< 25) to minimize PCR-induced errors. Overlapping-PCR utilizes two sets of primers, in which one of each set include complementary sequences at their 5′ termini. Two separate PCR reactions are performed. The products are then gel-purified to remove unincorporated primers, and PCR is performed a second time using only the outside pair of primers. An example is illustrated in *Figure 2*, in which a segment of the sequence from the P3 region of the genome of rhinovirus 14 (pT7HRV14) was introduced into a poliovirus replicon (pT7FLC/REP). PCR primers NP13 and NP14 are specific for the poliovirus and HRV sequence, respectively, but contain 5′ extensions which are complementary and represent the recombination junction. Two primary PCR products were joined by a further PCR reaction using outside primers 1201 and GYTO. The product of that reaction is cloned into pT7FLC/REP to produce plasmid pT7FLC/REPrhP3. A method for PCR is given in *Protocol 9*. A procedure for PCR ligation is given in *Protocol 6*.

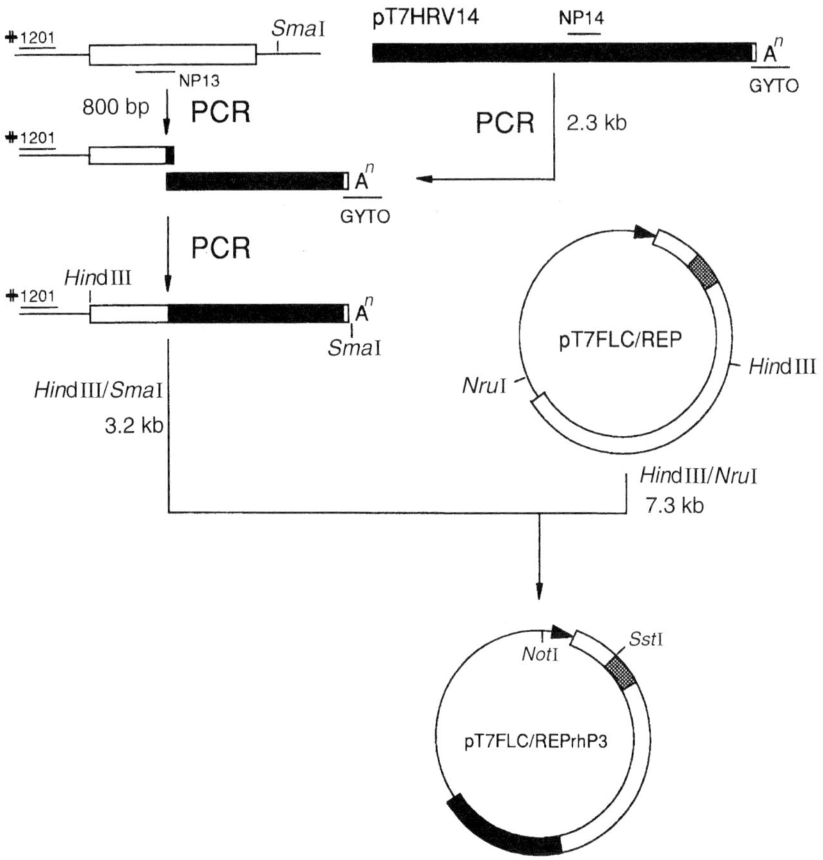

Figure 2 PCR mutagenesis by overlapping-PCR.

Protocol 6

PCR ligation

Equipment and reagents

- Thermal cycler (PCR) machine
- 2 pairs of PCR primers, appropriate to the target sequences
- 1 mM ATP
- Nucleotides, polymerase, buffer, and PCR template as in *Protocol 9*
- T4 polynucleotide kinase
- T4 DNA ligase

Method

1. Perform two PCR reactions as described in *Protocol 9* using cDNA as template in place of the RT reaction.
2. Purify the PCR products after agarose gel electrophoresis.

Protocol 6 continued

3. To each PCR product add 1 mM ATP and the supplier's reaction buffer. Heat the mix to 70 °C then add 50 U of T4 polynucleotide kinase. Incubate for 30 minutes at 37 °C in a 50 μl volume.
4. Combine 5 μl of each phosphorylated primer, representing approximately equimolar quantities, with 400 U of T4 DNA ligase at room temperature for 15 minutes.
5. Use 1 μl ligation reaction as the template in a PCR using external primers as in *Protocol 9*.

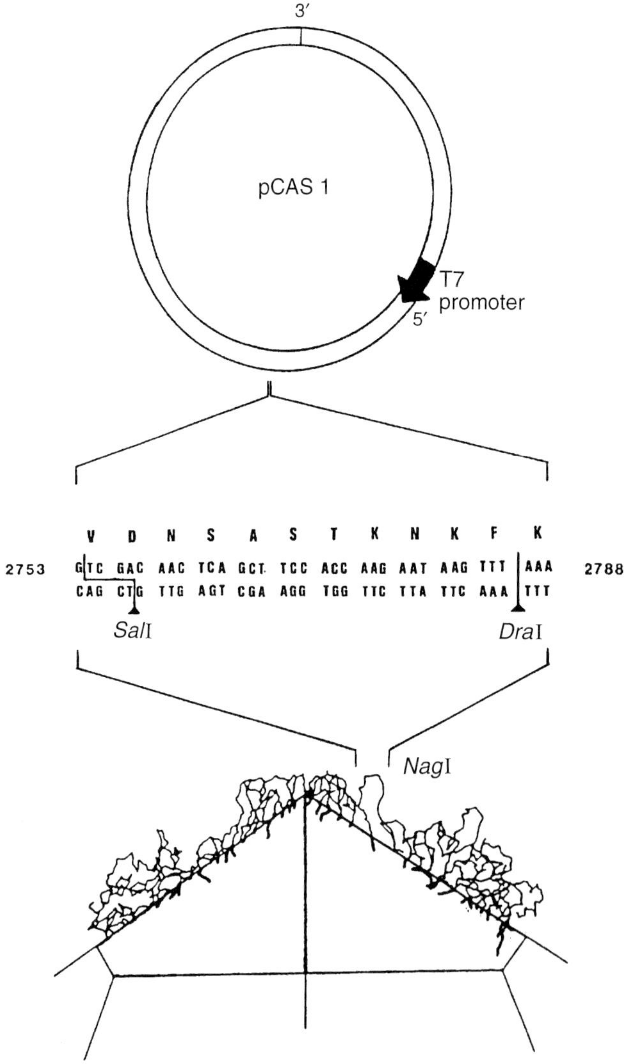

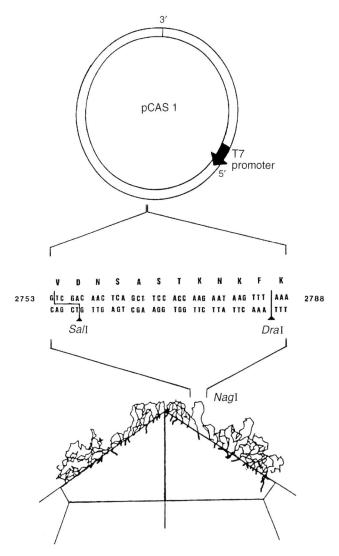

Figure 3 Cassette for generating chimeric poliovirus containing an insert in antigenic site 1.

When a series of many mutations are to be generated within the same region of the genome, it is worthwhile engineering convenient restriction enzyme sites into the area if they do not already exist. For example, when creating chimeric polioviruses in which foreign sequences were introduced into the VP1 antigenic site of poliovirus type 1, two unique restriction enzyme sites were first engineered into the sequences flanking the antigenic loop (see *Figure 3*) (11). The foreign sequences were then synthesized as complementary oligonucleotides that anneal to produce overhangs at their termini, which are compatible with the engineered restriction sites. The annealed oligonucleotides were simply ligated into digested vector.

2.5.2 Recovery of virus from infectious cDNA

The RNA genome of positive-strand RNA viruses is infectious. Introduced into cells it can serve as a messenger RNA and direct the synthesis of all the virus proteins required for replication. Genome-like RNAs can be transcribed *in vitro* from the wild-type or mutated cDNA, usually via a T7 promoter inserted in the cDNA upstream of the virus 5′ terminal nucleotide. A method for the synthesis and transfection of poliovirus synthetic RNA is given in *Protocol 7*.

Protocol 7

Synthesis and transfection of poliovirus RNA

Equipment and reagents

- T7 RNA polymerase
- cDNA of full-length poliovirus genome
- Appropriate restriction enzyme and buffer
- Phenol/chloroform (1:1)
- Ethanol
- Ribonucleotides and supplier's transcription buffer (Promega)
- DEPC-treated water (see *Protocol 5*)
- RNase inhibitor (e.g. RNA guard, Amersham Pharmacia Biotech)
- 1.5% agarose gel (1 × TAE buffer) and electrophoresis equipment and reagents
- 100 mM dithiothreitol (DTT)
- 1 × TAE buffer: glacial acetic acid 57.1 ml 0.5 M EDTA pH 8.0, 100 ml; Tris base 242 g, make up to 7850 ml with distilled water.
- Ethidium bromide
- 2 × HBS: 0.1 g Hepes, 0.16 g NaCl /litre in DEPC-treated water adjusted to pH 7.1
- 10 mg/ml DEAE Dextran
- Ohio HeLa cells
- Sterile cell-culture plates
- Growth maintenance and agar overlay DMEM (see *Protocols 1* and *2*)

Method

1. Digest 5 μg of the cDNA template with the restriction enzyme beyond the 3′ terminus of the poliovirus sequence.
2. Phenol:chloroform extract the digested template and ethanol-precipitate.
3. Resuspend the cDNA in 10 μl of the DEPC-treated water.

Protocol 7 continued

4. To 2 μl of the digested cDNA add:

5 mM rNTPs	10.0 μl
5 × supplier's transcription buffer	10.0 μl
100 mM DTT	5.0 μl
RNase inhibitor	1.0 μl
T7 polymerase	1.0 μl
DEPC-treated water	21.0 μl

5. Incubate at 37 °C for 1 hour.
6. Analyse 2 μl of the reaction on a 1.5% agarose gel stained with ethidium bromide (1 × TAE buffer).
7. To the remaining RNA add 500 μl 2 × Hepes, 20 μl DEAE Dextran, and 432 μl DEPC-treated water.
8. Make 10-fold dilutions of the RNA in 1 × transcription buffer.
9. Wash freshly seeded 80%-confluent Ohio HeLa monolayers on a 6-well plate with PBS.
10. Inoculate each well with 100 μl RNA dilutions.
11. Incubate at 37 °C for 1 hour.
12. Overlay with agar overlay as in *Protocol 1*.
13. Incubate for 3 days at 37 °C.
14. Visualize plaques[a] by eye or by staining cells with crystal violet stain.

[a] In excess of 10^5 plaques can be obtained from 1 μg of poliovirus RNA.

2.5.3 Transfection of RNA replicons

As mentioned in Section 2.4, the study of defective RNAs reveals the location of *cis*-acting RNA signals within the RNA virus genome (12). Using this information, an RNA replicon of poliovirus was designed in which a reporter gene, chloramphenicol acetyl transferase (CAT), was inserted in place of sequences in the genome which are not essential for replication or gene expression (10). These RNAs are generated by *in vitro* transcription from cDNAs mutated as described above. They are transfected into cells as described in *Protocol 7*, but rather than using a solid agar overlay, a liquid overlay of maintenance medium is used. After overnight inoculation, the cell lysates can be analysed for production of the reporter using a conventional CAT assay. The ability of the replicon to be packaged into capsids supplied *in trans* can be assayed by superinfection of the transfected cells with a helper virus. Following overnight incubation, progeny virus is released into the cell supernatant and can be passaged on to fresh monolayers. These are then analysed for reporter gene expression in the same way.

3 Determining the genotype of RNA virus mutants

The genotype of RNA virus mutants can be determined by sequencing either directly from RNA extracted from purified virions, or by amplification using reverse transcription and polymerase chain reaction, RT-PCR, with appropriate primers. The latter technique is described in *Protocols 8* and *9*. For mutants obtained as described in Section 2.1, determination of the site of mutation within the genome that confers resistance to selective pressure can be very informative. For example, when monoclonal antibodies are used to select escape mutants, the mutations selected will usually lie within the epitope of the antibody and thus indicate the location of antigenic sites on the virus structure. It is important to sequence several independent virus plaques at step 5 in *Protocol 1*, because there may be several different mutations which can confer resistance to a particular neutralizing monoclonal antibody. In some instances the mutations will not map to a contiguous region of the genome, indicating secondary and tertiary interactions with different virus proteins or domains within a single virus protein.

Mutations that give rise to drug resistance will usually be located within the virus protein which is the target for the drug, and therefore give clues as to the mechanism of action of the anti-viral activity. Again, by analysing several independently generated mutants, it has sometimes been shown that changes in more than one virus protein can overcome drug inhibition. Resistance to the anti-influenza drug amantadine can be mediated by changes in the M2 ion-channel protein and also by changes in the haemagglutinin (HA) protein for some virus strains. It is now known that M2 acts as a chaperone for HA during intracellular transport (13).

Protocol 8

Extraction of virus RNA from purified virus

Equipment and reagents

- Ultracentrifuge and Beckman SW55 rotor with polycarbonate tubes
- TNE buffer: 10 mM Tris-HCl pH 7.4, 100 mM NaCl, 1 mM EDTA
- 30% sucrose in TNE buffer
- TKM: 10 mM Tris-HCl pH 7.4, 10 mM KCl, 1.5 mM $MgCl_2$
- RNase inhibitor
- 20% SDS
- 5 mg/ml proteinase K
- 56 °C water bath
- SNL: 5% SDS, 100 mM Na acetate pH 7.0, 1.5 M LiCl
- DEPC-treated water (see *Protocol 5*)
- Phenol saturated with DEPC-treated water
- Chloroform
- Phenol/chloroform (1:1)
- 3 M Na acetate pH 5.2
- 100% and 70% ethanol
- Vacuum-drier

Protocol 8 continued

Method

1. Layer 3 ml of a cell culture supernatant containing virus onto 7 ml 30% (w/v) sucrose in TNE buffer. Centrifuge at $>$100 000 g for 2 hours at 4 °C.
2. Resuspend the virus pellet in 300 μl TKM. Leave the pellet overnight at 4 °C if it is difficult to resuspend.
3. Add 4 μl of the RNase inhibitor, 4.5 μl 20% SDS, and 15 μl 5 mg/ml proteinase K. Heat at 56 °C for 10 minutes.
4. Add 35 μl SNL. Mix. Add 300 μl phenol saturated with DEPC-treated water. Heat at 56 °C for 5 minutes inverting every 20 seconds.
5. Cool to room temperature. Add 300 μl chloroform. Mix and microcentrifuge at 12 000 g for 5 minutes.
6. Remove the upper layer to a fresh tube and perform a phenol/chloroform extraction twice.
7. Add 30 μl 3 M Na Acetate pH 5.2 and 1 ml 100% ethanol. Place at −70 °C for 30 minutes.
8. Microcentrifuge for 15 minutes at 12 000 g. Wash the RNA pellet by adding 500 μl 70% ethanol. Microcentrifuge for 15 minutes at 12 000 g.
9. Vacuum-dry the RNA pellet. Resuspend in 20 μl of DEPC-treated water.
10. Quantify the RNA by measuring absorbance at 260 nm.

Protocol 9

Reverse-transcription and PCR from virus RNA

Equipment and reagents

- Dry ice
- Purified virus RNA (see *Protocol 8*)
- Appropriate primers
- DEPC-treated water (see *Protocol 5*)
- 1 M Tris-HCl pH 8.5
- 1 M KCl
- 0.2 M $MgCl_2$
- 20 mM DTT
- 10 mM of each dNTP
- AMV reverse transcriptase (Gibco, Promega, NEB, etc.)
- RNase inhibitor (Gibco, Promega, NEB, etc.)
- 42 °C water bath
- PCR equipment and reagents
- Proof-reading polymerase and buffer (Vent, Stratagene)
- Agar gel electrophoresis equipment and reagents

A. Reverse transcription

1. Mix 2 μl of the purified virus RNA (see *Protocol 8*) with 2 μl of a complementary primer diluted to a concentration of 10 mM. Make the volume up to 12 μl by adding 8 μl DEPC-treated water. Heat to 95 °C for 1 minute, then snap-cool by placing the tube into a small beaker containing dry ice.

Protocol 9 continued

2. Assemble the components of the reverse transcription (RT) reaction on ice as follows. To 12 μl RNA plus annealed primer add:

1 M Tris-HCl pH 8.5	1.0 μl
1 M KCl	1.0 μl
0.2 M $MgCl_2$	1.0 μl
20 mM DTT	1.0 μl
10 mM dNTPs	2.0 μl
RNase inhibitor	0.5 μl

3. Move the tube to a 42 °C water bath. Add 1 μl AMV reverse transcriptase. Incubate at 42 °C for 45 minutes.
4. Heat to 95 °C for 5 minutes.

B. Polymerase chain reaction

1. Remove 10 μl of the RT reaction to a fresh tube suitable for use in a PCR heating block. Add:

distilled water	49 μl
10 × buffer for the proof-reading polymerase (e.g. Vent)	10 μl
2 mM dNTPs	10 μl
Primer 1 (may be the same as the RT primer)	10 μl
Primer 2	10 μl
Proof-reading polymerase (Vent)	1 μl

2. Perform PCR for 25 cycles (must be optimized for each primer set).
3. Analyse the PCR products by agarose gel electrophoresis, e.g. 1.5% agarose gel.

4 Analysis of the frequency of RNA virus mutants

As explained in Section 2.1, the frequency at which RNA virus mutants with single point mutations arise is approximately 1 in 10 000 (3). If a mutant cannot be isolated when selection pressure is applied, it is likely that more than one nucleotide change, or an insertion/deletion event, is required to generate the desired phenotype.

Sometimes, a single point mutation gives rise to a virus with an unstable phenotype, and a second site mutation is rapidly selected to compensate. This is the case for most influenza viruses that escape inhibition by the anti-viral agent GG167, which is an inhibitor of viral neuraminidase (NA). Escape can be mediated by mutations within the NA enzyme, which decrease its affinity for the substrate since the drug is a substrate analogue. However, in cell culture, the most readily isolated mutants have mutations in HA; this results in a decrease in the affinity of the haemagglutinin for the sialic acid receptor, thus reducing the efficiency required from the NA enzyme. Many GG167 escape mutants contain mutations in both NA and HA genes (14).

The location of the second site mutation can be revealing about the interactions that occur between proteins or RNAs within the virus. When reverse genetic techniques are used to design site-specific mutations, three outcomes are possible:

(a) The mutant is recovered at the same frequency as wild-type virus.

(b) A virus is recovered following transfection, but at a lower frequency than for wild-type virus.

(c) No mutants are recovered.

In the second instance, the most likely explanation is that the introduced mutation renders the virus non-viable, but that a second site mutation—which occurs at a frequency of 1 in 10 000 or less—may restore viability. Careful sequence analysis of the mutant virus genome will identify the site of the second mutation. In this way, it was first demonstrated that poliovirus protein 3C interacts with the 5′ end of the genomic RNA, since mutations introduced into the non-coding region resulted in second site changes within the 3C protein (15). The recovery of poliovirus mutants exemplifying the first two instances is shown in *Figure 4*. Synthetic RNA for wild-type poliovirus (top) or two mutants

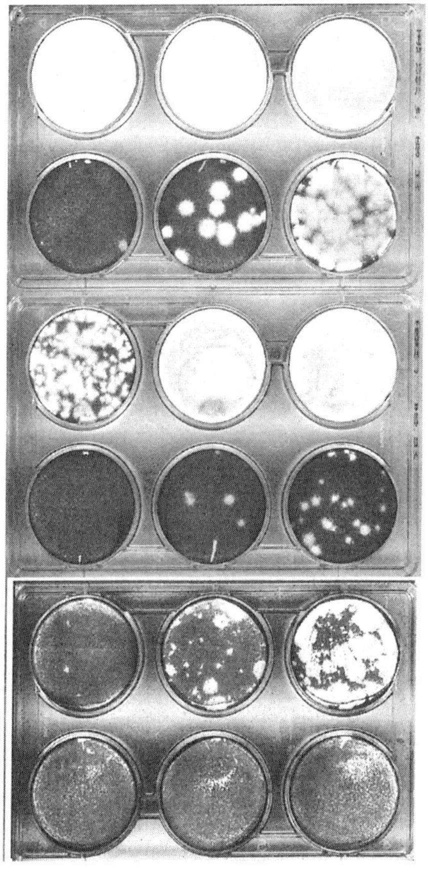

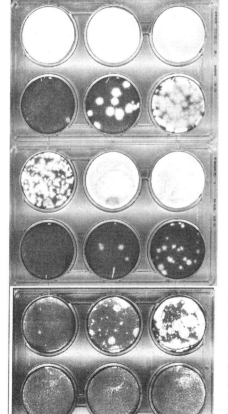

Figure 4 Recovery of poliovirus mutants by direct plaque assay of synthetic RNA (see text for details). This type of analysis can only be applied when the recovery efficiency of infectious virus from cDNA is high, as for the positive-sense RNA viruses. The reverse genetics techniques for negative-sense viruses, described in Chapter 9, will require much refinement before second site mutants can be recovered.

(below) was serially diluted in 10-fold steps before transfection on to cells in 6-well plates. 1 μg RNA was transfected into the top right-hand well and the lower left-hand well received only 10 pg RNA. The first mutant is viable, but the mutation confers a small plaque phenotype. The second mutant is recovered 1000-fold less efficiently and in some of the plaques, a second site mutation has restored a large plaque phenotype.

References

1. Burnet, F. M. (1959). In *The viruses 3* (ed. F. M. Burnet and W. M. Stanley), p. 255. Academic Press, New York.
2. Cooper, P. D. (1969). In *Biochemistry of viruses* (ed. H. B. Levy), p. 177. Marcel Dekker, New York.
3. Domingo, E., Martinez Salas, E., Sobrino, F., de la Torre, J. C., Portela, A., Ortin, J., and Holland, J. (1985). *Gene*, **40**, 1.
4. Cooper, P. D. (1964). *Virology*, **22**, 186.
5. Almond, J. W., Mcgeoch, D., and Barry, R. D. (1977). *Virology*, **81**, 62.
6. Schulman, J. L. and Palese, P. (1977). *J. Virol.*, **24**, 170.
7. Gehard, W., Yewdell, J., and Frankel, M. (1980). In *Structure and variation in influenza virus* (ed. W. G. Laver and G. M. Air), p. 273. Elsevier, New York.
8. Minor, P. D., Schild, G. C., Bootman, J., Evans, D. M., Ferguson, M., Reeve, P., Spitz, M., Stanway, G., Cann, A. J., Hauptmann, R., Clarke, L. D., Mountford, R. C., and Almond, J. W. (1983). *Nature*, **301**, 674.
9. Liu, C. G. and Air, G. M. (1993). *Virology*, **194**, 403.
10. Percy, N., Barclay, W. S., Sullivan, M., and Almond, J. W. (1992). *J. Virol.*, **66**, 5040.
11. Burke, K. L., Evans, D. J., Jenkins, O., Meredith, J., D'Souza, E. D. A., and Almond, J. W. (1989). *J. Gen. Virol.*, **70**, 2475.
12. Hagino Yamagishi, K. and Nomoto, A. (1989). *J. Virol.*, **63**, 5386.
13. Hay, A. J. (1992). *Semin. Virol.*, **3**, 21.
14. Blick, T. J., Sahasrabudhe, A., McDonald, M., Owens, I. J., Morley, P. J., Fenton, R. J., and McKimm-Breschkin, J.L. (1998). *Virology*, **246**, 95.
15. Andino, R., Rieckhof, G. E., Trono, D., and Baltimore, D. (1990). *J. Virol.*, **64**, 607.

Chapter 3
Analysis of transcriptional control in RNA virus infections

S. Makino
Department of Microbiology and Immunology, The University of Texas Medical Branch at Galveston, Galveston, Texas 77555-1019, U.S.A.

1 Introduction

This chapter will illustrate several commonly used methods for the analysis of mRNAs and their transcriptional control by RNA viruses. I will first describe the conventional methods for characterizing mRNAs in virus-infected cells. Subsequently, methods utilizing genetically engineered virus RNA or cDNA with or without reporter genes will be described. These methods enable the analysis of *cis*- and *trans*-acting RNA sequences to be determined.

Most RNA viruses (except retroviruses) undergo RNA-dependent RNA synthesis. They must synthesize virus genomic RNA, which is destined to be packaged into the progeny virion particles. This process is termed RNA replication. In addition, they also synthesize mRNAs (often smaller than the genomic RNA), which are used for translation of virus proteins. This process is termed transcription. For some RNA viruses, genomic RNA or its complementary RNA serves as mRNA for virus protein synthesis. In these cases, the replication of the RNA genome is synonymous with the transcription of mRNAs. As a result, the terms transcription and replication are often used indiscriminately. Very often, mRNAs have a different structure from those of genomic RNA, even though mRNAs may be nearly equivalent to the genomic size. In still other cases, mRNAs have structures clearly discernible from the genomic RNA. Thus, transcription and replication of virus RNA are separable and usually under different regulation. In this chapter, we will focus mainly on the analysis of mRNA transcription, even though many of the techniques are applicable to genomic RNA replication.

2 Analysis of virus mRNAs in virus-infected cells

2.1 General considerations

For successful extraction and characterization of virus mRNAs, it is crucial to avoid degradation by RNases. Cellular RNases, other RNases that may be present in solutions and glassware, and those present human skin, saliva, and hair can degrade the RNA of interest.

Disposable gloves should always be worn and talking should be avoided during extraction or characterization of virus RNAs; these procedures prevent contamination of the samples by RNase from skin and saliva. The activity of cellular RNases can be reduced by keeping the samples at a low temperature (0–4 °C). Samples containing RNA can be incubated at a higher temperature (room temperature to 37 °C) once proteinase or phenol is added to the samples.

The purchase of new chemicals for the preparation of solutions for RNA studies is advisable. Chemical bottles used for RNA studies should be stored separately from those used for other purposes. For dispensing chemicals from a bottle, a baked spatula should be used. More practically, tilt a chemical bottle and pour the contents into a weighing dish. If an excess amount of chemical is poured on to the weighing dish, collect the extra amount of chemical using a baked spatula or a sheet of weighing paper. The excess chemical should be discarded and not returned to the stock bottle.

Contamination of buffers with RNases can be eliminated by using properly prepared RNase-free solutions. It is crucial that all glassware is baked at 180–200 °C for at least 6 h; the majority of RNases should be inactivated by this treatment. Filter-sterilized distilled water (or filtered ultrapure water) is suitable for preparing the solutions; 0.22 μm disposable filters are used. All solutions should be prepared in baked glassware or disposable plasticware. Stirring bars used for dissolving solutions should also be baked. For adjusting pH, a pH meter probe should be extensively rinsed with filter-sterilized distilled water prior to use. Residual water should be removed by blotting. After adjusting the pH, the solutions should be kept in baked bottles. Bottle caps are usually not resistant to baking. In our laboratory, bottle caps are washed with detergent, rinsed with distilled water, sterilized by autoclaving. Since the caps may not be completely RNase-free, avoid touching the cap with the solution; do not shake bottles upside down. If solutions are prepared carefully as described above, sterilization by autoclaving is usually unnecessary. If solutions for RNA studies are prepared properly, treatment of solutions with diethyl pyrocarbonate (DEPC), which is known to be a strong RNase inhibitor, is unnecessary. We have successfully extracted and characterized a large RNA, the 32 kb-long coronavirus genomic RNA, the biggest virus RNA currently known, without DEPC treatment of solutions.

2.2 Radiolabelling and extraction of virus RNAs

RNA viruses, in general, replicate in the cytoplasm of the infected cells (with a few exceptions, such as influenza and Borna viruses). Their RNA replication and transcription are usually considered to be independent of DNA-dependent RNA synthesis. Thus, virus RNA can be metabolically labelled with radioisotopes in the presence of actinomycin D, which inhibits host-cell RNA synthesis. Also, virus RNAs can be isolated from the cytoplasm after removal of cell nuclei. This can substantially reduce the background because of the removal of cellular DNA.

The amount of virus mRNA that can be obtained varies greatly, depending on the kinetics of virus infection. Later in infection, more virus RNA will accumulate, but the cytopathic effects evident late in the infection often affects the quality of the RNA. Thus, the optimum timepoint for harvesting virus RNA should be determined individually for each virus. Various methods have been employed to extract virus mRNAs. *Protocol 1* gives a high yield of RNA of satisfactory quality in our hands. *Protocol 2* describes ^{32}P-radiolabelling of murine coronavirus mouse hepatitis virus (MHV) RNA in DBT (delayed brain tumour) cells. If [^{3}H]uridine is used for radiolabelling, the phosphate-free MEM incubation step in *Protocol 1* should be omitted.

Protocol 1

Radiolabelling and extraction of virus RNA from virus-infected cells

Equipment and reagents

- 60-mm tissue-culture dish
- Sterilized rubber policeman (e.g. Fisher)
- ^{32}Pi (Amersham or NEN) 400–800 mli ml^{-1}
- Sterilized, chilled, phosphate-buffered saline (PBS) pH 7.4
- 2 × proteinase K buffer: 0.2 M Tris–HCl pH 7.5, 25 mM Na_2EDTA, 0.3 M NaCl, 2% SDS
- 1 mg/ml actinomycin D stock (actinomycin D dissolves in water after overnight incubation at room temperature)
- 5 × phosphate-free Eagle's solution: dissolve 17 g NaCl, 1 g KCl, 0.73 g sodium citrate, 0.51 g $MgSO_4{\cdot}7H_2O$ (or 0.24 g $MgSO_4$), 2.5 g glucose, and 15 mg Phenol Red in 500 ml of water. After all chemicals are completely dissolved, add 0.66 g $CaCl_2{\cdot}2H_2O$ (or 0.5 g $CaCl_2$), and then autoclave. Store at 4 °C.
- Proteinase K solution: 10 mg/ml proteinase K in 10 mM Tris–HCl pH 7.5, 14 mM NaCl
- Phosphate-free, minimal essential medium (MEM): mix 100 ml of 5 × phosphate-free Eagle's solution, 10 ml 50 × MEM amino acid solution (Gibco-BRL), 5 ml 100 × MEM vitamin solution (Gibco-BRL), 5 ml 100 × antibiotics solution, e.g. kanamycin sulfate (Gibco-BRL), and 380 ml of water. Then, adjust pH using sterilized sodium bicarbonate. Store at 4 °C.
- Low-phosphate MEM: 9 vol. of phosphate-free MEM and 1 vol. of Eagle's MEM
- Dialysed heat-inactivated fetal calf serum
- Phenol/chloroform mixture: 50:50 (v/v)
- Lysis buffer: 0.5% NP-40, 0.1 M NaCl, 10 mM Tris–HCl pH 7.5, 1 mM EDTA

Method

1. Change the culture medium to low-phosphate medium containing FCS (98% low-phosphate MEM and 2% dialysed heat-inactivated FCS) and 2.5 μg/ml of actinomycin D at 2.5 h postinfection (p.i.) for MHV.
2. Change the culture medium to phosphate-free medium containing FCS (98% phosphate-free MEM and 2% dialysed heat-inactivated FCS) at 6 h p.i.

Protocol 1 continued

3. Add ^{32}Pi (250 μCi/ml) to the medium.
4. Incubate at 37 °C for 2 h.
5. Remove the culture medium.
6. Wash the cells with chilled PBS twice
7. Add 1 ml of chilled PBS to the culture
8. Use a rubber policeman to scrape cells from the dish and suspend them in PBS.
9. Transfer the cells in PBS into a 1.5 ml microcentrifuge tube.
10. Centrifuge for 5 sec at 16 000 g in a microcentrifuge at 4 °C.
11. Remove and discard the supernatant.
12. Add 200 μl of lysis buffer to the centrifuge tube and briefly vortex.
13. Keep on ice for 5 to 10 min.
14. Centrifuge for 30 sec at 16 000 g in a microcentrifuge at 4 °C.
15. Transfer the supernatant to a new tube.
16. Add 200 μl of 2 × proteinase K buffer and 8 μl of proteinase K to the 200 μl of supernatant.
17. Vortex gently.
18. Incubate for 30 min in a 37 °C water bath.
19. Add 400 μl of phenol/chloroform and vortex vigorously for 30 sec.
20. Centrifuge for 5 min at 16 000 g in a microcentrifuge at room temperature.
21. Transfer the supernatant to a new tube, and repeat steps 19–20.
22. Transfer the 400 μl upper layer to a new microcentrifuge tube and add 2.5 vol. of chilled ethanol.
23. Store at −20 °C or −80 °C.

2.3 Gel electrophoresis of RNAs

The size and species of radiolabelled intracellular virus mRNAs are determined by agarose gel electrophoresis under denaturing conditions. Two different agarose gel electrophoresis protocols are presented below.

2.3.1 Glyoxal gel electrophoresis

The glyoxal gel electrophoresis system employs the fact that glyoxal specifically binds to guanosine residues of RNA in neutral pH or slightly acidic pH. Glyoxalation sterically hinders G–C base-pair formation, thus preventing the renaturation of RNA secondary structure, resulting in denaturation of RNA. In glyoxal gels, virus RNA is first incubated with glyoxal and then separated by agarose gel electrophoresis in a phosphate buffer lacking denaturing agents (1).

Protocol 2

Glyoxal gel electrophoresis

Equipment and reagents

- Gel electrophoresis apparatus
- Peristaltic pump
- Power supply
- Agarose (electrophoresis grade)
- 30% glyoxal (aqueous solution): completely deionize by mixing with an ion-exchange resin (e.g. Amberlite mixed-bed resin, Sigma) for several hours—a glyoxal solution attains neutral pH after complete deionization. Aliquot the deionized glyoxal into microcentrifuge tubes and store at −80 °C. Glyoxal is stable for several months at −80 °C.
- 1 M sodium phosphate buffer pH 7.0
- Gel running buffer: 10 ml of 1 M sodium phosphate buffer, 990 ml of water
- 10 × buffer: 10 mM EDTA, 1% SDS, 100 mM sodium phosphate buffer pH 7.0
- DM buffer: 16.95 μl glyoxal, 75 μl dimethyl sulfoxide (DMSO), 15 μl 10 × buffer (make fresh)
- Dye buffer: 10 mM EDTA, 1% SDS (make fresh)
- Dye: 0.25% Bromophenol Blue, 0.25% xylene cyanol FF, 50% glycerol (make fresh)
- Dye/dye buffer (dye:dye buffer 1:1)

Method

1. Prepare the gel by dissolving 0.5–1 g of agarose in 100 ml of 10 mM sodium phosphate buffer in a microwave oven.
2. Wash and assemble the gel electrophoresis apparatus. Make sure the comb is about 5 mm above the plate.
3. Pour the agarose gel into the centre of the plate.
4. Denature the RNA samples as follows. Mix 7.13 μl of the DM buffer and 2.87 μl of RNA, then vortex briefly. Spin down by brief centrifugation. Incubate at 50 °C for 30–60 minutes. Add 2.5 μl of the dye/dye buffer.
5. Pre-run the gel while the RNA samples are denatured. Remove the end panels and comb. Pour 10 mM sodium phosphate running buffer into the gel apparatus and cover with a lid. Attach peristaltic-pump tubing to the electrophoresis apparatus and start buffer circulation. Connect the power supply and run at 16–20 mA.
6. Turn off the pump and power supply. Load the RNA sample carefully into a well on the gel.
7. Run at 20 mA with constant buffer circulation overnight.

2.3.2 Formaldehyde gel electrophoresis

In the formaldehyde agarose-gel electrophoresis system, gel and buffer both contain denaturing reagent, formaldehyde (2). Glyoxal gel electrophoresis requires circulation of buffer, while formaldehyde gel electrophoresis must be performed in a chemical hood. Although both electrophoresis systems are widely used, we feel that the formaldehyde gel electrophoresis system is easier to perform.

Protocol 3

Formaldehyde gel electrophoresis

Equipment and reagents

- 10 × RBS: 200 mM 3-(*N*-morpholino)-propanesulfonic acid (Mops), 10 mM EDTA, 50 mM sodium acetate pH 7.0
- Agarose (electrophoresis grade)
- Formaldehyde
- Running buffer: 100 ml of 10 × RBS buffer, 180 ml formaldehyde, 720 ml distilled water
- Sample buffer: 10 μl of 10 × RBS, 50 μl formamide, 17.8 μl formaldehyde.
- Formamide (molecular biology grade)
- 6 × dye solution: 0.25% (w/v) Bromophenol Blue, 0.25% (w/v) xylene cyanol, 40% (w/v)

Method

1. Prepare the gel. Mix 1–2 g of agarose, 10 ml of 10 × RBS, and 72 ml of distilled water. Dissolve the agarose in a microwave oven. Add 18 ml of formaldehyde when the temperature of gel drops to 50–60 °C.
2. Wash and assemble the gel electrophoresis apparatus.
3. Pour the agarose gel into the centre of plate.
4. Prepare the RNA sample by first dissolving the sample RNA in 2 μl of distilled water. Add 8 μl of sample buffer, mix gently by pipetting, and put on ice.
5. Denature the RNA at 65 °C for 5–10 minutes and then immediately chill on ice for 2–3 minutes.
6. Add 2 μl of 6 × dye solution, mix well, and carefully load the RNA sample into a well on the gel.
7. Run the gel at 50–100 V.
8. If gel is to be dried after electrophoresis, soak it first in 0.1 M ammonium acetate for 20 min and then in water for 30 min (twice) with gentle agitation.

3 Analysis of virus mRNA structure

3.1 General considerations

Virus mRNAs, in general, have very similar structures to that of cellular mRNAs, i.e. they have a 3′-poly(A) and a 5′-cap structure. However, many virus genomes have unique structures, such as 5′-linked virus proteins. This chapter will not discuss the characterization of these specific structures. Instead, it will deal with the questions of how to determine which regions of the genomic RNA a virus mRNA represents, and what is the structural relationship between the different mRNA species. These are central questions regarding virus mRNAs.

The advance of cloning and sequencing technologies has facilitated the characterization of mRNAs. However, simpler experiments may reveal many essential features of mRNAs without going though cloning, which may be limited by the availability of a sufficient quantity of virus RNA. This section will describe some of the basic techniques for characterizing mRNAs.

3.2 Separation of virus RNA in preparative agarose gels

To characterize virus mRNA structure, radiolabelled virus RNA is usually separated by gel electrophoresis. Purified RNAs are then extracted from the gel for further characterization. Where unlabelled RNAs are to be extracted, the RNA is usually spiked with a small amount of labelled RNA. For the isolation of specific virus RNA from gels, virus RNAs are separated by preparative gel electrophoresis. We use partially denatured gel electrophoresis; 6 M urea is used as a denaturing reagent. Alternatively, virus RNA is first denatured by heating, and then separated under non-denaturing conditions using a low melting-point agarose gel.

Protocol 4

Urea–agarose gel electrophoresis (3)

Equipment and reagents

- Agarose (electrophoresis grade)
- 10 × TSE buffer: 400 mM Tris, 200 mM sodium acetate, 20 mM EDTA, 330 mM acetic acid, pH 7.4
- 9 M urea solution (9 M ultrapure urea solution is used for preparation of the gel, and 9 M demonized urea solution is used for electrophoresis buffer)
- 1 M sodium phosphate buffer pH 7.0
- 250 mM EDTA
- 10% SDS
- Running buffer: 120 ml of 10 × TSE buffer, 800 ml of deionized 9 M urea solution, 280 ml of distilled water
- Sample buffer: 10 μl of 1 M sodium phosphate buffer pH 7.0, 10 μl of 10% SDS, 4 μl of EDTA, 76 μl of distilled water
- 6 × dye solution: 0.25% (w/v) Bromophenol Blue, 0.25% (w/v) xylene cyanol, 40% (w/v) sucrose)

Method

1. Prepare the gel as follows. Mix 1–2 g of agarose, 10 ml of 10 × TSE buffer, and 23 ml of distilled water. Dissolve the agarose in a microwave oven. Add 66 ml of 9 M ultrapure urea solution when the temperature of the gel drops to 50–60 °C.
2. Wash and assemble the gel electrophoresis apparatus in a cold room (4 °C).
3. Pour agarose gel into the centre of plate.
4. Prepare the RNA sample by first dissolving the sample RNA in 9 μl of distilled water. Add 1 μl of sample buffer and 2.5 μl of 6 × dye solution, mix gently by pipetting, and put on ice.
5. Denature the RNA at 56 °C for 5–10 minutes and then immediately chill on ice for 2–3 minutes.

Protocol 4 continued

6. Load the RNA sample carefully into a well in the gel.
7. Run RNA in the gel at 50–100 V at 4 °C overnight.
8. If the gel needs to be dried after electrophoresis, first soak it in 0.1 M ammonium acetate for 2 h with a change of buffer and then in water for 30 min (twice) with gentle agitation.

Protocol 5

Low melting-point agarose gel electrophoresis

Equipment and reagents

- Low melting-point agarose (electrophoresis grade)
- 10 × TBE buffer: 890 mM Tris base, 890 mM boric acid, 20 mM EDTA
- 1 M sodium phosphate buffer pH 7.0
- 250 mM EDTA
- 10% SDS
- Running buffer: 120 ml of 10 × TBE buffer, 1020 ml of distilled water
- Sample buffer: 10 μl of 1 M sodium phosphate buffer pH 7.0, 10 μl of 10% SDS, 4 μl of EDTA, 76 μl of distilled water
- 6 × dye: 0.25% (w/v) Bromophenol Blue, 0.25% (w/v) xylene cyanol, 40% (w/v) sucrose)

Method

1. Prepare the gel as follows. Mix 1–2 g of agarose, 10 ml of 10 × TBE buffer, and 80 ml of distilled water. Dissolve the agarose in a microwave oven, and keep at 50–60 °C.
2. Wash and assemble the gel electrophoresis apparatus.
3. Pour the agarose gel into the centre of plate.
4. Prepare RNA sample by first dissolving the sample RNA in 9 μl of distilled water. Add 1 μl of the sample buffer and 2.5 μl of 6 × dye solution, mix gently by pipetting, and put on ice.
5. Denature the RNA at 100 °C for 2–3 minutes. And then immediately chill on ice for 2–3 min.
6. Load the RNA sample into a well on the gel carefully.
7. Run the RNA on the gel at 50–100 V (lower voltage is preferred for the purification of RNA).

3.3 Extraction of virus RNA from gel slices

After separation of virus mRNA in preparative gel electrophoresis, the location of virus mRNA is determined by exposing a wet-gel to X-ray film. After a gel slice that contains virus mRNA is cut from the preparative gel, the RNA in the gel slice is extracted. The following procedures are used in our laboratory.

Protocol 6

Isolation of RNA from a gel slice after urea–agarose gel electrophoresis (4)

Equipment and reagents

- 1-butanol
- Hexadecyltrimethylammonium bromide (Sigma)
- Antifoam A (Sigma)
- 0.2 M NaCl
- Chloroform
- Ethanol

A. Preparation of solutions

1. Mix 150 ml of 1-butanol and 150 ml of distilled water in a glass bottle.
2. Allow the butanol and water phases to separate, then transfer 100 ml of the upper butanol fraction to another glass bottle.
3. Transfer 100 ml of the water fraction to another glass bottle.
4. Add 1 g of hexadecyltrimethylammonium bromide to 100 ml of the butanol fraction.[a]
5. Dissolve hexadecyltrimethylammonium bromide by shaking the bottle.
6. Add 100 ml of the butanol-saturated water from step 3.
7. Add 50 μl of Antifoam A and shake well.
8. Leave the bottle at room temperature overnight.
9. Take the upper butanol fraction and lower water fraction into two separate bottles and keep them at 37 °C.

B. Isolation of RNA from urea–agarose gels

1. Transfer the urea-agarose gel slices into a microcentrifuge tube.
2. Incubate at 75 °C until the gel slices melt.
3. Estimate the total volume.
4. Cool the melted gel solution at 37 °C and add equal volumes of the equilibrated butanol and water phases (Part A, step 9).
5. Invert the tube 50 times and centrifuge at 10 000 g for 5 sec.
6. Transfer the upper butanol fraction into another microcentrifuge tube; the RNA is now in the butanol fraction.
7. Add an equal amount of saturated butanol to the remaining water phase.
8. Repeat steps 5–6, twice, for back extraction.
9. Pool the butanol fractions.
10. Add a one-quarter volume of 0.2 M NaCl to the combined butanol fraction.
11. Invert the tube 50 times and centrifuge at 10 000 g for 5 sec.

Protocol 6 continued

12. Transfer the lower water phase to another microcentrifuge; the RNA is now transferred to the water phase.
13. Add the same volume of 0.2 M NaCl solution used in step 10 to the remaining butanol fraction and repeat the salt extraction.
14. Add an equal volume of chloroform dropwise to the combined aqueous solutions.
15. Shake the tube and keep the tube at 0 °C for several minutes; the RNA is still in the water phase, while hexadecyltrimethylammonium bromide precipitates in this step.
16. Briefly centrifuge the tube and take the upper water phase.
17. Precipitate the RNA from the upper water phase with ethanol.

[a] If a large quantity of RNA (up to 150 μg RNA/ml of agarose gel) is used, use 3.67 g of hexadecyltrimethylammonium bromide per 100 ml of butanol.

Protocol 7

Isolation of RNA from a gel slice after low melting-point agarose gel electrophoresis

Equipment and reagents

- TE buffer: 10 mM Tris–HCl pH 7.4, 1 mM EDTA
- Pre-warmed (37 °C) saturated phenol
- Pre-warmed (37 °C) saturated phenol/chloroform (50:50)
- Chloroform

Method

1. Melt the gel slice at 50 °C in a microcentrifuge tube.
2. Add 3 volumes (or more) of pre-warmed (37 °C) TE buffer.
3. Add an equal volume of pre-warmed saturated phenol.
4. Vigorously mix the tube using a vortex mixer for 1 min.
5. Centrifuge in a microcentrifuge at 15 000 g for 5 min.
6. Take the upper aqueous phase to a new microcentrifuge tube.
7. Add an equal volume of pre-warmed phenol/chloroform and vortex mix the tube for 1 min.
8. Centrifuge in a microcentrifuge at 15 000 g for 5 min.
9. Transfer the upper aqueous phase to a new microcentrifuge tube.
10. Add an equal volume of chloroform and vigorously mix the tube using a vortex mixer for 1 min.
11. Centrifuge in a microcentrifuge at 15 000 g for 5 min.
12. Take the upper aqueous phase to a new microcentrifuge tube.
13. Concentrate the RNA by precipitation using ethanol.

3.4 One-dimensional oligonucleotide fingerprinting

Oligonucleotide fingerprinting of RNase T1-digested RNA was used extensively for characterizing RNA before the era of cloning and sequencing. This method has now fallen out of use and very few laboratories are equipped to perform two-dimensional fingerprinting. However, we have found that one-dimensional oligonucleotide fingerprinting is a very quick and easily performed approach, which gives a lot of information regarding mRNA structures. This procedure requires no equipment other than standard DNA-sequencing apparatus. For this procedure, ^{32}P-labelled RNA is digested with RNase T1, which generates a series of oligonucleotides ending at the G residues at the 3′ end. Depending on the distribution of Gs in the RNA, an array of oligonucleotides of different sizes characteristic of each RNA is generated. In conventional two-dimensional gel electrophoresis, these T1 oligonucleotides are separated by size and charge into 2-D fingerprints. This allows all the oligonucleotides to be separated and further characterized. However, this procedure is cumbersome. One-dimensional fingerprinting obviates the need for separation by electrophoresis in an acidic environment, which is the most difficult part.

Protocol 8

One-dimensional oligonucleotide fingerprinting

Equipment and reagents

- ^{32}P-labelled purified RNA (see *Protocols 6 & 7*)
- Vertical gel electrophoresis apparatus
- TE buffer: 10 mM Tris-HCl pH 7.4, 1 mM EDTA
- Loading solution: formamide containing 0.1% (w/v) Bromophenol Blue and 0.1% (w/v) xylene cyanol
- Gel stock solution: 172 g acrylamide and 12 g methylene-bisacrylamide in 700 ml water
- RNase T1
- Gel solution: 200 ml of the gel stock solution, 11 ml of 1 M Tris-borate buffer pH 8.2, 11 ml water, freshly prepared 1 ml of 10% ammonium persulfate, 100 μl of TEMED
- 22% polyacrylamide gel made up in gel solution
- Running buffer: 50 mM Tris-borate pH 8.2

Method

1. Incubate ^{32}P-labelled purified RNA with RNase T1 in TE buffer at 37 °C for 1 h.[a]
2. Add the same volume of the loading solution.
3. Load on to the 22% polyacrylamide slab gel.
4. Run at 650 V for 16 h. Alternatively, apply the sample on to a sequencing gel containing a high concentration of urea.

Protocol 8 continued

5. After electrophoresis, expose the gel exposed to X-ray film without drying. If the sample is separated on sequencing gels, fix the gels and then dry under a vacuum. Expose the dried gel to an X-ray film.

[a] Two units of RNase T1 completely digest 10 μg of RNA at 37 °C for 20 min, or 1 unit of RNase T1 completely digests 20 μg of RNA at 37 °C for 2–3 h incubation.

3.5 Northern blot hybridization

Preliminary structures for mRNAs can be derived from Northern blot hybridization using different RNA probes. Both cDNA probes labelled by random priming and RNA probes made by *in vitro* transcription are useful for this purpose. The use of oligonucleotides will further enhance the ability to detect minor nucleotide differences between different RNA species. Usually, the total cytoplasmic RNA can be used for Northern blot analysis. However, if the relative amount of virus RNA is not sufficiently high, the background may be very high. In this case, poly(A)$^+$ RNA should be used for hybridization.

Significant progress has been made using non-radiolabelled probes in nucleic acid hybridization procedures, including Northern blot analysis. We found that the sensitivity of Northern blot analysis using the DIG-labelling and detection system (Boehringer-Mannheim (now Roche Diagnostics)) is nearly 100 times more sensitive than conventional Northern blot analysis using random-primed probes. The protocol of sensitive Northern blot analysis is included in the commercially available DIG-labelling kit. The following section describes Northern blot hybridization using a radiolabelled oligonucleotide probe.

Protocol 9
Hybridization using oligonucleotide probes

Equipment and reagents

- Equipment and reagents for formaldehyde gel electrophoresis (see *Protocol 3*)
- UV-crosslinking apparatus (e.g. Stratagene, Stratalinker 2400)
- 50 mM NaOH
- 1 M Tris–HCl pH 7.2
- 20 × SSC: 3 M NaCl, 0.3 M sodium citrate
- [^{32}P]5′ end-labelled oligonucleotide
- Sheared salmon sperm DNA
- Hybridization buffer: 5 × SSC, 0.1% Ficoll, 0.1% polyvinylpyrrolidone, 0.1% bovine serum albumin, 50 mM sodium phosphate pH 7.0, 1% SDS
- Yeast tRNA (Sigma, type X)
- Nylon hybridization membrane (cut into 10 mm × 7 mm or 14 mm × 12 mm pieces) (e.g. Biodyne, ICN Radiochemicals)
- Membrane washing buffer: 2 × SSC
- 50 °C or 65 °C water bath

Protocol 9 continued

Method

1. Separate RNA molecules by formaldehyde gel electrophoresis as described in *Protocol 3*.
2. After the electrophoresis, soak the gel in 200 ml of 50 mM NaOH for 20 min with gentle agitation.[a]
3. Soak the gel in 200 ml of 100 mM Tris–HCl pH 7.2 for 10 min.
4. Soak the gel in 200 ml of 20 × SSC for 10 min.
5. Transfer the RNA samples from the gel to the nylon membrane in 20 × SSC and leave overnight at room temperature.
6. Crosslink the RNAs to the nylon membrane using a UV-crosslinker.
7. Prepare 15 ml (for 10 mm × 7 mm membrane) or 25 ml (for 14 mm × 12 mm membrane) hybridization solution containing 1 × hybridization buffer, 100 μg sheared salmon sperm DNA, and 50 μg/ml of yeast tRNA
8. Place the crosslinked membrane in a plastic bag and pour in the hybridization solution. Seal the bag and incubate the membrane in a 50 °C or 65 °C water bath for 4–5 hours.
9. Add 8–10 ml (10 mm × 7 mm membrane) or 10–15 ml (14 mm × 12 mm membrane) of [^{32}P]5′ end-labelled oligonucleotide directly into the hybridization solution in the bag.
10. Incubate overnight in the 50 °C or 65 °C water bath.
11. Prepare 600 ml of the washing solution (2 × SSC)
12. Pour off the hybridization solution and place the membrane in 200 ml of the washing solution. Wash the membrane on a shaker for 30 min at room temperature. Repeat this wash once more.
13. For the third wash, put the membrane in 200 ml pf the washing solution in a water bath, at the hybridization temperature.
14. Pour off the washing solution and air-dry the membrane for 3–5 minutes.
15. Wrap the membrane in a plastic wrap and expose it to an X-ray film.

[a] This treatment degrades large molecular weight RNAs into smaller RNAs, which are transferred more efficiently than very large RNA molecules to the hybridization membrane.

4 The use of RNA reporter constructs for transcriptional assays

To manipulate virus RNA sequences for transcriptional analysis, a full-length cDNA copy of virus RNA is frequently used. The cDNA vector typically contains a T7 promoter in front of the virus cDNA sequences, allowing a complete virus RNA to be transcribed *in vitro* by T7 RNA polymerase. This *in vitro* transcribed RNA can be infectious, i.e. when transfected into cultured cells it leads to RNA replication and the complete process of virus replication, sometimes resulting in the production of virus particles. Such an infectious RNA and its parent cDNA construct can be used for mutational analysis. These types of infectious RNAs and cDNA constructs are available for many viruses, e.g. poliovirus, Sindbis virus, Dengue virus, hepatitis C virus, etc.

This approach, however, is not yet feasible for all RNA viruses, particularly those with large, segmented or negative-strand RNA genomes. An alternative approach for these viruses is the use of a defective-interfering (DI) RNA construct, which allows analysis of the *cis*-acting sequences involved in RNA transcription, replication, and recombination. Even for those viruses for which an infectious cDNA (RNA) is available, such DI RNA constructs are powerful tools for the analysis of the various *cis*-acting RNA sequences, because these types of DI RNA constructs are smaller and thus easier to use for mutational analysis than a full-length virus RNA. Also, the DI RNA sequences will be separate from the virus genome, which provides all the gene products necessary for RNA synthesis. Thus, the *cis*-acting RNA sequences are independent of the virus protein, enabling unequivocal analysis of the RNA sequences.

The DI RNA typically contains the 5′- and 3′-ends of the virus genomic sequences. The minimum sequences required for its infectivity vary with the virus RNA. Sometimes, there are other sequences derived from the internal region of the virus genome. There are also other variations of the end sequence, such as the inverted 5′- or 3′-ends. Because these DI RNAs lack all the genes encoding the gene products required for RNA synthesis, they cannot replicate unless a full-length, wild-type virus RNA (termed 'helper RNA') is also present. The helper RNA provides the machinery and enzymes *in trans*, while the DI RNA provides the *cis*-acting signals for DI RNA replication. DI RNA has a significant replication advantage over the helper RNA in that it often becomes the predominant RNA species after a few passages. Thus, it can be amplified much faster than the wild-type RNA, an advantage in studies of RNA transcription. Another advantage of DI RNA is its small size, which makes it easier to perform genetic alterations of the RNA.

A reporter gene can also be inserted into the RNA to facilitate analysis of RNA replication or transcription. Chloramphenicol acetyltransferase (CAT), luciferase (Luc), or green fluorescent protein (GFP) are some of the most commonly used reporter genes. Some DI RNAs contain an endogenous ORF, which can be translated either directly or after transcription of a mRNA. In these cases, the reporter genes can be constructed as a fusion protein with the virus coding sequences, so that they can be expressed using the endogenous regulatory elements for transcription and/or translation of the virus proteins. Alternatively, the reporter gene can be placed behind an inserted transcription promoter, so that a separate mRNA can be transcribed from the DI RNA, allowing the expression of the protein. These reporter genes thus allow the determination of replication or transcription of the DI RNA.

Typically, the DI RNA sequence is placed behind a T7 or T3 RNA polymerase promoter. The RNA transcript is transfected into the virus-infected cells immediately after virus infection. The virus RNA serves as the source of all the virus *trans*-acting factors. The DI RNA synthesis is then analysed at various timepoints after infection by the methods described above or by assaying the reporter genes.

It should be noted that these DI RNAs allow relatively accurate determination of the *cis*-acting sequences required for virus RNA synthesis. However, the sequence requirement of RNA synthesis is very often affected by the overall RNA structure. Thus, the results obtained from the DI RNA study may not necessarily reflect those of the natural virus RNAs.

Some DI RNA constructs may require 5′- and 3′-end sequences that exactly match those of the natural virus RNAs. The presence of additional nucleotides at both ends may interfere with virus RNA synthesis. This is especially true for (–)strand RNA viruses. In these cases, the precise 3′-end can often be generated by adding the hepatitis delta virus ribozyme to the 3′-end of the RNA. This ribozyme can cleave RNA at the site precisely upstream of the ribozyme domain.

Another commonly used method for RNA transcriptional analysis is to transfect the cDNA construct of DI RNAs, which contain a T7 RNA polymerase promoter. The transfected cells are then infected with a recombinant vaccinia virus encoding T7 RNA polymerase, which will transcribe the DI RNA *in vivo*. The cells are further infected with the wild-type virus of the DI RNA. This approach yields a large amount of DI RNA in the cells, allowing detection of a low level of RNA synthesis. This method has been used for various virus systems, particularly negative-stranded RNA viruses. However, this method is more suitable for analysing the virus gene products than studying the mechanism of RNA synthesis, since T7 polymerase-mediated transcription often causes RNA recombination and other unexplainable RNA artefacts.

4.1 RNA transcription *in vitro*

In this section, we describe *in vitro* RNA transcription and RNA transfection using cloned mouse hepatitis virus (MHV) DI RNA as an example. MHV is a prototype coronavirus, which contains a non-segmented, single-stranded, positive-strand RNA genome. The 5′-end and 3′-end of virus RNA have a cap structure and poly(A) sequence, respectively. The entire sequence of MHV DI RNA is inserted downstream of the T7 promoter in a plasmid. A unique restriction site is placed downstream of the poly(A) sequence of MHV DI cDNA. To produce RNA transcripts *in vitro*, plasmid DNA is first digested with a unique restriction enzyme downstream of the poly(A) sequence. MHV DI RNA transcripts are synthesized *in vitro* using bacterial T7 polymerase from the linearized plasmid. It is expected that *in vitro* synthesized RNA is capped, as an RNA cap analogue is included in the transcription reaction. Extra non-MHV sequence is present at the very 3′-end of the transcripts, yet these extra nucleotides do not affect MHV DI RNA replication (5). If precise very 3′-end termini are required for RNA transcription *in vitro*, the virus cDNA sequence is inserted upstream of the hepatitis delta virus ribozyme sequence (6).

Protocol 10

In vitro transcription of a DI RNA construct

Equipment and reagents

- Transcription-optimized 5 × buffer (Promega)
- 100 mM DTT
- 40 U/μl recombinant RNasin® ribonuclease inhibitor (Promega)
- 10 mM each ATP, CTP, UTP, and GTP
- 0.5 mM GTP
- 1 mg/ml BSA
- 2.5 mM RNA cap analog (New England BioLabs)
- 1–5 μg/3 μl linearized template DNA (approximately)
- 20 U/μl T7 RNA polymerase (Promega)
- 1 U/μl RQ1 RNase-free DNase (Promega)
- Phenol:chloroform:isoamyl alcohol (50:50:1)
- Chloroform
- 5 M ammonium acetate
- Ethanol (100% and 70%)
- RNase-free distilled water

Method

1. Mix the following components in the order listed below:

10 μl transcription-optimized 5 × buffer	
100 mM DTT	5 μl
recombinant RNasin® ribonuclease inhibitor1	1 μl
10 mM ATP	5 μl
10 mM UTP	5 μl
10 mM CTP	5 μl
0.5 mM GTP	5 μl
1 mg/ml BSA	5 μl
2.5 mM RNA cap analog	5 μl
~ 1–5 μg/3 μl linearized template DNA[a]	3 μl
20 U/μl T7 RNA polymerase	1 μl
Water	to 50 μl

2. Incubate at 37 °C for 1 h.
3. Add 5 μl of 10 mM GTP and 1 μl of T7 RNA polymerase.
4. Incubate at 37 °C for 60 minutes.
5. Add 2 μl of RQ1 RNase-free DNase to digest the template DNA.
6. Incubate at 37 °C for 15 minutes.[b]
7. Add 100 μl of RNase-free distilled water.
8. Extract with an equal volume (150 μl) of phenol:chloroform:isoamyl alcohol. Vortex vigorously for 1 minute and centrifuge in a microcentrifuge at 12 000 g for 2 minutes.

Protocol 10 continued

9. Transfer the upper, aqueous phase to a new tube and add an equal volume (150 μl) of chloroform. Vortex for 1 minute and centrifuge in a microcentrifuge at 12 000 g for 2 minutes.
10. Transfer the upper, aqueous phase to a new tube.
11. Add 100 μl of 5 M ammonium acetate and 625 μl of 100% ethanol. Mix well and place at −70 °C to −80 °C for 20 minutes or at −20 °C overnight.
12. Centrifuge at 12 000 g for 15 minutes.
13. Carefully pour off the supernatant and wash the pellet with 70% ethanol.
14. Dry the pellet under vacuum.
15. Dissolve the RNA in 20–40 μl of distilled water or TE buffer.

[a] DI RNA expression vector should be linearized by digestion with an appropriate restriction endonuclease. Do not use restriction enzymes that produce 3′-protruding ends, because extraneous transcripts have been reported to appear in addition to the expected transcript when such templates are transcribed. After restriction enzyme digestion, examine complete linearization of the DNA by agarose gel electrophoresis.

[b] RNA can be examined by non-denaturing agarose gel electrophoresis.

4.2 RNA transfection of DI RNA construct and helper virus infection

RNA transfection is commonly used to introduce MHV DI RNA transcripts into cells. Although various RNA transfection procedures are described in the literature, the lipofection procedure is used most commonly. The lipofection procedure was initially described for DNA transfection (7), and later it was shown that lipofection is also useful for RNA transfection. In lipofection, nucleic acids are mixed with liposomes in a test tube. Negatively charged nucleic acids bind to the surface of liposomes. When the mixture of liposomes and nucleic acid is added to the cells, they attach to the cell surface and nucleic acids are taken up by endocytosis. In many cases, serum inhibits the attachment of the complex of lipofection reagent and nucleic acid to the cell surface. Accordingly, cells are usually washed extensively to remove serum prior to lipofection. Many lipofection reagents are commercially available. RNA transfection efficiency is affected by various factors, including the density and types of cells, type and amount of lipofection reagent, and amount of RNA. The best transfection condition needs to be determined from several pilot experiments, in which various parameters, e.g. cell density, source, amount, and concentration of reagent, are changed. Electroporation is also used for high-efficiency transfection. Again, the condition of electroporation that provides high-efficiency needs to be tested for each cell line.

We describe a typical RNA transfection procedure using transfection of MHV DI RNA as an example. As described above, replication of helper virus is necessary to provide all *trans*-acting elements for DI RNA replication. We usually transfect MHV DI RNA into MHV-infected cells, although DI RNA replication also occurs efficiently when cells are transfected with DI RNA first and then infected with the helper virus.

Protocol 11
Lipofection

Equipment and reagents

- *In vitro* transcribed DI RNA
- 60 mm plate of subconfluent (approximately 80%) cells
- MEM minimal essential medium (Gibco-BRL) supplemented with 2 mM L-Glutamine 2.2 g/l sodium bicarbonate and 0.1 mg/ml kamanyian sulphate (Gibco-BRL), pH 7.2
- Helper virus
- Polystyrene tubes
- 29.5 g/l tryptose phosphate broth (TPB, Difco)
- Fetal calf serum (FCS)
- 1 mg/ml lipofectin® reagent (Gibco-BRL)
- Complete culture medium: 88% MEM, 10% TPB, 2% FCS

Method

1. Wash the cells with pre-warmed MEM and infect helper virus at 5–10 m.o.i.
2. Rock the cell plate every 15 minutes during the virus adsorption period.
3. During helper virus adsorption, prepare the transfection mixture as followings. Resuspend *in vitro* transcribed DI RNA in 20–40 μl of distilled water and add pre-warmed MEM to the RNA solution up to 250 μl (Mix A) in a polystyrene tube.
4. Prepare Lipofectin®reagent mixture. Gently mix 7.5 μl of Lipofectin®reagent with 242.5 μl of pre-warmed MEM (Mix B) in a polystyrene tube.
5. At 15 minutes prior to transfection, add Mix A to Mix B (Mix C) and incubate at room temperature.
6. After 60 minutes' adsorption of helper virus to the cells, aspirate the supernatant from the cells and wash the cells with pre-warmed MEM two or three times.
7. Overlay Mix C on to the cells and spread it well on to the cell surface.
8. Incubate at 37 °C for 60 minutes. During incubation, rock the cells every 15 minutes.
9. Add 1.5 ml of pre-warmed MEM and incubate at 37 °C for 60 minutes.
10. Aspirate the supernatant and add 5 ml of complete culture medium.
11. Incubate at 37 °C until virus harvesting time.

4.3 DNA transfection and vaccinia virus infection

This section describes the expression of MHV DI RNA using DNA transfection and infection of recombinant vaccinia virus expressing T7 polymerase. In this system, MHV DI RNA is transcribed from transfected plasmid DNA, in which the MHV DI cDNA sequence is placed downstream of the T7 promoter, by expressed

T7 polymerase, which is encoded in recombinant vaccinia virus vTF7-3 (8). Because vaccinia virus carries an RNA capping function, synthesized MHV DI RNA is capped. Usually plasmid DNA is not linearized for DNA transfection. Instead, the T7 terminator is inserted downstream of the virus cDNA sequence; transcription is terminated at the T7 terminator in transfected cells. Insertion of the hepatitis delta virus ribozyme downstream of virus cDNA is also used to produce virus RNA with a precise 3′-end. (Recombinant vaccinia virus, vTF7-3 is available from Dr Bernard Moss, at the National Institutes of Health, Bethesda, MD, USA.)

Protocol 12

DNA transfection and expression of DI RNA by recombinant vaccinia virus

Equipment and reagents

- DMEM (high glucose)
- FCS
- MEM/FCS mix: 97.5% MEM (*Protocol 11*), 2.5% FCS
- Lipofectin reagent (Gibco-BRL)
- Vaccinia virus (> 5×10^8 p.f.u./ml titre
- 40 mg/ml cytosine-β-D-arabinofuranoside (Sigma)
- Clear-capped polystyrene tubes
- Humidified 3% CO_2 incubator

Method

1. Pre-warm the cell growth media to 37 °C.
2. Aspirate the media and wash the cells once with 5 ml of pre-warmed MEM.
3. Dilute the vTF7-3 and inoculate the cells. Use an m.o.i. of vTF7-3 higher than 5.
4. Incubate the infected cells in a 37 °C CO_2 incubator for 1.5–2 hours, rocking the plates every 15–20 min.
5. Pre-warm the MEM and DMEM (high glucose) in a 37 °C water bath
6. Prepare the DNA transfection mixture in clear-capped polystyrene tubes. Mix 20 μl of the Lipofectin reagent, 8–10 μg of DNA in DMEM (total volume 400 μl). Let the tubes stand at room temperature for 10–15 min.
7. While the transfection mixture incubates, aspirate the inoculum from each plate and wash with pre-warmed MEM three times.
8. After the third wash, aspirate the MEM completely and add the transfection mixture.
9. Let the plates sit for 10 min at room temperature, and add 1.6 ml of pre-warmed DMEM to each plate.
10. Incubate the plates in the 37 °C CO_2 incubator for 2–4 hours.
11. Aspirate the inoculum and wash the cells with pre-warmed MEM.
12. Add the helper virus (in this case MHV) to each plate.

Protocol 12 continued

13. Incubate the plates in the 37 °C CO_2 incubator for 1 hour. Rock the plates every 15 min.
14. Prepare the virus infection media (5 ml per plate).
15. After 1 hour, aspirate the inoculum and add MEM/FCS mix. Also add 5 μl of AraC to each plate.
16. Incubate the plates in the 37 °C CO_2 incubator for 10–12 hours.
17. At 10–12 hours' post-helper virus infection, extract intracellular RNA or harvest virus from the supernatant.

4.4 CAT reporter assay system

In some cases, replicating virus RNA, DI RNA, or virus mRNA encode reporter genes. Quantitation of a reporter-gene expression indicates the amount of virus RNA encoding the reporter gene. A CAT reporter assay system is described in this section.

Protocol 13

CAT reporter assay

Equipment and reagents

- PBS, pH 7.4
- NTE buffer: 0.1 M NaCl, 0.01 M Tris–HCl pH 7.2, 1 mM EDTA
- 0.25 M Tris–HCl pH 8.0
- [^{14}C]Chloramphenicol (0.05 mCi/ml) (1 CN)
- 5 mg/ml n-butyryl CoA
- Thin-layer chromatography chamber, etc.
- Ethyl acetate
- Rubber policeman (e.g. Fisher)
- Dry ice/ethanol bath
- SpeedVac (Sorvall)

Method

1. Remove the medium from the cells. Wash the cells three times with PBS buffer. After the final wash, remove as much PBS as possible.
2. Add 1 ml of the NTE buffer/6 cm culture dish, and incubate the cells for 5 min at room temperature.
3. Scrape off the cells using a rubber policeman and transfer the cells to a microcentrifuge tube.
4. Centrifuge at 15 000 g for 1 minute at 4 °C.
5. Remove the supernatant and resuspend the pellet in 100 μl of 0.25 M Tris–HCl pH 8.0.

Protocol 13 continued

6. Subject the extracts to three rapid freeze/thaw cycles, vortexing the tube vigorously after each thaw cycle. Use a dry ice/ethanol bath for freezing, and a 37°C water bath for quick thawing the sample.
7. Heat only those lysates for use in the CAT assay at 60°C for 10 min to inactivate endogenous deacetylase activity.
8. Centrifuge the extracts at 15 000 g for 2 min. Transfer each supernatant to a separate fresh tube. (Store the extracts after this step at −70°C, if desired.)
9. Prepare the following reaction mixture in a 1.5 ml microcentrifuge tube: 50 μl of the cell extract, 3 μl of [^{14}C]chloramphenicol, 5 μl of n-butyryl Coenzyme A and 47 μl of 0.25 M Tris–HCl pH 8.0.
10. Incubate the reaction at 37°C for 30 minutes to 20 hours, depending on expected activity.
11. Briefly spin the tubes in a microcentrifuge.
12. Add 500 μl of ethyl acetate to each tube and mix by vortexing, then centrifuge at 15 000 g for 5 min at 4°C.
13. Transfer the upper solution to fresh microcentrifuge tubes and dry completely in a SpeedVac.
14. Add 20 μl of cold ethyl acetate/dry tube and wash down the tube sides.
15. Analyse by thin layer chromatography.

4.5 Detection and quantitation of minute amount of virus RNAs

4.5.1 Quantitative RT–PCR

Sometimes, Northern blot analysis is not sensitive enough to demonstrate or quantify minute amounts of virus RNAs. In the case of MHV, it is very difficult to detect MHV negative-strand RNAs by Northern blot analysis. We used quantitative RT–PCR to determine the amount of negative-strand MHV DI RNA (9). In quantitative RT–PCR internal control RNA (ICR) is used; although this has a similar structure as the virus RNA of interest, it contains a small deletion within the amplified region. ICR is synthesized by *in vitro* transcription of PCR products that contain the T7 promoter sequence at the 5′-end, or *in vitro* transcription of a linearized plasmid DNA that encodes ICR downstream of the T7 promoter. A known amount of ICR is mixed with an intracellular RNA sample containing a small amount of the virus RNA of interest. Usually, serially diluted ICR is mixed with the same volume of an aliquot of the intracellular RNA sample. The cDNA of the virus RNA of interest and that of the ICR are produced by incubating the RNA sample with a specific primer and reverse transcriptase. After cDNA synthesis, PCR products of both virus RNA and ICR are produced. Virus RNA-specific PCR products and ICR-specific products are co-amplified and the amount of both PCR products are independent of cycle number, or of the concentration of primers or dNTPs. After amplification, PCR products are separated by agarose gel electrophoresis and the amount of PCR produced from the virus RNA is compared with the amount of PCR products of ICR. Because the amount of ICR prior to RT–PCR is known, the amount of virus RNA prior to RT–PCR is easily estimated by comparing the amount of PCR products from virus RNA and ICR. In the following example, an approximately 80 nt-long deletion is introduced in ICR. Although we used an *in vitro* transcription kit and total RNA isolation kit from a company (Ambion) here, similar results are obtained by using the *in vitro* RNA transcription system and RNA extraction procedures described above.

Protocol 14

Quantitative RT–PCR

Equipment and reagents

- Linearized plasmid
- Phenol:chloroform (50:50)
- Intracellular RNA containing the virus RNA of interest
- 50–200 mM oligonucleotide
- 42 °C water bath
- 100 °C heat block
- UV spectrophotometer
- *In vitro* transcription kit (Ambion)
- PCR reagents including *Taq* DNA polymerase (Perkin-Elmer)
- Total RNA isolation kit (Ambion, Totally RNA™)
- 1 M KCl
- 1 M Tris–HCl pH 8.3
- 62.5 mM methyl mercury
- 100 mM $MgCl_2$
- 700 mM β-mercaptoethanol
- 100 mM DTT
- 12.5 mM dNTP mix (12.5 mM each of ATP, GTP, CTP, TTP)
- RNasin (Promega)
- AMV reverse transcriptase (Promega)
- RQ1 RNase-free DNase (Promega)
- 1 × PCR buffer: 10 mM Tris–HCl, 1.5 mM $MgCl_2$, 50 mM KCl, 0.1 mg/ml gelatin, pH 8.3

Method

1. Using the T7 *in vitro* transcription kit, synthesize ICR *in vitro* from 1 μg of a linearized plasmid.
2. Digest the plasmid DNA with RQ1 RNase-free DNase by incubating the sample at 30 °C for 2 hours.
3. Extract ICR by phenol/chloroform and precipitate RNA with ethanol. Suspend ICR in water and quantitate the amount of ICR using a UV spectrophotometer.
4. Make a series of two- to fivefold dilutions of ICR in microcentrifuge tubes.
5. Add the same amount of intracellular RNA, containing the virus RNA of interest, to all microcentrifuge tubes prepared in step 4. Ensure the total RNA in each tube is 6.72 μl.
6. Denature the RNAs by incubating the tubes on a 100 °C heat block for 4–5 min.

Protocol 14 continued

7. Immediately transfer to ice and keep for 2 min. Spin in a microcentrifuge 10 seconds at 16 000 g to collect all the liquid.
8. Add 1.28 μl of 62.5 mM methyl mercury to the supernatant. Mix and incubate at room temperature for 10 min.
9. Add the following reagents to the tube:

Distilled H_2O	22.5 μl
100 mM $MgCl_2$	5.0 μl
1 M KCl	5.0 μl
1 M Tris–HCl (pH 8.3)	2.5 μl
700 mM β-mercaptoethanol	2.0 μl
Oligonucleotide (50–200 mM)	1.0 μl
100 mM DTT	5.0 μl
12.5 mM dNTPs	5.0 μl
RNasin	1.0 μl
AMV reverse transcriptase	1.0 μl

10. Incubate the mix in a 42 °C water bath for 1 hour.
11. Heat in a 100 °C heat block for 2–3 min to inactivate reverse transcriptase.
12. Briefly centrifuge to collect all the liquid.
13. To clean up the cDNA, perform a phenol/chloroform extraction followed by ethanol precipitation. Alternatively, remove excess oligonucleotide and heat-inactivated enzyme by spin column chromatography.
14. Transfer a small volume (1–10 μl) of the sample containing single-strand cDNA to another tube that contains 1 × PCR buffer, 2.5 U *Taq* DNA Polymerase, and two primers (10 pmol). Ensure the final volume of the sample is 100 μl.
15. Perform PCR with the desirable temperature cycle profile.

4.5.2 RNase protection assay

In RNase protection assays, a labelled RNA probe (usually radiolabelled at one end of RNA or internally) is hybridized with the virus RNA of interest. After hybridization, the RNA–RNA hybrid is treated with single-strand specific RNases, e.g. RNase T1 and RNase A. These RNases digest single-strand RNA, but do not digest double-stranded RNA hybrids. The RNase-resistant portion of the probe is detected by gel electrophoresis. The RNase protection assay is significantly more sensitive than a conventional Northern blot assay; although we have not extensively quantified the sensitivity of the RNase protection assay, this assay appears to be at least 10 times more sensitive than a conventional Northern blot assay. We failed to detect minute amounts of MHV DI RNAs in early infection, yet we successfully demonstrated very low amounts of positive-strand MHV DI RNAs early in infection by the RNase protection assay described below (9).

Protocol 15
RNase protection assay

Equipment and reagents

- 6% sequencing gel
- Sequencing gel solutions and apparatus
- 80% formamide
- 40 mM Pipes
- 400 mM NaCl pH 6.4 and 100 mM NaCl
- 1 mM EDTA and 5 mM EDTA
- Solution A: 80% formamide, 40 mM Pipes, 400 mM NaCl, 1 mM EDTA
- 10 mM Tris–HCl pH 7.5
- 15 μg/ml RNase A
- 1 μg/ml RNase T1
- 10% SDS
- Solution B: 100 mM NaCl, 10 mM Tris–HCl pH 7.5, 5 mM EDTA, 15 μg/ml of RNase A, 1 μg/ml of RNase T1
- 5 mg/ml proteinase K
- Phenol saturated with 0.1 M Tris–HCl pH 8.0
- Phenol:chloroform (50:50)
- Ethanol
- Loading buffer: 90% formamide, 10 mM Tris–boric acid pH 7.0, 2 mM EDTA, 0.01% (w/v) xylene cyanol, 0.01% (w/v) Bromophenol Blue

Method

1. Prepare radiolabelled riboprobe (e.g. ^{32}P-labelled RNA probe by *in vitro* transcription of RNA from plasmid DNA or PCR products).
2. Extract intracellular RNA containing RNA molecules of the interest.
3. Incubate the extracted RNA at 100 °C for 2 min in microcentrifuge to denature RNAs.
4. Immediately place the tube on ice.
5. Incubate heated denatured RNAs with riboprobe in 30 μl of Solution A at 60 °C.
6. After 8 h incubation, add 300 μl of Solution B.
7. Incubate the sample for 10 min at 15 °C.
8. Terminate the RNase reactions by adding 10 μl of 10% SDS and 2 μl of 5 mg/ml proteinase K, followed by incubation at 37 °C for 15 min, phenol extraction, phenol/chloroform extraction, and ethanol precipitation.
9. Redissolve precipitated RNA in 5 μl of loading buffer and apply to a 6% sequencing gel.

References

1. Carmichael, G. G. and McMaster, G. K. (1980). *Methods in enzymology*, Vol. 65 (ed. L. Grossman and K. Moldave), p. 47. Academic Press, London.
2. Sambrook, I., Fritsh, E. F., and Maniatis, T. (ed.) (1989). *Molecular cloning: a laboratory manual* (2nd edn), p. 7.34. Cold Spring Harbor Laboratory Press, NY.
3. Makino, S., Taguchi, F., and Fujiwara, K. (1984). *Virology*, **133**, 9.

4. Langridge, J., Langridge, P., and Bergquist, P. L. (1980). *Anal. Biochem.*, **103**, 264.
5. Makino, S., Joo, M., and Makino, J. K. (1991). *J. Virol.*, **65**, 6031.
6. Joo, M., Banerjee, S., and Makino, S. (1996). *J. Virol.*, **70**, 5769.
7. Felgner, P. L., Gadek, T. R., Holm, M., Roman, R., Chan, H. W., Wenz, M., Northop, J. P., Ringgold, G. M., and Danielson, M. (1987). *Proc. Natl Acad. Sci. USA*, **84**, 7413.
8. Fuerst, T. R., Niles, E. G., Studier, F. W., and Moss, B. (1986). *Proc. Natl Acad. Sci. USA*, **83**, 8122.
9. An, S., Maeda, A., and Makino, S. (1998). *J. Virol.*, **72**, 8517.

Chapter 4

Analysis of RNA virus-encoded proteinases

M.D. Ryan, M. Flint, M.L.L. Donnelly, E. Byrne, and V. Cowton
School of Biology, University of St. Andrews, Centre for Biomolecular Sciences, North Haugh, St. Andrews KY16 9ST, U.K.

1 Introduction

An extremely common strategy of positive-strand RNA viruses is to encode some, if not all, of their proteins in the form of polyproteins. Picorna-, flavi-, and potyviruses, for example, encode all their proteins in a single, long, open reading frame (ORF). Other viruses may have monopartite genomes encoding more than one polyprotein (e.g. togaviruses). Multipartite genomes are very common in plant viruses and each of the different RNA strands may also encode a polyprotein (e.g. bymo-, como-, nepoviruses).

The proteolytic processing of polyproteins may be mediated by virus-encoded proteinases alone, or in combination with host-cell proteinases. Virus-encoded proteolytic activities may be present as the sole function of a discrete processing product, or may be an activity of a proteinase domain within a protein that possesses multiple biochemical activities. The proteolytic processing of the polyprotein may, therefore, result in the generation of:

- the proteinase as a discrete product (N- and C-termini both defined);
- a product with the proteinase domain at either the N- or C-terminus (one terminus of the proteinase defined); or
- a product in which the proteinase domain is internal (neither terminus defined).

Many techniques are available for studying the biochemical and structural properties of virus-encoded proteinases, some of which (e.g. high-throughput screening techniques) require equipment and resources not normally available in academic laboratories. Whilst it is not absolutely essential that a complete nucleotide sequence of the region suspected of containing a proteinase domain or that cDNA clones are available, progress is severely hampered if this is not the case. Many of the techniques described below make the assumption that these data and reagents are available.

This chapter will describe a broad range of techniques that can be used to characterize, clone, express, and assay virus-encoded proteinases. Excellent reviews are available that discuss the organization of virus genomes, virus-encoded proteinases, and the techniques that have been used in their analysis (1–12).

2 Defining the proteinase type

2.1 Inhibitor studies

Proteinases have four major catalytic mechanisms classified on the identity of the nucleophilic species: serine, cysteine, and acidic or metalloproteinases. The classical method of identifying which type of proteinase is being studied is that of inhibitor 'profiling'. The activity of proteinases purified from virus-infected cells or heterologous expression systems can be assayed in the presence of inhibitors of each class of proteinases. This presupposes, however, that a suitable assay system for the proteinase is available. The substrate for the assay may be a region of a polyprotein containing a suitable cleavage site or a synthetic peptide. Polyprotein-type substrates may be produced by purification of cloned/(bacterially) expressed material, or be produced using an *in vitro* translation system (see Section 3.2 and *Protocol 2*). In the latter system very small amounts of substrate are synthesized, but they can be labelled to produce a sensitive assay (see *Protocol 2*). An important note here is that experiments using a region of a polyprotein as the substrate may proceed with only an approximate knowledge of the polyprotein cleavage site.

In contrast, assays based on the (inhibition of) proteolysis of peptidic substrates require prior identification of the amino acid sequence of the cleavage site, and the flanking sequences, to design suitable peptide substrates. The quantities of proteinase required for these types of characterization demand purification from virus-infected cells, or expression in bacteria (Section 4). The cleavage of peptides can be monitored by standard chromatographic techniques, or peptidic substrates may be modified to produce chromogenic, fluorometric, or radiochemical assays (13–16).

2.2 Sequence analysis

An important tool at the outset of such a programme of work is a desktop computer able to access databases. This is not a trivial point, since not only was the presence of many virus-encoded proteinases/proteolytic domains predicted by sequence analysis and alignments, but in many cases the identities of specific residues involved in catalysis were also accurately predicted (reviewed in refs 7, 9–11).

2.2.1 Sequence motifs

Unlike many other enzymes (e.g. helicases, polymerases), in general, proteinases possess very few sequence 'motifs' to facilitate their identification from an inspection of the primary sequence data. A number of short motifs are found that are characteristic for certain proteinase types (e.g. -DT/SG-aspartyl proteinases; -GxSG-serine proteinases) and can be found by the use of a simple text editor. Superb resources in this respect are the MEROPS database of peptidases (http://www.bi.bbsrc.ac.uk/world/Labs/peptidase/merops.htm) and the sequence alignments given in ref. 12.

2.2.2 Sequence alignments

Alignments may be performed between sequences thought to encode a virus proteinase and the rapidly expanding database of virus-encoded proteinase sequences. Modern desktop computers are able to perform quite large multiple sequence alignments using public domain software such as CLUSTAL X, PHYLIP, MACAW, and GENEDOC (available for both Macintosh and Windows operating systems). When aligning a proteinase domain with other virus or cellular enzyme sequences it is recommended that alignments are performed varying the lengths and regions of the test sequence as some algorithms produce 'sub-optimal' alignments if extensive N- or C-terminal extensions of the test sequence are present. MACAW is particularly useful in searching for shorter 'blocks' of similarities between sequences. It should be noted, however, that it may prove difficult to achieve a 'meaningful' alignment, particularly for proteinases such as the thiol or cysteine proteinases which may show very low overall similarities.

The programs we routinely use (10, 11) for determining the relatedness of aligned sequences are PROTDIST/PROTPARS (PHYLIP suite of programs) and TREECONW, although many other programs are available. If an alignment with the sequence of another (characterized) proteinase can be made, this may indicate which residues may be involved in catalysis or substrate-binding, the importance of which may be confirmed by site-directed mutagenetic experiments.

3 Delimiting the proteinase (domain)

RNA virus polyprotein processing may involve a number of different proteinases or proteinase domains. The potyvirus polyprotein, for example, is processed by three different proteolytic activities; the NIa proteinase (cysteine nucleophile in a serine proteinase chymotrypsin-like fold), the HC proteinase (cysteine nucleophile in a papain-like fold), and the P1 proteinase (serine proteinase; reviewed in refs 7 and 11). Polyprotein cleavages may occur in *cis* (co-translational, intramolecular) or in *trans* (post-translational, intermolecular). This section describes the methods that may be employed to differentiate between and characterize proteolytic activities.

3.1 Deletion/truncation analysis

Proteolytic activities can be mapped to regions of the polyprotein using a variety of methods. cDNA encoding the polyprotein may be cloned into transcription vectors and restriction enzyme (RE) sites throughout the cDNA used to produce linear template cDNA for 'run-off' transcription (see *Figure 1*, Panel E). The series

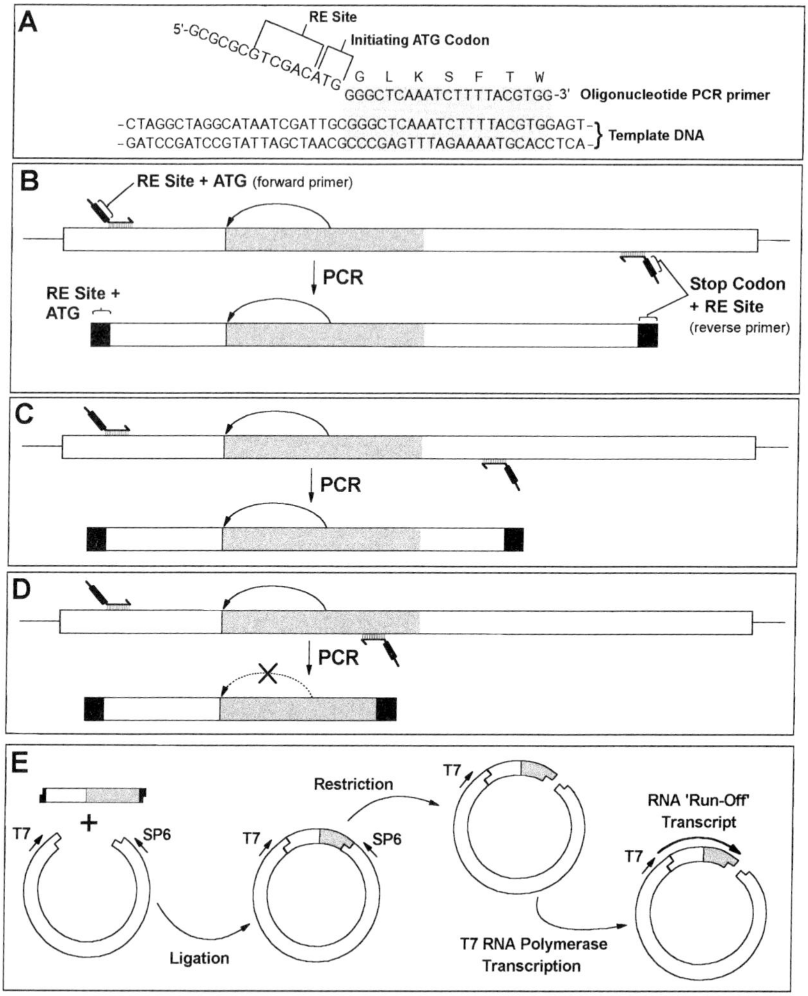

Figure 1 Proteinase analysis by truncation. The sequence of the 'forward' oligonucleotide PCR primer is shown. Sequences not complementary to the template contain a region that will allow subsequent restriction. A non-template translation-start codon is included. The region of the primer complementary to the template is highlighted (Panel A). The virus coding region (boxed area) and the proteinase domain (grey area) are shown. The curved arrow represents the N-terminal cleavage in *cis*. The position of the oligonucleotides relative to the proteinase domain are shown (arrows) together with the PCR product (Panel B). Successive truncations of the sequences (Panel C) will result in a proteolytically inactive product (Panel D). The PCR product is restricted and ligated into a similarly restricted transcription vector (opposing T7 and SP6 RNA polymerase promoters). The plasmid clone is linearized by restriction prior to transcription using T7 RNA polymerase (Panel E).

of truncated RNA transcripts may be used to programme *in vitro* translation systems (Section 3.2). The endogenous proteolytic processing properties of the truncated polyproteins is determined by the analysis of the translation profile by SDS-polyacrylamide gel electrophoresis (PAGE) and autoradiography (17, 18).

The use of the polymerase chain reaction (PCR) enables the (relatively) rapid construction of clones encoding defined regions of polyproteins into transcription vectors. The experimental strategy for the cDNA subcloning is described in *Protocol 1* and the analysis of the polyproteins is described in Sections 3.2 and 4. This strategy, shown in *Figure 1*, has the advantage that RE sites are created at desired positions, without relying upon the distribution of sites within the template sequence. Appropriate (naturally occurring) RE sites may also be used to produce other truncated forms of the protein. Successively shorter regions of the cDNA encoding the polyprotein are amplified, with a fixed N-terminal position, producing C-terminal truncations (see *Figure 1*, Panels B, C). The hypothetical proteinase domain illustrated cleaves at its own N-terminus in *cis*. As deletions extend into the proteinase domain its activity is destroyed (see *Figure 1*, Panel D). If the proteinase cleaves C-terminally, the cloning strategy would be reversed: successively shorter PCR products would be amplified, with a fixed C-terminal position, such that the polyprotein would be N-terminally truncated. Detailed protocols describing the individual molecular biology experimental steps are published elsewhere (19).

Protocol 1

Subcloning into transcription vectors

Equipment and reagents

- DNA sequence analysis software
- Thermocycling block
- Oligonucleotide PCR primers
- Template DNA
- Nucleoside triphosphate stock solutions (10 mM each)
- Horizontal agarose gel electrophoresis equipment and reagents
- DNA restriction/modification enzymes/thermostable polymerase (and buffers supplied by manufacturer)
- DNA purification kit
- Phenol:chloroform
- Ethanol:acetate
- Transcription vector(s)

A. Design of oligonucleotide primers

1. Inspect the DNA sequence for the presence and absence of restriction enzyme (RE) sites present in the transcription vector(s).[a] Include a translation start codon into the region of the forward primer which is not complementary to the template (*Figure 1*, Panel A). Similarly, include a translation stop codon between the last template codon and the RE site in the reverse primer.
2. Determine the length of the primer that is complementary to the template by the A/T G/C content of this region.[b]

B. Amplification of the cDNA

1. Amplify the cDNA using an appropriate PCR protocol (18), during which the non-template bases are added to the termini of the amplified product (see *Figure 1*,

Protocol 1 continued

Panel B). Use a single forward primer and a nested set of reverse primers to make a truncation series by amplifying successively shorter regions of the cDNA (see *Figure 1*, Panels B–D).

2. Resolve the PCR product from the template DNA and unincorporated primers by agarose gel electrophoresis.
3. Purify the PCR product from the agarose using a commercially available DNA purification kit. This step will remove the template DNA (usually a plasmid construct) which could give rise to 'background' clones.

C. Cloning the PCR product

1. Digest the gel-purified PCR product with the appropriate restriction enzymes.
2. Extract the doubly-restricted PCR product with phenol/chloroform (1:1) to remove protein, and then precipitate with 2.5 volumes of ethanol and 1/20 volume of 2M sodium acetate pH 5.6.[c]
3. Ligate the restricted PCR product into an appropriately restricted transcription vector (see *Figure 1*, Panel E).

[a] RE sites that are not within the region of cDNA to be amplified can be used in the oligonucleotide design to facilitate cloning. The choice of restriction endonuclease sites to be created will determine the orientation of the insert with respect to the T7 or SP6 RNA polymerase promoter sites within the transcription vector. The nucleotides located 5′ of the RE site are required for the endonuclease to cleave at the end(s) of the amplified DNA. Restriction enzymes differ in their requirements for the length of sequences flanking the RE site for it to be cleaved.

[b] It should be borne in mind that the expense of synthesis of extra bases (to increase complementarity) at this stage of the experimental design may save valuable time and consumables later on. In particular, check the primer for internal complementarity—particularly in the 3′ region

[c] It is not necessary to remove the very short, terminal restriction fragments by further gel electrophoresis and DNA purification.

3.2 Screening for proteolytic activity using translation systems *in vitro*

A rapid method of analysing the endogenous proteolytic properties of regions of polyproteins is by the use of translation systems *in vitro*. Two types of *in vitro* translation protocols may be employed; 'uncoupled', in which RNA is transcribed *in vitro* and subsequently used to programme an *in vitro* translation reaction (Section 3.2.1 and Protocol 2), or a 'coupled' transcription–translation reaction system (Section 3.2.2).

In vitro translation systems may be used to synthesize labelled polyprotein substrates, which can be used to assay proteinases expressed by other means (Section 3.2.3). The very small quantities of enzyme or substrate synthesized in these systems means that monomolecular reactions (cleavages in *cis*), which are not sensitive to dilution, are much more easily detected than bimolecular reactions (cleavages in *trans*). It is preferable, therefore, to analyse a polyprotein region comprising both the proteinase (domain) and a *cis* cleavage site. Suitable regions of the virus genome may be amplified by PCR and cloned into transcription vectors as described in *Protocol 1*. Purified virus RNA or infected-cell mRNAs may also be used to programme *in vitro* translation reactions—some types of vRNA translating very well, others very poorly. These systems may also be used for 'inhibitor profiling' experiments (Section 2.1).

Although of great utility, translation profiles derived from these *in vitro* systems should be interpreted with care. Translation products derived from internal initiation may be confused with proteolytic processing products. Internal initiation may be reduced by lowering the amount of transcript RNA used to programme the translation reaction (only possible using an uncoupled translation system) and by increasing the ion concentrations (20). The apparent absence of proteolytic activity may be due to a number of reasons. The active form of the enzyme may be a homodimer (e.g. HIV proteinase), the assembly being disfavoured under conditions where very little monomer is synthesized. A number of virus proteinases require Zn^{2+} as a structural, rather than catalytic, component, and insufficient levels of this ion in the translation system have been shown to severely affect activity. The redox potential of the system may also affect the activity of the enzyme: processing mediated by thiol proteinases may be enhanced by the addition of reducing agents (2.5 mM dithiothreitol). The proteinase domain of the NS3 protein of hepatitis C, flavi- and pestiviruses requires an 'activating' protein encoded elsewhere in the genome (reviewed in ref. 11). The absence of such activating co-factor sequences from cDNA constructs will severely depress proteinase activity.

3.2.1 Uncoupled translation systems

For the uncoupled *in vitro* translation systems, the transcription vector clone is restricted at the desired site (see *Figure 1*, Panel E) and the linearized template is used to programme the *in vitro* translation system (see *Protocol 2*). A critical step in this procedure is preparing RNase-free template DNA. Although the use of uncoupled systems requires additional steps (*in vitro* transcription/gel analysis of transcription reaction) this method allows the translation reaction to be programmed with defined amounts of transcript RNA. It is recommended that fresh transcription reactions are performed prior to translation since storage of transcription mixtures at $-20\,°C$ results in transcript degradation.

The results from individual steps involved in an *in vitro* transcription/translation experiment are shown in *Figure 2*. The region of the HRV14 genome-length cDNA encoding the P1 and P2 regions of the polyprotein was amplified using PCR. Unique restriction enzyme sites were created at each terminus together with a translation stop codon (see *Protocol 1*). The PCR product was doubly restricted and the restriction fragment ligated into a transcription vector. The plasmid DNA was linearized by restriction at the RE site created at the 3′ end of the construct (see *Figure 2*, Panel A). The purified, restricted, template was transcribed *in vitro* and the reaction analysed by agarose gel electrophoresis (see *Figure 2*, Panel B). An aliquot of the transcription reaction was used to programme an *in vitro* translation system and the translation products analysed by SDS-PAGE. The P2 region of the polyprotein contains a proteinase (2A), which performs a single polyprotein cleavage at its own N-terminus producing the P1 and P2 proteolysis products (see *Figure 2*, Panel C).

Protocol 2

Uncoupled transcription and translation *in vitro*

Equipment and reagents

- Linearized template DNA
- Phenol:chloroform
- Ethanol:acetate
- T7/SP6 RNA Polymerase
- 10 × concentration polymerase buffer (supplied by enzyme manufacturer)
- Stock solution of rNTPs (10 mM each)
- RNase-free water
- RNase-free 0.5 ml or 0.2 ml plastic reaction tubes
- RNase inhibitor
- Constant temperature apparatus (or use thermocycler block as in *Protocol 1*)
- Horizontal agarose gel apparatus and reagents (ethanol-sterilize the gel tank)
- Ethidium bromide
- Translation reaction components (supplied by manufacturer)
- Label ([^{35}S]methionine or -cysteine, [^{14}C]- or [^{3}H]amino acids, or biotinylated lysine)
- Protein gel apparatus and reagents
- Autoradiography/phosphorimaging equipment and reagents

A. Transcription *in vitro*

1. Keep all reaction components on ice. Make the final reaction volume to 20 μl. Perform all reactions with RNase-free plasticware and wear gloves. If possible, use 'dry' heating blocks for incubations as water baths rapidly become contaminated.
2. Phenol/chloroform extract the linearized template DNA twice and ethanol/acetate precipitate (using sterile tubes) as in Protocol 1. Redissolve in RNase-free water.
3. Add 1 μg of template DNA to the transcription reaction mixture containing:
   1 × transcription buffer (T7 or SP6, as appropriate)
   1 mM (each) rNTPs
   1 unit RNase inhibitor
   Make up to a volume of 19 μl with RNase-free water.
   Add 1 μl of the RNA polymerase.
4. Mix thoroughly by repeated pipetting (avoid frothing) and transfer to the heating block. Incubate at 30 °C for 1 h.
5. Transfer the reaction back on to ice during the gel analysis of the transcription reaction.

B. Analysis of the transcription reaction

1. Prepare a thoroughly clean, ethanol-sterilized, agarose gel tank and an agarose gel containing ethidium bromide (0.5 μg/ml) made using RNase-free water (prepare immediately prior to use). Prepare the sample loading buffer with equal care.

Protocol 2 continued

2. Load an aliquot of the transcription reaction (5 μl) on to the gel alongside an equivalent amount of linearized template DNA (control).
3. Illuminate the gel with UV.[a]

C. Translation *in vitro*

The transcription reaction may be used to programme an *in vitro* translation mixture directly, without purification of the transcript RNA.

1. Keep all the reaction components on ice and make sure the same precautions are taken against RNase contamination as described in Part A above.
2. Set up the translation reaction mixture according to the manufacturer's instructions. Programme the individual translation reactions with the volume of the *in vitro* transcription mixture containing 1 μg, 500 ng, and 250 ng of RNA transcript.
3. Transfer the translation reaction mixture to a heating block and incubate at 30 °C for 1 h.
4. Analyse the translation reaction by SDS-PAGE and determine the distribution of label using the detection/visualization techniques appropriate for the label chosen (see *Figure 2*, Panel C).

[a] The transcript RNA should be clearly visible as a discrete (not 'smeared') product in 5–10-fold molar excess over the template DNA (*Figure 2*, Panel B).

3.2.2 Coupled transcription/translation systems

Coupled *in vitro* translation systems (e.g. rabbit reticulocyte lysate, wheatgerm extract) combine the RNA transcription and translation reactions in one step. Plasmid DNA (1 μg) is used to programme the translation mixture without prior restriction. It is possible to use linear template DNA in these coupled systems, although the transcription/translation of this type of template DNA is noticeably better in wheatgerm extract than in the rabbit reticulocyte lysate. Detailed experimental protocols for both systems are supplied by the manufacturer (e.g. Promega).

It is strongly recommended that plasmid DNA purified using proprietary affinity columns should be subsequently phenol/chloroform-extracted twice, ethanol/acetate-precipitated, and dissolved in RNase-free water using the same precautions (RNase-free plasticware, gloves, etc.) as if it were RNA, rather than DNA.

3.2.3 Translation systems and interactions *in trans*

Although a very small quantity of translation product is synthesized in these systems, it can be labelled to (relatively) high specific activities. A single translation reaction (50 μl) may yield enough labelled substrate for 20 proteinase assays. If both proteinase and substrate are to be synthesized in these systems two methods may be used.

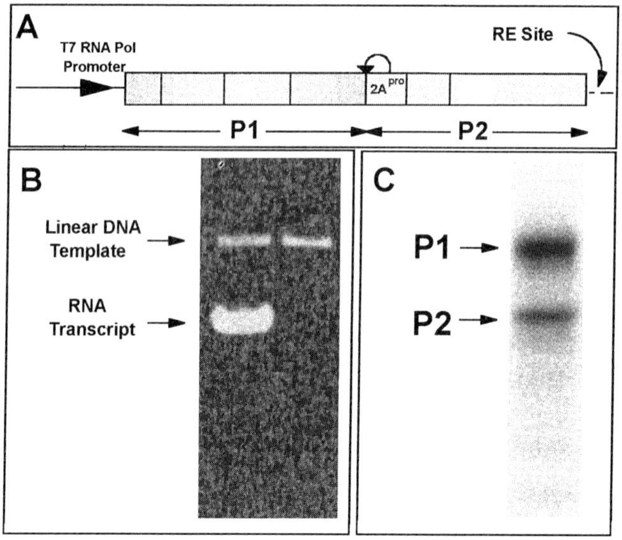

Figure 2 *In vitro* transcription and translation. Sequences encoding the P1–P2 region of the human rhinovirus 14 (boxed area) are cloned into a transcription vector. The curved arrow indicates the N-terminal polyprotein cleavage mediated by $2A^{pro}$. The plasmid is linearized downstream of the HRV sequences (Panel A) and used to programme an *in vitro* transcription reaction. The transcription reaction is analysed by agarose gel electrophoresis (left lane) and compared to an equal amount of template DNA alone (right lane). The discrete RNA transcript is in ~8-fold molar excess over the template DNA (Panel B). The transcript RNA is used to programme an *in vitro* translation reaction and the products analysed by SDS-PAGE and autoradiography. Only two major translation products are observed—the $2A^{pro}$ mediated cleavage products P1 and P2 (Panel C).

First, cDNA clones encoding the proteinase and the substrate (individually) are both used to programme a single translation reaction—a co-translation. Since all translation products become labelled, it should be ensured (by computer prediction of the molecular masses of the anticipated products) that any co-migration of the translation/processing products on SDS-PAGE gels does not hamper the analysis. Second, the cDNA clone encoding the proteinase is used to programme a translation mixture that is not supplemented with label. A closely similar reaction should be run in parallel, supplemented with label, to give a reasonable indication if the (unlabelled) reaction was successful. The cDNA clone encoding the substrate is used to programme another translation reaction, supplemented with label.

In each case, following the period of synthesis (1 hour), translation is arrested by the addition of cycloheximide to a final concentration of 800 μg/ml. The unlabelled proteinase translation reaction is then mixed with the labelled substrate reaction. Titration of one reaction against the other will determine the optimal *trans* processing mixture. The co-translated or combined reaction mixtures are overlaid with mineral oil (50 μl) and incubated at 30 °C for 16 hours to increase the yield of product from the bimolecular interactions. The translation/processing products are subsequently analysed as described in *Protocol 2*.

3.2.4 Rapid analysis of site-directed mutants

The analysis of the proteolytic properties of site-directed mutagenetic forms of the proteinase may be greatly accelerated by *in vitro* transcription using PCR products as the template DNA without recourse to molecular cloning (19) (see *Protocol 3* and *Figure 3*). Since the translation analysis is performed on a large population of molecules rather than a molecular clone, errors introduced by three PCRs is less problematic than one might think. One may, however, be unlucky enough to suffer from the 'founder-effect' (a mutation introduced early in the amplification process), which may affect the translational analysis. It is recommended, therefore, that the nucleotide sequence of the 'overlap' PCR product (from PCR-3, see *Figure 3*) be determined.

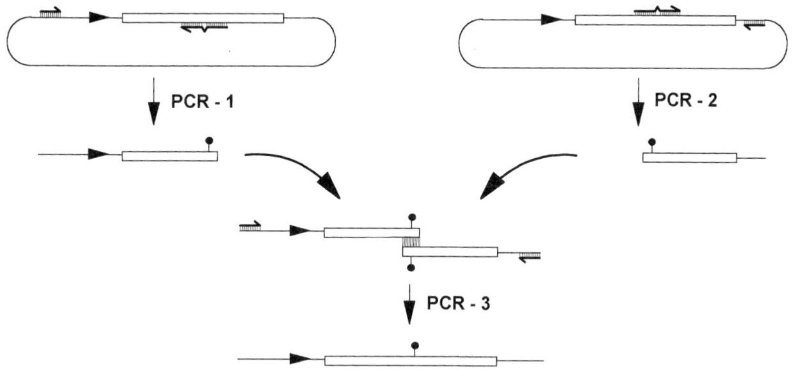

Figure 3 Overlap polymerase chain reaction. The positions of the oligonucleotide primers are shown (single arrow heads) in relation to the RNA polymerase promoter (arrow) and the region encoding the virus sequences (boxed area). The mutagenetic mis-match between the primers (annealing within the virus sequences) and the template are shown. The positions of the mutation introduced into the coding sequences by the PCR reactions are shown (circle). The (purified) products from reactions 1 and 2 are mixed and amplified using primers annealing to the vector sequences.

Protocol 3

Analysis of site-directed proteinase mutants

Equipment and reagents

- Plasmid clone of proteinase in a transcription vector
- PCR reagents and thermocycling block
- Oligonucleotide PCR primers
- Horizontal agarose gel electrophoresis equipment and reagents
- DNA restriction enzymes
- DNA purification kit
- Transcription reagents (see *Protocol 2*)
- Translation reaction mixture and reagents (see *Protocol 2*)

Method

1. In the first PCR reaction, design the forward oligonucleotide primer to anneal to the template DNA some 50 bp upstream of the RNA polymerase promoter site (e.g.

Protocol 3 continued

within the vector sequences). Design the reverse primer to contain 20 bases either side of the mutation site (PCR-1; see *Figure* 3).

2. In the second reaction, ensure the forward PCR primer is the complement of the reverse primer used in the first reaction. The reverse primer anneals to the template DNA immediately downstream of the proteinase insert sequences (PCR-2; see *Figure* 3).
3. Purify the PCR products from reactions PCR-1 and PCR-2 by agarose gel electrophoresis, followed by purification using a DNA purification kit.[a]
4. Mix equal quantities of PCR products (1 μg each) with the forward primer from PCR-1 and the reverse primer from PCR-2.[b]
5. Purify the full-length PCR product by agarose gel electrophoresis and a DNA purification kit.
6. Use the purified product (1 μg) to programme an *in vitro* transcription reaction.
7. Analyse the transcript by agarose gel electrophoresis and use an aliquot of the transcription reaction to programme an *in vitro* translation system (see *Protocol* 2).[c]

[a] Purification is essential since unincorporated primers must be removed.

[b] The two templates anneal (via the complementarity designed into the mutagenetic primers) and, following strand elongation primed by the other template, the full-length product is amplified (PCR-3; *Figure* 3).

[c] The purified product (1 μg) may be used to programme a coupled wheatgerm translation directly.

4 Bacterial expression

Expression of virus proteinases in bacterial systems offers many advantages over the *in vitro* systems, but has potential pitfalls. If the proteinase is required in quantities sufficient for more extensive biochemical (e.g. kinetic studies, screening for inhibitors, etc.) or biophysical analyses, then this would normally be the method of choice. Proteinases are, however, potentially toxic expression products, especially when expressed to the desired (high) levels. The toxicity of the expressed proteinase can be inferred by monitoring the OD_{650} throughout the bacterial growth/induction phases of the expression experiment (21). If the proteolytic processing is to be studied within the bacterium, monospecific antibodies directed against the anticipated cleavage products may be essential. An alternative is to create 'artificial' reporter protein or proteinase polyproteins (see Section 6).

4.1 Expression of inactive proteinase

Expression of an inactive form of the proteinase may overcome toxicity problems. The inactive enzyme expression product may be used to raise antibodies, or may be used to study the binding (rather than proteolysis) of other virus or cellular proteins. Certain virus proteinases also bind specific RNA structures within the virus genome (reviewed in ref. 10) and proteinase–RNA binding studies may be performed with a proteolytically inactive form of the protein.

The proteinase may be inactivated by the insertion or the deletion of coding sequences (whilst maintaining the open reading frame), although this may affect any other binding properties. A finer lesion may be introduced into the proteinase by the substitution of one of the active site residues (see Section 5). The literature contains numerous examples of this approach (reviewed in refs 7, 9–11), and has proved particularly valuable in the expression of virus proteinases designed to lead to protein structural studies. Commonly the active-site cysteine or serine residue is substituted by an alanine. The highly conserved stereochemistry of the active sites of these enzymes permits accurate modelling of the original nucleophile on to the resolved structure.

4.2 Affinity 'tagging'

Many systems are available to express proteins in *Escherichia coli* in the form of fusion proteins, or with N- or C-terminal peptide 'tags'. This has two potential benefits: an increase in the solubility of the expressed product; and it also provides a method whereby the expressed product may be purified via affinity chromatography. Many proteinases that function as part of a larger protein may tolerate C- or N-terminal fusion partners whilst retaining proteolytic activity.

The fusion partner used for proteinases that liberate either their own N- or C-terminus (cleavage in *cis*) should be chosen with care, since autoproteolysis of the fusion protein or 'trimming' of the peptide tag will defeat the purpose of the experiment. For proteinases that perform a C-terminal cleavage, the (N-terminal) fusion partners most widely used are glutathione S-transferase (GST) and maltose binding protein (MBP). Expression vectors are available for the construction of either N- or C-terminal histidine tags.

4.3 Insolubility of expressed proteinases

Picornavirus 3C proteinases when expressed in *E. coli* form inclusion bodies. The solving of the atomic structure of two 3C proteinases, HRV14 and hepatitis A virus, illustrates two methods of how this problem may be solved.

In the case of the rhinovirus 3C proteinase, the wild-type proteinase was expressed and this resulted in the formation of inclusion bodies. This insolubility 'problem' was used in a positive sense, in that purification of the inclusion bodies (a relatively straightforward procedure) leads to a considerable purification of the expressed product. The inclusion bodies were extracted with 6 M guanidine-HCl and 20 mM β-mercaptoethanol, the denaturant dissolving the proteinase. The denatured proteinase was then refolded by dialysis against 20 mM phosphate, 20 mM β-mercaptoethanol, 1 mM EDTA and purified using a number of chromatographic methods (22).

Expression of the hepatitis A virus proteinase followed a different strategy. In this case, a doubly-mutated form of the proteinase was created: Cys^{24}(Ser) and Cys^{172}(Ala), the latter mutation replacing the active site nucleophile. In this case a soluble product was expressed to high levels (10% of cellular protein) and could be easily purified (23).

5 Substrate specificity

Proteinases expressed in sufficient quantities to perform assays in *trans* may be analysed using two main methods. The purified proteinase is used to cleave synthetic peptide substrates corresponding to cleavage sites. The importance of residues flanking the scissile bond can be determined by residue substitutions and comparison of the K_m and V_{max} values obtained for the cleavage of each synthetic peptide (reviewed in ref. 5). Alternatively, site-directed mutagenesis may be performed on a suitable (labelled) protein substrate (see Section 3.2).

6 Artificial 'reporter' polyproteins

An alternative to raising monospecific antibodies against virus proteins, for monitoring proteolysis, is to create an 'artificial' polyprotein system using reporter proteins. This has the advantage that the proteinase activity can be monitored via the activity/property of the reporter protein(s) and antibodies against the reporter proteins may be commercially available. Copy DNAs encoding such polyproteins are assembled using overlap-PCR (see *Protocol 3*). Factors that must be taken into account in the design of oligonucleotide primers are:

(a) removal of stop codons (if present) from the N-terminal protein (e.g. GFP in the construct encoding GFP-$2A^{pro}$; see *Figure 4*, Panel A);

(b) inclusion of sequences flanking the proteinase cleavage site required to maintain its integrity as a suitable substrate (4–8 residues, depending somewhat upon proteinase type);

(c) there is sufficient complementarity between the initial PCR products to permit the 'overlap' reaction (see *Protocol 3* and *Figure 3*);

(d) creation of a stop codon at the end of the coding region if one is not present in the template sequence (e.g. $2A^{pro}$ in the construct encoding GFP–$2A^{pro}$; see *Figure 4*, Panel A);

(e) creation of RE sites at the ends of the clones (multiple sites may be created) to permit easier cloning into the target vector(s);

(f) construction of the clone using DNA sequence-analysis software and the need to check that a single open reading frame is maintained.

The L proteinase from foot-and-mouth disease virus (C-terminal cleavage in *cis*) was linked to green fluorescent protein (GFP) in precisely this manner. Similarly, the 2A proteinase from HRV14 (N-terminal cleavage in *cis*) was linked to

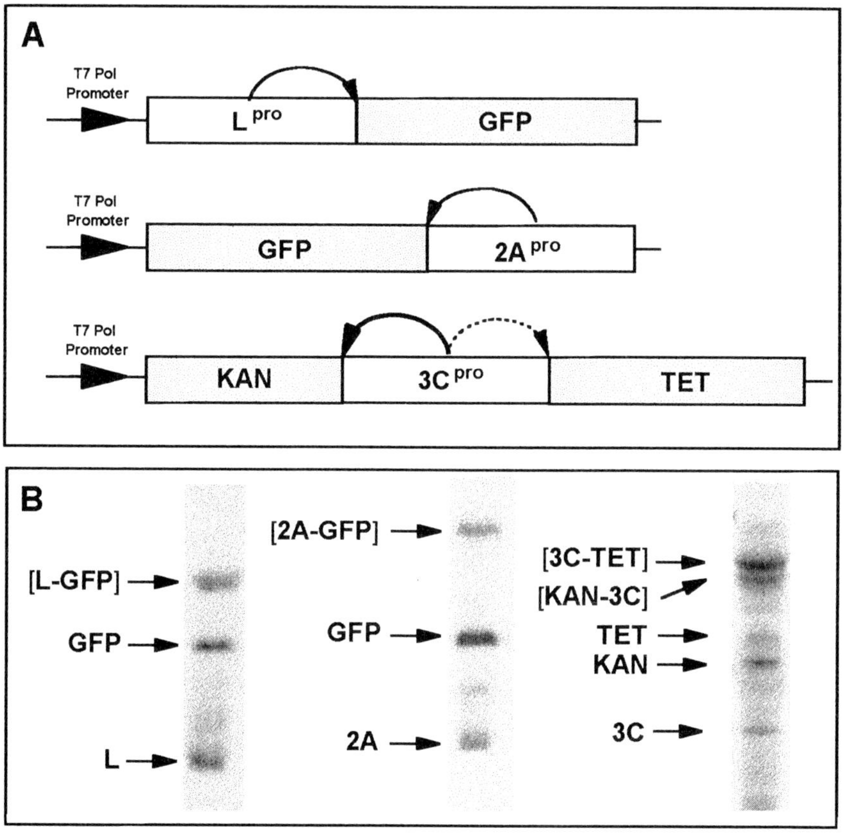

Figure 4 Reporter protein systems. Coding sequences within transcription vectors are shown (boxed areas). The cleavage activities of the virus proteinases L, 2A, and 3C are indicated by curved arrows. The reporter proteins green fluorescent protein (GFP), kanamycin resistance (KAN), and tetracycline resistance (TET) are used to create reporter protein–proteinase fusions (Panel A). Translation profiles derived from these constructs demonstrate the proteolytic properties of the proteinases in these systems. The identities of the cleavage products are shown, with uncleaved products included in square brackets (Panel B).

GFP (see *Figure 4*, Panel A). The 3C proteinase from HRV14 may cleave at both N- and C-termini, and this proteinase was linked to the kanamycin resistance protein (KAN) and the tetracycline resistance protein (TET; see *Figure 4*, Panel A). These constructs were analysed using coupled transcription/translation systems and the translation products analysed by SDS-PAGE and autoradiography (see *Figure 4*, Panel B). The L and 2A proteinases are cleaving themselves (in *cis*) from GFP as in their native polyprotein context, although uncleaved products are observed. The 3C proteinase is cleaving predominantly at its own N-terminus (uncleaved [3C-TET] is a major translation product, although the products of C-terminal cleavage are observed (uncleaved [KAN-3C]) together with the products from cleavage at both sites (KAN, 3C, and TET).

References

1. Goldbach, R. and Wellink, J. (1988). *Intervirology*, **29**, 260.
2. Wellink, J. and van Kammen, A. (1988). *Arch. Virol.*, **98**, 1.
3. Hellen, C. U. T., Krausslich, H-G., and Wimmer, E. (1989). *Biochemistry*, **28**, 9881.
4. Harris, K. S., Hellen, C. U. T., and Wimmer, E. (1990). *Semin. Virol.*, 1, 323.
5. Kay, J. and Dunn, B. M. (1990). *Biochem. Biophys. Acta*, **1048**, 1.
6. Palmenberg, A. C. (1990). *Ann. Rev. Microbiol.*, **44**, 603.
7. Dougherty, W. G. and Semler, B. L. (1993). *Microbiol. Rev.*, **57**, 781.
8. Koonin, E. V. and Dolja, V. V. (1993). *Crit. Rev. Biochem. Mol. Biol.*, **28**, 375.
9. Gorbalenya, A. E. and Snijder, E. J. (1996). *Perspect. Drug Discov. Design*, **6**, 64.
10. Ryan, M. D. and Flint, M. (1997). *J. Gen. Virol.* **78**, 699.
11. Ryan, M. D., Monoghan, S., and Flint, M. (1998). *J. Gen. Virol.*, **79**, 947.
12. Barrett, A. J., Rawlings, N. D., and Woessner, J. F. (ed.) (1998). *Handbook of proteolytic enzymes*. Academic Press, London.
13. Darke, P. L., Nutt, S. F., Brady, S. F., Garsky, V. M., Ciccarone, T. M., Leu, C. T., Lumma, P. K., Freidinger, R. M., and Sigal, I. S. (1988). *Biochem. Biophys. Res. Commun.*, **156**, 297.
14. Tomaszek, T. A., Magaard, V. W., Bryan, H. G., Morre, M. L., and Meek, T. D. (1988). *Biochem. Biophys. Res. Commun.*, **168**, 274.
15. Tamburini, P. P., Dreyer, R. N., Hansen, J., Letsinger, J., Elting, J., Willse, A. G., Dally, R., Hanko, R., Osterman, D., Kamarck, M. E., and Warren, H. Y. (1990). *Anal. Biochem.*, **186**, 363.
16. Hyland, L. J., Dayton, B. D., Moore, M. L., Shu, A. Y., Heys, J. R., and Meek, T. D. (1990). *Anal. Biochem.*, **188**, 408.
17. Ryan, M. D., Belsham, G. J., and King, A. M. Q. (1989). *Virology*, **173**, 35.
18. Harris, B. D. and Rickwood, D. (ed.) (1990). *Gel electrophoresis of proteins: a practical approach*. Oxford University Press, Oxford.
19. Glover, D. M. and Hames, B. D. (ed.) (1994). *DNA cloning 1: a practical approach—core techniques*. Oxford University Press, Oxford.
20. Dasso, M. C. and Jackson, R. J. (1989). *Nucl. Acids Res.*, **17**, 3129.
21. Glover, D. M. and Hames, B. D. (ed.) (1996). *DNA cloning 2: a practical approach—expression systems*. Oxford University Press, Oxford.
22. Mathews, D. A., Smith, W. W., Ferre, R. A., Condon, B., Budahazi, G., Sisson, W., Villafranca, J. E., Janson, C. A., McElroy, H. E., Gribskov, C. L., and Worland, S. (1994). *Cell*, **77**, 761.
23. Malcom, B. A., Chin, S. M., Jewell, D. A., Stratton-Thomas, J. R., Thudium, K. B., Ralston, R., and Rosenberg, S. (1992). *Biochemistry*, **31**, 3358.

Chapter 5

Detection and analysis of host gene targets for oncogenic retroviruses

J.C. Neil and A. Terry
Molecular Oncology Laboratory, University of Glasgow, Department of Veterinary Pathology, Bearsden, Glasgow G61 1QH, U.K.

1 Introduction

Unlike the DNA viruses, where oncogenic representatives are found in disparate families, only two RNA virus families, the Retroviridae and the Flavoviridae, have so far been implicated in neoplasia. The mechanisms by which the latter operate are largely unknown and the prototypic example, hepatitis C virus, appears to be an indirect cause of cancer, for which the roles of specific gene products are as yet undetermined. In contrast, the Retroviridae have received much attention as causes of cancer. Animal retroviruses have been studied for many years, mainly as model carcinogens; their chief contribution has been through the identification of oncogenes that play a role in cancers of diverse aetiology. This field of work is the main focus of this chapter. However, it should be noted that these agents can be important causes of disease in their natural host species, and that since the discovery of human retroviruses there is much wider interest in retrovirus pathogenesis and methods of their control.

Many retroviruses and related elements have become incorporated into chromosomal DNA of host germline cells and inherited in Mendelian fashion. Most of these are replication-defective viruses, which are incapable of spread within the host and are generally tolerated as harmless passengers. However, new integrations will occur in somatic cells if these agents are reactivated or spread to new hosts as infectious viruses, and the possibility exists that host cellular gene functions will be disrupted. If damage occurs to a sufficient number of critical host genes, cancer may develop. In short, this is a summary of our current view of the mode of action of the 'simple retroviruses' listed in *Table 1*. In the light of this model, an explanation for the variable oncogenic efficiency of different retrovirus families can be advanced. The factors favouring tumour development by this mechanism include constitutive replication to high levels and a lack of significant cytopathic effect on the infected cell. These features are shared by the potently oncogenic simple retroviruses in *Table 1*.

In contrast, the complex retroviruses of the HTLV/BLV group are relatively inefficient carcinogens and owe their oncogenic potential to virus regulatory

Table 1 Directly oncogenic retroviruses and known modes of action

Virus	Mode of action			Genome structure	Tumour type
	Insertional mutagenesis	Oncogene transduction	Transactivation		
Avian leukosis virus	+	+	–	Simple	Bursal lymphoma
Reticulo-endotheliosis virus	+	+	–	Simple	Lymphoma
Mouse mammary tumour virus	+	–	–	Simple*	Mammary carcinoma
Murine leukaemia viruses	+	+	–	Simple	Lymphomas
Feline leukaemia virus	+	+	–	Simple	T-lymphoma, other
Human T-cell leukaemia virus	–	–	+	Complex	Adult T-cell leukaemia
Bovine leukaemia virus	–	–	+	Complex	B-cell leukaemia

*Carries *sag* (superantigen gene) in addition to simple retrovirus *gag–pol–env* structure.

proteins (e.g. Tax). These proteins serve to stimulate cell proliferation in the early stages of infection, but may be selectively downregulated along with virus expression in cells progressing to malignancy. Although HTLV/BLV proviruses are generally found as clonally integrated elements in tumour DNA, there is only fragmentary evidence that this results in activation of genes in *cis*, and this mechanism does not appear to be a significant feature of their oncogenic action. Finally, the lentiviruses are omitted from this list. Although immunosuppressive lentiviruses such as HIV clearly increase the cancer risk for infected individuals, they are generally absent from the tumour cells and appear to act mainly as co-factors for other, normally weak, oncogenic viruses. For a comprehensive coverage of retrovirus biology, see refs 1 and 2.

2 Insertional mutagenesis

While some retrovirus-like elements (e.g. Ty retrotransposons of yeast) integrate at specific sites in host DNA, the retroviruses of higher eukaryotes show little, if any, preference for a particular primary DNA sequence or chromosomal domain in host DNA (3). The finding of a common region of virus integration in a series of independent tumours is therefore a likely indication that the integration event has played a determining role in tumour outgrowth. This assumption has been amply corroborated by analyses of common insertion sites, which have revealed the presence of numerous oncogenes or tumour suppressor genes in close proximity (4).

2.1 Analysis of virus integration patterns

The presence of somatically acquired provirus integrations in tumour DNA is a first clue that the agent in question has played a role in the initiation or maintenance of the transformed state. The detection of such virus sequences is therefore the first step to be undertaken in analysing tumours of suspected virus aetiology. For this exercise, knowledge of the primary sequence and restriction map of the provirus is helpful, though not essential. However, as some oncogenic retroviruses have closely related endogenous counterparts present at multiple copies in host DNA, it is important to be able to distinguish these to gauge the number and pattern of new insertions. This problem is most acute for the murine and feline leukaemia viruses, where it is necessary to generate subgenomic probes to distinguish endogenous from exogenous viruses. As shown in *Figure 1*, the two regions of the MLV and FeLV genomes that show the greatest divergence from their related endogenous viruses are the enhancer region of the long terminal repeat (U3) and the 5′ end of the *env* gene which encodes the receptor-binding domain. Probes from either region have proved useful for analysing the new somatically acquired integrations by these viruses.

Southern-blot hybridization analysis remains the definitive tool for examining provirus integrations in tumours, although PCR methods might be considered if the amounts of material are limited. The benefits of blot analysis are that it can establish the complexity of the tumour population as well as the number of newly acquired proviruses, two important indicators of the scale of the task ahead in analysing integration sites. If the tumour cell can be shown to be clonal with respect to new virus integrations, this is also a useful (though not definitive) sign that at least some of the insertions have played a role in determining the tumour phenotype and promoting cell growth. Another use of blot analysis is in the demonstration of new integrants in tumour cell clones in association with novel phenotypic properties (e.g. invasiveness)—this has proved to be a fruitful means of locating those genes capable of influencing tumour progression (5).

Though many technical variations have been introduced making blot hybridization simpler, faster, and more sensitive, detecting single-copy sequences in higher eukaryotic DNA still requires high-specific activity probes and low back-

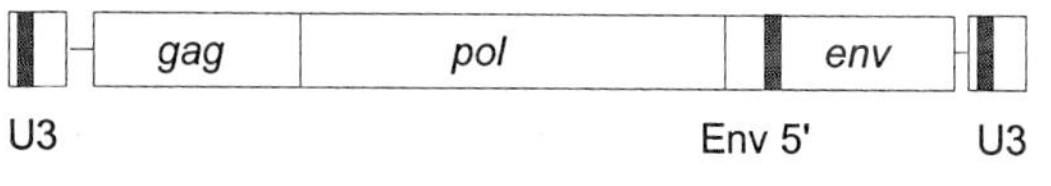

Figure 1 Outline genetic structure of the provirus form of murine and feline leukaemia viruses and the location of sequences suitable for use as specific probes for exogenous provirus insertions. The virus sequences at these positions are sufficiently mismatched with endogenous proviruses to allow discrimination by high-stringency hybridization. Useful probes of this type have been derived from the enhancer domain of the LTR (U3) (25, 26) or from the 5′ portion of the *env* gene (27, 28). For other retroviruses listed in *Table 1*, specific probes are easier to derive due to the lack of closely related endogenous viruses (REV) or their presence at relatively low copy numbers (ALV, MMTV).

ground. The method described in *Protocol 1* is the one in current use in our laboratory. A useful adjunct to this analysis is the use of a standard reference DNA with a clonal virus integration pattern, which can be created, if necessary, by *in vitro* infection and cell cloning.

Protocol 1

Analysis of integrated virus sequences in tumour DNA

Equipment and reagents

- TE buffer: 10 mM Tris-HCl, 1 mM EDTA pH 8.0
- 7.5 M Ammonium acetate
- TAE buffer: 40 mM Tris–acetate, 2 mM EDTA pH~8.5
- 20 × SSC: 3 M sodium chloride, 0.3 M trisodium citrate
- 5 × Denhardt's solution: 0.1% (w/v) Ficoll type-400, 0.1% (w/v) polyvinyl pyrrolidone, 0.1% (w/v) bovine serum albumin (fraction V)
- Vacuum drier
- 0.6–0.8% agarose gel
- Agarose gel electrophoresis equipment and reagents
- Bromophenol Blue
- Ethidium bromide
- Gel denaturing solution: 1.5 M NaCl, 0.5 M NaOH
- Gel neutralizing solution: 1.5 M NaCl, 0.5 M Tris–HCl pH 7.4
- Pre-hybridization solution: 5 × SSC, 0.5% SDS, 5 × Denhardt's solution, 20 μg/ml sonicated salmon sperm DNA
- Hybond N membrane (Amersham Pharmacia Biotech)
- Nick columns (Amersham Pharmacia Biotech)
- High Prime labelling kit (Roche Diagnostics)
- [α-^{32}p] dCTP 3000 Ci/mmol, 10 mCi/ml

Method

1. Extract DNA by a gentle method that generates a high average molecular weight preparation (> 50 kb) and redissolve in TE buffer.
2. Digest 20 μg aliquots overnight with 80–100 units of the appropriate restriction endonucleases (e.g. *Eco*RI, *Bam*HI) in a convenient volume (50–200 μl). Precipitate the DNA with 0.5 volumes salt and 2 volumes ethanol, pellet in a microcentrifuge 22,000 g, 10 minutes, briefly dry under vacuum, and resuspend (>4 h) in 35 μl TE buffer at 37 °C.
3. Separate by agarose gel electrophoresis (0.6–0.8%) in TAE buffer at 26 V overnight, or until the tracker dye (Bromophenol Blue) has run to two-thirds of the gel length. Stain the gel with ethidium bromide (0.5 μg/ml) to visualize the DNA under UV light. Denature the gel for 30 min in the gel denaturing solution then neutralize for 30 min in the gel neutralizing solution.
4. Transfer by capillary blotting on to Hybond N membrane in 20 × SSC overnight then UV-crosslink.
5. Pre-hybridize (>4 h) in 15 ml of the pre-hybridization solution at 65 °C.

Protocol 1 continued

6. Radiolabel the probe by random priming 10–30 ng of the probe DNA fragment using 'High Prime' and [α-^{32}P]dCTP. Separate the probe from the unincorporated label using a 'Nick Column'. Add 1×10^6 c.p.m. of the probe per millilitre of prehybridization solution. Allow hybridization to take place at 65 °C overnight.
7. If the probe is highly homologous to the target DNA, wash the blot three times (20 min each time) in $0.1 \times$ SSC, 0.5% SDS at 60 °C, place in a sealed plastic bag and expose to X-ray film overnight.

2.2 Cloning virus integration sites

For obvious reasons, most investigators choose to select tumours with a low number of newly acquired proviruses for further analysis. In spite of improvements in cloning technology, few will wish to map an unnecessarily large number of insertion sites, particularly if many of these prove to be solo integrations that played no role in tumour development or progression.

Methods for the initial cloning of a panel of integrated proviruses are generally a compromise, between the desire to obtain the maximum number of clones with greatest ease and the need for adequate genetic material for further analysis of the chromosomal region. Conventional lambda bacteriophage vectors remain our favoured option, as these yield large enough DNA fragments to ensure that sufficient host DNA sequences will be available to allow the next analytical steps to proceed without further primary cloning.

A requirement for cloning is a virus-specific probe capable of efficiently selecting recombinants from a background of host DNA. As a general rule, if a probe is useful for single-copy sequence detection on Southern blot analysis, it will be adequate for phage library screening.

The protocol will vary according to the vector and the specific restriction sites available for cloning. Various commercial suppliers provide replacement vectors with large insert capacity (up to 23 kb, e.g. Lambda EMBL 3, 4, Lambda FIX II, Lambda GEM-11,12). Vectors such as the Zap Express vector have a smaller cloning capacity (up to 12 kb), but have the advantage of easy retrieval of the insert by *in vivo* excision and phagemid replication. The major choices relate to the restriction enzyme to be used, and whether to size-select the DNA or to use partial digestion of the target DNA. If a fragment or a series of fragments of defined size are to be cloned, then size-selection of the digested DNA will increase the representation of these sequences in the library. Also, size-selection reduces the number of small unlinked fragments that may be ligated in tandem into the vector and thus complicate subsequent analysis (Lambda FIX II or Lambda GEM-12 are designed to reduce this problem). Partial digestion of the target DNA is a useful strategy for obtaining larger segments of flanking DNA, particularly where a common 6-base cutter is to be used to cut the genomic DNA (e.g. *Bam*HI). To devise the ideal strategy, the target fragments should initially be characterized.

Possible pitfalls of this method of cloning include the presence of sequences

Protocol 2

Bacteriophage cloning of integration sites

Equipment and reagents

- Lambda vector arms, e.g. Lambda EMBL 3, 4, Lambda FIX II, or Lambda Zap Express Stratagene) or Lambda GEM-11,12 (Promega)
- Packaging mixture (Stratagene or Promega)
- Denaturing solution: 1.5 M NaCl, 0.5 M NaOH
- Neutralising solution: 1.5 M NaCl, 0.5 M Tris-HCl pH 8.0
- Rinse solution: 0.2 M Tris-HCl pH 7.5, 2 × SSC
- Hybond N membrane (Amersham, Pharmacia, Biotech)
- Pre-hybridisation solution: 5 × SSC, 0.5% SDS, 5 × Denhardt's solution, 20 μg/ml sonicated salmon sperm DNA
- SM buffer: 10 mM sodium chloride, 10 mM magnesium sulphate heptahydrate, 50 mM Tris-HCl pH 7.5, 0.01% (w/v) gelatin
- Appropriate bacterial strain
- Large agar plates (24 × 24 cm)
- Hybond N membrane (Amersham, Pharmacia, Biotech)

Method

1. Digest a 10–100 μg aliquot of genomic DNA carrying the virus insertion with the chosen restriction endonuclease. If required, select for the appropriate size range by gel electrophoresis and DNA extraction from the agarose gel.
2. Ligate digested DNA into the phage arms.
3. Package the ligated DNA using packaging mixes.
4. Titre the packaged phage by plating on the appropriate bacterial strain.
5. Plate approximately 2×10^5 p.f.u. per large agar plate (24 × 24 cm). Ideally, screen a total of 10^6 or more plaques for a full representation of the genome.
6. Using the Hybond N membrane, lift duplicate filter impressions from each plate, carefully marking the orientation of the filters with respect to the plates.
7. Lyse phage particles on the filter by immersion in denaturing solution for 2 minutes, bind phage DNA to the filter by soaking for 5 minutes in neutralising solution then rinsing for up to 30 seconds in rinse solution. UV crosslink and hybridize with the virus-specific probe, in pre-hybridisation solution using similar conditions to those described for Southern blotting in *Protocol 1*.
8. Identify areas of the plates that show positive plaques on duplicate lifts; excise from the plate, elute in SM buffer, titrate, and plate at a lower density.
9. Repeat steps 6 to 8 for each positive area until clearly isolated positive plaques are obtained for each positive spot seen on the original filter.[a]
10. Test plaque-purified clones for homogeneity by plating approximately 100 plaques, lifting phage impressions, and hybridizing with the virus-specific probe. Check that every plaque on the agar plate is positive. If not carry out a further round of purification.
11. Plate each positive isolated plaque at high density and make a plate lysate of phage particles.
12. Lyse the phage and purify the phage DNA as described in ref 5A.

[a] Note that some positive plaques may be very small due to reduced replication efficiency.

Reference 5A. Sambrook, J., Fritsch, E. F., Maniatis, T. (1989) Molecular cloning: A laboratory manual, second edition. Cold Spring Harbor Laboratory Press.

incompatible with cloning in phage vectors. An example is the 5′ end of the mouse mammary tumour virus (MMTV), which proved unstable in phage vectors, hampering the analysis of integration sites and the derivation of infectious molecular clones of MMTV (6).

To expedite cloning in phage vectors, viruses have been generated containing selectable markers. For example, a *supF* gene was inserted into the long terminal repeat of Moloney MLV as a potential cloning aid (7). Although this approach can undoubtedly accelerate the cloning process, we have chosen not to use it—the modified Moloney MLV LTR proved to be unstable *in vivo* in infected mice, with evidence of accumulating deletions of the supF element in tumours (our unpublished results). This method may therefore be unsuitable where significant virus replication has preceded cloning.

An alternative approach to the cloning of virus junction fragments is to use the inverse PCR (8), a method that has been used to good effect by a number of laboratories. The method requires knowledge of the primary sequence of the infecting virus. The basic principle is that tumour DNA is digested to generate fragments which will include virus–host junctions. The digested DNA is then religated at low concentration, favouring the formation of DNA circles. These circles are then linearized by digestion with an enzyme known to cut within the virus sequences, generating a linear template in which the (unknown) host-cell DNA sequences are now flanked by virus sequences. A library of integration sites can therefore be generated by PCR amplification between the virus termini. This is depicted in diagrammatic form in *Figure 2*. We have found that this method works with acceptable efficiency using 4- or 5-base cutting enzymes as enzyme A (e.g. *Sau*3AI, *Dde*I for FeLV), although the junction fragments generated are often too small to allow informative analysis of the locus. While the use of 6-base cutters for enzyme A would generate larger fragments they will be less efficiently amplified by PCR, and will have a greater chance of containing a site for enzyme B. Enzyme B is preferably a relatively rare cutter, which also contains sites in the body of the provirus (e.g. *Sma*I or *Kpn*I for FeLV, MLV) to ensure that the PCR reaction does not amplify internal genomic fragments.

Protocol 3

Inverse PCR cloning of provirus integration sites

Equipment and reagents

- Appropriate restriction enzymes
- PCR primers and reagents (see *Protocol 4*, Chapter 6)
- Ethanol
- Agarose gel electrophoresis equipment
- TBE buffer: 0.089 M Tris base, 0.089 M boric acid, 2 mM EDTA pH 8.0

Method

1. Identify a suitable enzyme that cuts within virus sequences close to, but outside, the LTR (enzyme A in *Figure 2*).

Protocol 3 continued

2. Digest DNA with enzyme A.
3. Religate at a range of concentrations from 10^{-2} to 10^{-6} mg/ml.
4. Concentrate DNA by ethanol precipitation.
5. Cut DNA with enzyme B.
6. PCR-amplify using primers with opposite orientation in the unrearranged provirus (see *Figure 2*). Check the utility of these primers by their ability to amplify undigested DNA or plasmid constructs containing an internally deleted provirus. Choose the primers with care where closely related endogenous proviruses are present, with at least one primer specific for exogenous virus.[a]
7. Analyse the products by gel electrophoresis on 1% agarose TBE or 4% polyacrylamide and ethidium bromide staining, and by cloning and sequencing.

[a] For FeLV and MLV, the U3 domain of the LTR yields ideal primers.

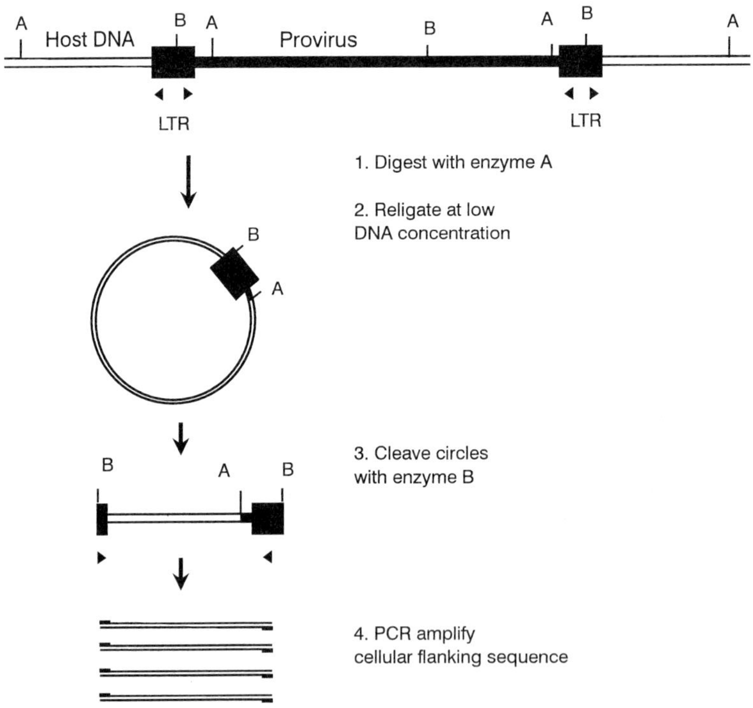

Figure 2 Inverse polymerase chain reaction. This method is a general means of recovering unknown DNA sequences by virtue of their linkage to a known sequence. In the example illustrated, tumour DNA containing integrated proviruses is digested with a restriction enzyme (A) which cleaves at a known location within the provirus and at an indeterminate location in the cellular flanking sequences. Religation of this DNA at low concentration favours the formation of DNA circles. The DNA can then be linearized by cleavage at another known restriction enzyme site (B) within the provirus and amplified by PCR using primers based on the virus sequence.

Other PCR-based methods have been developed to solve this problem. An example is vectorette PCR, where DNA is restriction-enzyme digested and ligated to add a PCR primer to the fragment ends. Using this primer in combination with a second primer within the known sequence (in this case the provirus), it is therefore possible to amplify the unknown junction fragment (9).

2.3 Analysis of host DNA flanking the provirus integration sites

After obtaining the cellular flanking sequence, the next step is to assess the significance of the cloned integration site(s). The presence of even two integrations in the same small region of DNA in independently derived tumours is highly improbable in the absence of a selective process. Analysis of a panel of such tumours for integrations in the same chromosomal domain is therefore an important step towards sorting incidental events from those that have played a determining role. However, even solo integration events could conceivably have played such a role, and could have been under-represented if the panel of tumours available is not large. It may therefore be useful to analyse these integrations for the presence of genes of interest.

To locate the retrovirus integration site to a specific host chromosomal region and check for redundancy with known virus insertion loci, it is helpful to identify a stretch of sequence that can be used as a single-copy probe. We have used a shotgun cloning/hybridization screening method to short-cut this search.

Protocol 4

Rapid cloning of single-copy sequences from host flanking DNA

Equipment and reagents

- Appropriate restriction enzymes
- Appropriate hybridization probes
- Hybond N membrane (Amersham Pharmacia Biotech)

Method

1. Digest the recombinant phage DNA containing the provirus and flanking sequences with an enzyme that generates 300 bp–1 kb fragments (e.g. *Sau*3AI, *Hin*fI) and clone these fragments into a suitable high copy-number plasmid vector. Generate four replica filters by plating a series of recombinants and lysing the bacterial cells on the filter by standard methods (10)
2. Hybridize filter 1 with a probe generated by labelling complete phage lambda DNA or preferably the cloning vector arms (if available): exclude positive clones.
3. Hybridize filter 2 with a probe generated by labelling a clone (e.g. plasmid insert) containing an entire retrovirus provirus: exclude positive clones.

Protocol 4 continued

4. Hybridize filter 3 with labelled total DNA from the host (e.g. mouse, cat, rat): exclude positive clones (to screen out abundant repetitive elements).
5. Hybridize filter 4 with a probe derived by labelling the original phage recombinant: select positive clones.
6. Test clones passing this screening process for suitability in Southern blot hybridization. Those with larger inserts (> 300 bp) are more likely to be effective probes for this purpose.

When a single-copy probe has been identified it can be used to screen a panel of tumours for integrations at the same locus. It is desirable to establish a partial restriction map of surrounding DNA to ensure that the screening process encompasses as large a domain of cellular DNA as possible. By standard Southern blot procedures it should be possible to screen a domain of 30–40 kb for the presence of rearrangements suggestive of provirus insertion. *Figure 3* shows a

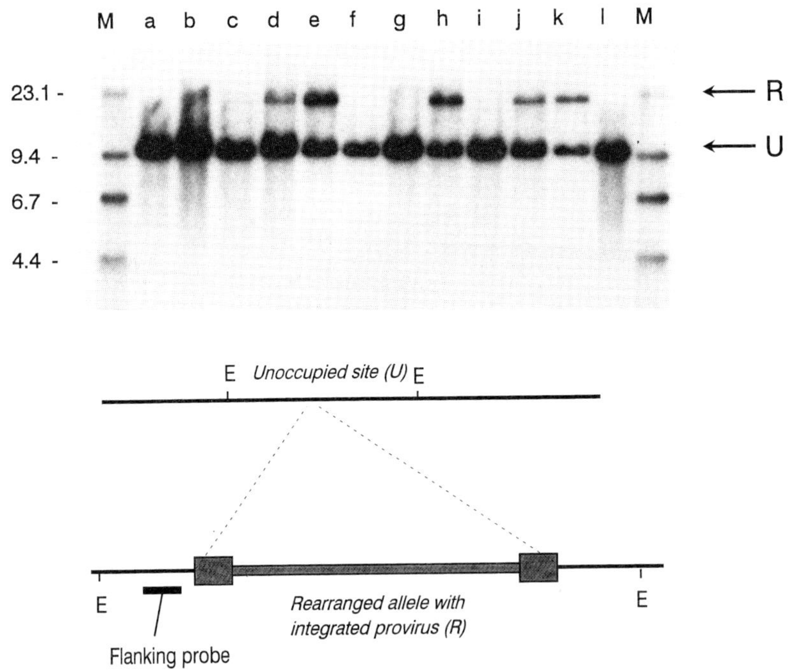

Figure 3 Southern blot analysis of a common insertion site for murine leukaemia virus in virus-accelerated tumours of CD2-myc transgenic mice. DNA from individual tumours was digested with *Eco*RI (E), separated on an agarose gel, blotted, and hybridized with a probe derived from the cellular sequences flanking an integrated provirus obtained by lambda cloning. The probe detects the unoccupied site (U) and a series of rearranged fragments (R) corresponding to the presence of MLV proviruses within the *Eco*RI fragment. As *Eco*RI does not cut within typical MLV proviruses, the rearranged fragments are around 8.5 kb larger than the unoccupied allele (adapted from ref. 11).

typical pattern produced by a series of integrations at an established common insertion site (the *til-1* locus (11).

To check for redundancy with known retrovirus insertion sites and cellular genes, the insertion locus can be mapped to its chromosomal location. This exercise is relatively easily performed if the DNA is of human or murine origin, as the Human Genome Mapping Project has led to the availability of large sequence databases and numerous aids to gene mapping. Chromosomal location of genes in the mouse has been greatly aided by the development of large panels of interspecific backcross mice. A selected laboratory mouse strain is interbred with a wild-mouse relative (*Mus spretus*) and the F_1 is then backcrossed to either parental strain, generating offspring which are either heterozygous or homozygous for the backcross parent. As *Mus spretus* has many polymorphisms relative to *Mus musculus*, typing the offspring can be achieved by using RFLPs or PCR-based analyses. The European Interspecific Backcross (EUCIB) represents a resource of almost 1000 progeny that have been typed for anchor markers on each chromosome. It is therefore possible to establish linkage to any new marker by analysing DNA samples from a selection of the backcross panel, and then locating it more precisely by comparison with a subset of mice known to be recombinant for the relevant chromosome. Extensive linkage maps have now been generated for EUCIB (www.hgmp.mrc.ac.uk/MBx/)

Protocol 5

Chromosomal location of insertion loci in the mouse

Equipment and reagents

- Reference DNA samples from backcross parental mice (*Mus spretus* and a standard laboratory mouse strain)
- Backcross DNA samples (obtained from HGMP Resource Centre)

Method

1. Identify a stretch of unique sequence that can be readily amplified from the laboratory mouse parental DNA (C57 strain for EUCIB).
2. Amplify and compare these sequences from C57 and *Mus spretus* DNA to identify a PCR primer/restriction enzyme combination that will distinguish alleles of the parental mouse strains, e.g. by amplification across a polymorphic repeat sequence or by digestion of PCR products with a polymorphic restriction enzyme site.
3. Obtain a panel of recombinant inbred mouse DNAs, and type alleles present in individual samples by PCR analysis for the new marker.[a]

[a] The data can then be submitted for linkage analysis, which will provide a tentative location by comparison to a series of anchor markers. A further round of analysis with selected mice recombinant for the relevant chromosome will allow fine mapping relative to known linked markers. Together with comparisons with other backcross databases, this analysis can identify whether any known genes map close to the insertion locus.

For other species, where genome maps are much less complete, it may be more valuable to search for a region of DNA conserved in evolution and then map the homologous sequence to a human and/or mouse chromosomal location. The single-copy probes identified in the screening exercise above should be hybridized to a filter containing a range of animal DNA samples (a zooblot). Detection of sequences on a zooblot will require reduced stringency of hybridization or washing, which may require optimization. For an initial test we would recommend the procedure outlined in *Protocol 1*, but with 2 × SSC instead of 0.1 × SSC during the washing process (step 7).

If a zooblot positive probe can be identified it can be used for direct mapping to a human or mouse chromosome, or used to isolate homologous sequences from human or mouse genomic libraries (Genome Project Resource Centres supply academic laboratories with filters of YAC (Yeast artificial chromosome), PAC (P1-derived artificial chromosome), and BAC (Bacterial artificial chromosome) libraries and of chromosome-specific cosmid libraries). It is important to check that the probe is unique in the species of origin before attempting zooblot analysis—we have found that some repetitive sequences can give a misleading low copy-number hybridization pattern on heterologous DNA (e.g. certain human Alu sequences on mouse DNA). A further advantage of identifying zooblot positive sequences is that these may lead to the identification of adjacent genes: coding sequences and critical regulatory sequences flanking expressed genes are generally more highly conserved than introns or intergenic DNA.

DNA sequence analysis of the integration locus can, on occasion, lead directly to its identification if a match is found to a known database entry. It can also be useful to identify and exclude repetitive elements from further analyses. Further, the definition of primers for PCR amplification aids chromosomal localization of the sequence if mouse backcross or cell-hybrid panels are to be used.

2.4 Location of genes at integration sites

If mapping studies do not implicate a known gene or locus, the next step in the analysis of a novel common insertion site is to search for the target gene(s). This is not necessarily a straightforward exercise as the true target gene may be at a considerate distance from the insertions (up to 300 kb), and the effects on gene expression may be subtle and may be negative rather than positive. The location and orientation of the cluster of provirus insertions can be a useful clue, particularly if the common modes of gene activation (or inactivation) are taken into account. *Figure 4* is a summary diagram of the known modes of gene activation by retrovirus insertion. In the upstream enhancer mode, the cluster of insertions is often close to, but in the opposite orientation from, the target gene's promoter, and in this case the transcription unit structure is unaffected—only the levels of expression are altered. These effects appear to be able to operate over some distance, as several well-documented examples of long-range activation show. In the cases of promoter insertion or 3′ enhancer insertion the provirus insertions are within the transcription unit and will alter

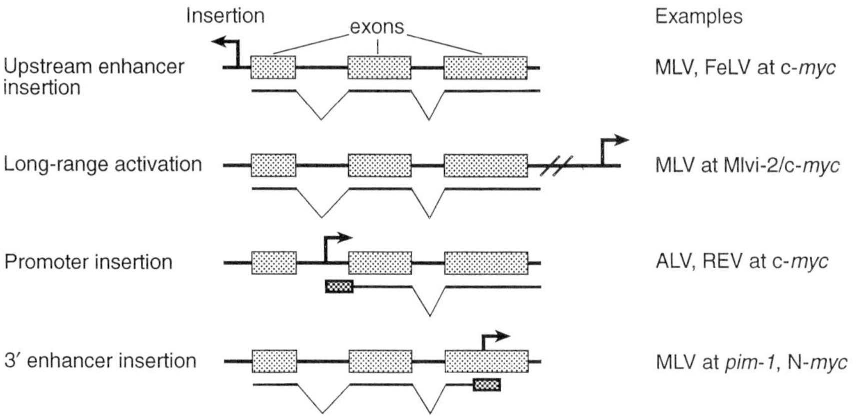

Figure 4 A series of possible relationships of clustered retrovirus integrations to the target gene. Filled boxes depict the exon structure of the cellular target gene with lines underneath showing RNA transcripts. Dark boxes indicate virus sequences inserted within the gene transcript, while large arrows indicate the site and orientation of the cluster of provirus insertions. Although many common insertion sites occur within or close to the target gene, with obvious disturbances of transcript structure (e.g. promoter insertion, 3′ truncation) or expression (nearby enhancer insertion), integrations can be at some considerable distance from the likely target gene and have subtle effects on expression.

its structure. In the former case, the virus R-U5 sequence becomes the 5′ end of the transcript, while in the latter case it is the U3-R sequence. Analysis of tumour cell RNA is a useful analytical tool in these cases.

High-throughput, DNA sequence analysis is a useful adjunct to screening for the affected transcription unit, although it is recognized that not all laboratories have access to such facilities. It should also be noted that most genes at insertion sites have been identified without large-scale genomic sequence analysis. For insertion sites in mouse or human DNA, the generation of a YAC or BAC contig surrounding the insertion site is now a routine matter with the aid of public domain resources as well as commercial libraries. There are a number of websites from which input sequences can be compared with database entries and analysed for the hallmarks of cellular genes (e.g. BLAST at www.ncbi.nlm.nih.gov, or NIX, an integrated package of multiple algorithms at www.hgmp.mrc.ac.uk). Despite advances in ease of use and computational power, sequence analysis and database comparison do not reliably detect all transcription units. The algorithms can give only probability estimates on features such as putative exons, promoters, and open reading frames and may miss non-standard genetic elements (12).

A combination of sequence analysis with methods capable of identifying transcription units is preferable in searching for target genes at insertion loci. A simple but laborious approach is to generate a panel of single-copy sequences from the flanking DNA and screen these against Northern blots for evidence of expression. This approach has been successful where the affected gene is close to the provirus insertion cluster and the effects on gene expression are striking (13). However, where analysis of larger stretches of DNA becomes necessary, then more rapid screening methods are desirable. Transcribed sequences can be selected from large genetically defined regions by techniques such as solution hybrid capture (14) or exon trapping (15). Commercial kits are available for the latter (Life Technologies Ltd).

Where long-range gene activation is suspected, larger scale analytical methods must be employed. YAC, PAC, and BAC clones corresponding to the locus can be isolated from libraries obtainable from Genome Resource Centres. For species other than human and mouse, library resources are more limited but are becoming more widely available. The larger inserts of these high-capacity vectors are analysed most conveniently by pulse-field gel electrophoresis, which can resolve fragments of greater than 1 Mb. The proximity of a CpG-rich island has been used as a useful clue to the presence of target genes in some cases (16, 17). Cleavage with restriction enzymes with CpG within their recognition sequence

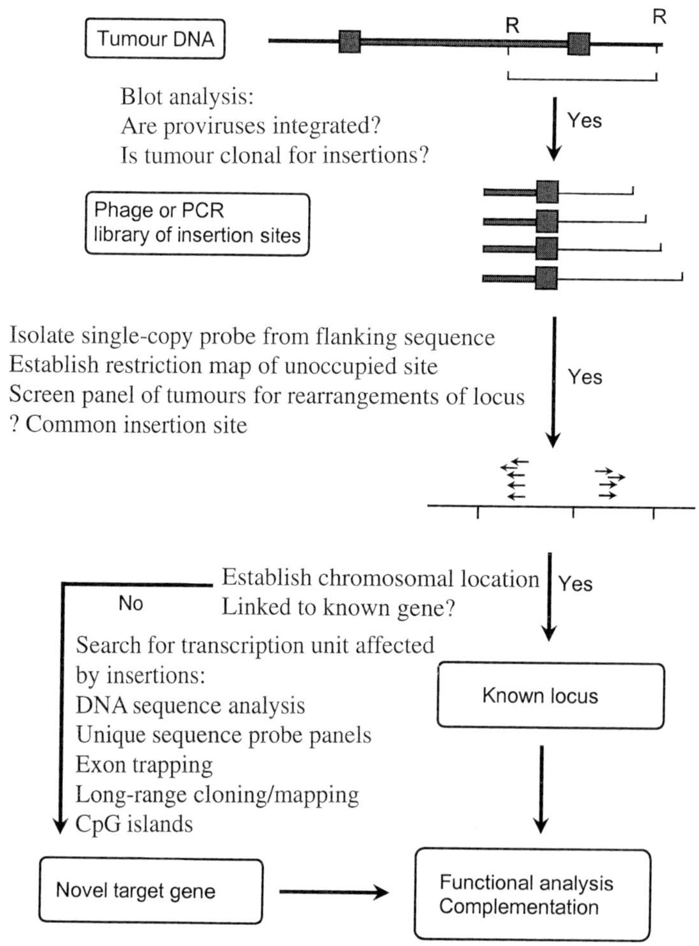

Figure 5 Overview of the steps involved in the analysis of insertional mutagenesis, from the initial demonstration of a clonal integration pattern in the tumour to the identification of the host gene targets of insertional mutagenesis.

can be used to screen for such elements: examples include *Nae*I, *Sma*I, *Eag*I, *Not*I, *Sac*II, *Asc*I, *Eco*47III.

As any transcription unit in the vicinity of the insertion locus may conceivably be a target for the oncogenic effects of insertion, identification of these is just the first step in characterizing the locus and the selective advantage it confers on the tumour cell. If the effects on gene expression are very strong, then there may be little difficulty in deciding where the focus of further investigation should lead. However, the example of the *pal-1/gfi-1* locus indicates the potential complexity of such cases. At this locus, four discrete clusters of mutually exclusive insertions have been identified within a 60 kb stretch of mouse DNA (18). While the cluster at *Evi-5* revealed a transcription unit encoding a product with homology to the *Tre2* oncogene (19), there is as yet no evidence that this transcription unit is upregulated by the insertions. It is possible that the common target for all these clustered insertions is *Gfi-1*, a zinc-finger transcription factor originally characterized by provirus insertions which gave rise to factor (IL-2) independent growth of T-cells (20).

The ultimate demonstration of relevance of any gene identified in this way is to show that it contributes to the transformed cell phenotype. This can be achieved either by introducing the rearranged allele into the murine germline as a transgene or by an *in vitro* assay for its effects on cell proliferation, survival, or differentiation. *Figure 5* is an overview of the steps involved in characterizing a retrovirus as a possible insertional mutagen and in identifying the target genes in this process.

3 Transduction of virus oncogenes

3.1 Properties of transducing viruses

The process of oncogene transduction involves a rare recombination between the retrovirus genome and host cell DNA, which results in the replacement of virus sequences with host genetic information and the formation of an acutely oncogenic variant. Several practical points should be raised when considering this mode of retrovirus oncogenesis.

- The process is rare, although in some types of tumour this can be a prevalent mechanism (e.g. c-*erbB* transduction in erythroblastosis of line 15 chickens, c-*myc* transduction in T lymphomas of domestic cats).
- Most isolates have been derived from naturally occurring tumours selected from large populations of infected animals (commercial chicken flocks, pet cats, wild mice).
- Examples come exclusively from the 'simple' retroviruses (see *Table 1*)—those which replicate to high titre *in vivo*, display broad tissue tropism, and lack complex autoregulation at the level of transcription and splicing.
- The transducing viruses are replication-defective and require a helper virus for their propagation (with one notable exception: Rous sarcoma virus).

3.2 Biological methods of detection

The basic approach to detecting acutely oncogenic viruses, which dates back to the earliest pioneering work on avian retroviruses, is the *in vivo* passage of cell-free extracts from tumours. This process can confirm the presence of an acutely oncogenic agent that, by definition, is distinguished from the more common slow-acting helper virus by the speed with which tumours develop. For example, FeLV typically takes around 2 years to induce neoplastic disease in its host, but the latent period is compressed to around 3 months when the inoculum contains a Myc-transducing virus (21). Some acutely oncogenic viruses operate even more rapidly, inducing tumours within 2 weeks of infection. Another hallmark of many transducing viruses is the transformation of an unusual cell type (e.g. muscle fibroblasts, endothelial cells) distinct from those typically affected by the associated helper virus.

In vitro transformation assays were developed mainly as analytical tools long after the initial discovery of acutely oncogenic viruses, such as Rous sarcoma virus (1911) and Fujinami sarcoma virus (1914). However, they may also be of use in primary screening for acute transforming viruses if an appropriate target cell type can be propagated in cell culture. Thus, focus-formation in primary chick-embryo fibroblasts reveals the transforming properties of a wide range of avian sarcoma viruses, as well as agents that transform macrophages such as MC29 avian myelocytomatosis virus. *Table 2* lists a variety of acutely oncogenic viruses and the *in vitro* transformation systems that have been used to demonstrate and analyse their transforming potential.

Focal transformation assays are useful not only for quantitating transforming viruses by titration, but they also allow the defective viruses to be separated from their associated helper virus by end-point cloning. Increasing dilutions of the transforming virus stock are used to infect target cells, and clones of transformed cells are analysed to find rare cells that have been infected with the transforming virus without concomitant infection with the helper virus. How-

Table 2 Some transformation assays for oncogene transducing retroviruses

Assay system	Phenotype	Helper virus	v-*onc* gene	Reference
Primary chick embryo	Foci or anchorage-independent colonies	ALV	v-*src*, v-*fps*, v-*yes*, v-*ros*, v-*myc*, v-*jun*	29, 30
Yolk-sac macrophages	Foci	ALV	v-*myb*, v-*myc*, v-*erb*-B	31
Mouse fibroblast, e.g. 3T3	Foci or anchorage-independent colonies	MLV	v-*mos*, v-*ras*, v-*abl*	32, 33
Long-term, bone-marrow culture	Colonies	MLV	v-*abl*	34
Feline embryo fibroblast	Foci	FeLV	v-*fes*, v-*sis*, v-*abl*, v-*fgr*, v-*kit*	35

ever, for a reasonable chance of success, high-titre stocks of transforming virus are required. Before the advent of molecular cloning techniques, the derivation of non-producer transformed cells provided a method for identifying the transforming virus and characterizing the defective virus genome and its products.

3.3 Molecular methods of detection

Some transducing viruses have been identified without the aid of an *in vitro* transformation assay, e.g. FeLV-derived transducing viruses carrying c-*myc*, T-cell receptor β-chain, or *Notch2* (22–24). The key feature of these discoveries is that molecular analysis and cloning were carried out directly on primary tumour material. This brings to the fore an important principle in the analysis of defective transforming virus complexes: passage under conditions that do not positively select for the transforming virus will lead to loss of titre and overgrowth of the replication-competent helper virus. *Figure 6* illustrates this general principle in diagrammatic form, and highlights important molecular clues to the possible presence of a transducing virus. The passage of a complex of helper and transforming virus in tumours *in vivo* or by selection of transformed cells *in vitro* will enrich the defective transforming component, which can generally be detected as a variant genome by Southern blot analysis.

Figure 7 is a diagrammatic overview of the steps that led to the discovery of various transducing retroviruses. Most of the pioneering work in this field occurred many years before the advent of molecular cloning, let alone PCR-based technologies. It is our view that the systematic analysis of tumours where transduction is a relatively common occurrence (e.g. on naturally occurring virus-associated lymphomas in pet cats or wild mice) may be a worthwhile

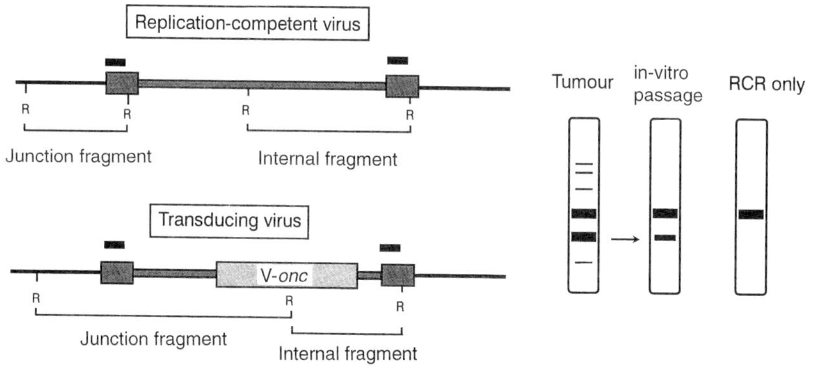

Figure 6 Detection of defective transducing retroviruses by Southern-blot hybridization analysis (right panel), and the effects of *in vitro* passage in conditions that do not select for retention of the defective virus. The black boxes (left panel) indicate U3 probe fragments. Restriction enzyme digestion of genomic DNA containing transducing virus yields a characteristic internal fragment detected by the U3 probe, which diminishes in intensity on *in vitro* passage and is absent from cells infected with a replication-competent helper virus (RCR). For an example see ref. 28.

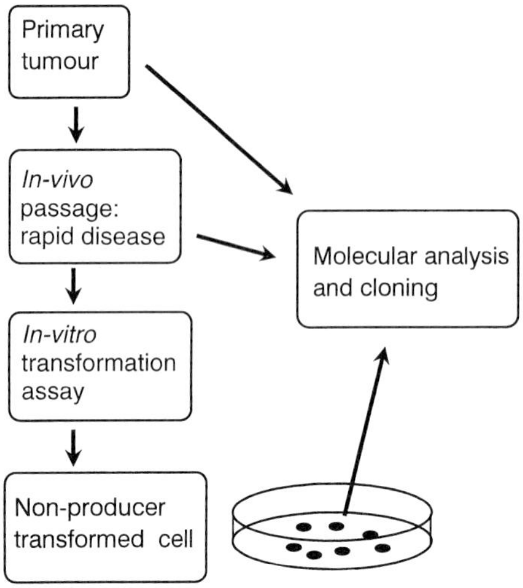

Figure 7 Overview of the discovery and characterization of transducing retroviruses. Most of the known cases were initially identified by virtue of their acute oncogenic potential, which was revealed on *in vivo* passage to new hosts. However, recent studies conducted on primary tumour material and cell lines have revealed further examples (e.g. v-*tcr*, v-*Notch*) for which *in vitro* transformation assays do not exist.

approach to the identification of further cellular genes that contribute to the neoplastic state.

References

1. (1997). *Retroviruses*. Edited by Coffin, J. M., Hughes, S. H., and Varmus, H. E. Cold Spring Harbor Laboratory Press, New York.
2. (1992). *The Retroviridae*. Editor J. Levy, Plenum Press, New York.
3. Withers-Ward, E. S., Kitamura, Y., Barnes, J. P., and Coffin, J. M. (1994). *Genes Dev.*, **8**, 1473–87.
4. Jonkers, J. and Berns, A. (1996). *Biochem. Biophys. Acta*, **1287**, 29–57.
5. Habets, G. G., Scholtes, E. H., Zuydgeest, D., *et al.* (1994). *Cell*, **77**, 537–49.

5a Sambrook, J., Fritsch, E. F., Maniatis, T. (1989) Molecular Cloning: A Laboratory Manual, Second Edition. Cold Spring Harbor Laboratory Press, New York.

6. Brookes, S., Placzek, M., Moore, R., Dickson, C., and Peters, G. (1986). *Nucleic Acids Res.*, **14**, 8231–45.
7. Reik, W., Weiher, H. and Jaenisch, R. (1985). *Proc. Natl Acad. Sci. USA*, **82**, 1141–5.
8. Silver, J. and Keerikatte, V. (1989). *J. Virol.*, **63**, 1924–8.
9. Arnold C. and Hodgson, I. J. (1991). *PCR Methods Appl.*, **1**, 39–42.
10. Grunstein, M. and Hogness, D. (1975). *Proc. Natl Acad. Sci. USA*, **72**, 3961–5.
11. Stewart, M. A., Terry, A., O'Hara, M., Cameron, E. R., Onions, D. E., and Neil, J. C. (1996). *J. Gen. Virol.*, **77**, 443–6.
12. Best, S., Le Tissier, P., Towers, G., and Stoye, J. P. (1999). *Nature*, **382**, 826–9.
13. Stewart, M., Terry, A., O'Hara, M., *et al.* (1997). *Proc. Natl Acad. Sci. USA*, **94**, 8646–51.
14. Hattier, T., Bell R., Shaffer, D., *et al.* (1995). *Mamm. Genome*, **6**, 873–9.
15. Duyk, G. M., Kim, S. W., Myers, R. M., and Cox, D. R. (1990). *Proc. Natl Acad. Sci. USA*, **87**, 8995–9.

16. van Lohuizen, M., Verbeek, S., Scheijen, B., Wientjens, E., van der Gulden, H., and Berns, A. (1991). *Cell*, **65**, 737–52.
17. Lammie, G. A., Smith, R., Silver, J., Brookes, S., Dickson, C., and Peters, G. (1992). *Oncogene*, **12**, 2381–7.
18. Scheijen, B., Jonkers, J., Acton, D., and Berns, A. (1997). *J. Virol.* **71**, 9–16.
19. Liao, X., Morse, H. C. I., Jenkins, N. A., and Copeland, N. G. (1997). *Oncogene*, **14**, 1023–9.
20. Gilks, C. B., Bear, S. E., Grimes, H. L., and Tsichlis, P. N. (1993). *Mol. Cell Biol.*, **13**, 1759–68.
21. Onions, D., Lees, G., Forrest, D., and Neil, J. (1987). *Int. J. Cancer*, **40**, 40–5.
22. Neil, J. C., Hughes, D., McFarlane, R., *et al.* (1984). *Nature*, **308**, 814–20.
23. Fulton, R., Forrest, D., McFarlane, R., Onions, D., and Neil, J. C. (1987). *Nature*, **326**, 190–4.
24. Rohn, J. L., Lauring, A. S., Linenberger, M. L., and Overbaugh, J. (1996). *J. Virol.*, **70**, 8071–80.
25. Cuypers, H. T., Selten, G., Zjilstra, M., de Goede, R., Melief, C., and Berns, A. (1986). *J. Virol.*, **60**, 230–1.
26. Overbaugh, J., Donahue, P. R., Quackenbush, S. L., Hoover, E. A., and Mullins, J. I. (1988). *Science*, **239**, 906–10.
27. Mucenski, M. L., Taylor, B. A., Ihle, J. N., *et al.* (1988). *Mol. Cell. Biol.*, **8**, 301–8.
28. Terry, A., Fulton, R., Stewart, M., Onions, D., and Neil, J. C. (1992). *J. Virol.*, **66**, 3538–49.
29. Vogt, P. K. (1969). In *Fundamental techniques of virology* (ed. K. Habel and N. P. Salzman). Academic Press, New York.
30. Wyke, J. A. and Linial, M. (1973). *Virology*, **53**, 152–61.
31. Graf, T. (1975). *Z. Naturforsch.*, **30c**, 847–9.
32. Hartley, J. L. and Rowe, W. P. (1966). *Proc. Natl Acad. Sci. USA*, **55**, 780–6.
33. Bassin, R. H., Tuttle, N., and Fischinger, P. J. (1970). *Int. J. Cancer*, **6**, 95–107.
34. Dexter, T. M., Scott, D., and Teich, N. M. (1977). *Cell*, **12**, 355–64.
35. Sarma, P. S. and Law, M. J. (1977). *Proc. Soc. Exp. Biol. Med.*, **156**, 480–4.

Chapter 6
Analysis of virus quasispecies

Jonathan Ball
Department of Microbiology, University Hospital, Queens Medical Centre, Nottingham NG7 2UH, U.K.

1 Virus quasispecies

A virus quasispecies can be thought of as a complex population of genetically related yet distinct variants (1). The term has been used to describe populations of a number of RNA viruses, e.g. human immunodeficiency virus type 1 (HIV-1) and hepatitis C virus (HCV). Although this definition of a quasispecies appears simple in theory, in practice it can be rather ambiguous (2). For example, the term has been applied to describe unrelated viruses present in different individuals, unrelated viruses arising from multiple infections within the same individual, and to genetically related but distinct variants co-circulating within the same individual and arising from a single infection. It is probably less confusing to use the term to describe the virus population co-existing within a single individual.

The major contributory factor in generating the high genetic heterogeneity observed in RNA virus populations is the error-prone nature of the virus RNA-dependent polymerases, which, unlike DNA polymerases, lack a proof-reading ability. Although the low fidelity of RNA virus polymerases has a key role in generating population diversity, other factors (such as the replicative rate of the virus and environmental selective pressures) are also extremely important. For example, the reported error rates of the HIV and human T-cell leukaemia virus-type 1 (HTLV-1) reverse transcriptases are comparable, yet the level of genomic heterogeneity observed in HTLV-1 is much lower than that observed in equivalent genomic regions of HIV-1 (3, 4).

An important consequence of virus quasispecies is that the virus population is able to respond very readily to environmental changes by selective outgrowth of pre-existing variants (1). For this reason some RNA viruses, such as HIV and HCV, are capable of setting up and maintaining an active and dynamic chronic infection, during which the quasispecies is constantly fluctuating and evolving in response to a variety of selective pressures such as host immune responses. In addition, the quasispecies nature of HIV and HCV also renders effective therapeutic intervention difficult. Some studies have shown that the failure of interferon treatment is often observed in individuals who harbour a diverse HCV population before therapy is initiated. Studies of the effects of treatment of HIV infection with a number of different classes of antiretrovirus agents have shown that acquisition of drug resistance, particularly during the administration of single or dual therapies, occurs readily *in vivo*. This is due to the outgrowth of pre-existing drug-resistant variants, and often the drug-resistant phenotype is conferred by the acquisition of a limited number of defined point mutations. Another well-documented consequence of virus quasispecies evolution is the production of virus variants with altered growth characteristics, such as replication kinetics or an ability to utilize different cellular receptors and therefore infect different cell types. Again this is particularly evident in HIV-1, where point mutations within a hypervariable region of the envelope gene, the so-called *V3* region, determine the virus's ability to utilize a variety of co-receptors (chemokine receptors), so determining which cell types are permissive for that particular variant of HIV. This property is very important in disease pathogenesis, where progression to symptomatic HIV-1 infection and the concomitant accelerated loss of immune function is associated with a switch in virus phenotype from macrophage-tropic/CCR5-utilizing to T-cell tropic/CXCR4-utilizing strains (5, 6). In addition, acquisition of specific mutations within the envelope gene also enable some variants to infect non-lymphoid cells, such as brain fibroblasts and epithelial cells, leading to disease in these tissues.

In the context of quasispecies analysis, retroviruses, for example HIV and HTLV, are worthy of special consideration because of their unique life cycle. During virus infection and replication, the virus RNA genome is reverse-transcribed to yield a double-stranded DNA genome. The double-stranded DNA is then transported to the nucleus, where essentially random integration into the host chromosomes takes place. Once integrated, the virus genome is known as the provirus. The provirus is maintained in a latent state until a variety of host cell and virus factors interact with control regions within the provirus genome, resulting in the synthesis and release of new progeny viruses. Therefore retrovirus quasispecies exist as two, often distinct, populations: one population comprising the latent proviruses (DNA) and the other the circulating virus particles (RNA) (7).

When carrying out analyses of virus quasispecies, it is important to remember that different regions of the virus genome will demonstrate varying degrees of complexity, reflecting the nature and degree of selection exerted upon them. One other consideration is that genetic changes observed in one genomic region might have occurred because of a selective pressure exerted elsewhere—a so-called population bottleneck or sweep. Finally, it is also necessary to appreciate that genomic regions are exposed to multiple, and often opposed, selective pressure. Again this can be illustrated by reference to the *V3* region of the HIV envelope gene—which is exposed to positive selective pressures to escape host immune responses, yet at the same time is under restraining pressure to maintain it's function in co-receptor interaction.

Therefore, from this brief introduction we can see that analyses of virus quasispecies composition and evolution can provide important insights into the underlying biology of the virus population.

2 Choice of analytical method

Undoubtedly the most informative method of analysing virus quasispecies would be to determine the nucleotide sequence of the entire genome of all the variants present in the population. However, considering that the HCV and HIV genomes are approximately 9 kilobases (kb) in length and an 'average' HIV-infected individual harbours well over 10^{12} variants at any one time, this approach is unsuitable! Therefore, the methods used in quasispecies analysis are a compromise between obtaining sufficient information to draw adequate conclusions, but without an exhaustive amount of effort. The methods routinely used can be crudely divided into those which simultaneously analyse multiple variants (termed population analyses) and those which analyse representative individual variants. The latter almost exclusively involves separate sequence analysis of multiple variants representative of the population. This provides the maximum amount of information for each variant, but because this process is relatively labour-intensive it restricts the length of the subgenomic region and the number of variants that can be analysed. In contrast, population analyses are less laborious but provide less information about the quasispecies. These are typified by 'majority' or 'consensus' nucleotide sequencing, heteroduplex mapping and tracking analysis (HMA/HTA), single-stranded conformation polymorphism (SSCP), length polymorphism analysis, sequence discriminatory PCR, and probe extension and line probe assays. Obviously, the choice of assay is of paramount importance when studying virus quasispecies, and this will vary between viruses and the study objectives. The relative merits of each method will be included in the subsequent introductory paragraphs relating to each of the techniques described.

3 Nucleic acid extraction

Unfortunately, nucleic acids within virus particles and infected cells are complexed to nuclear proteins and other macromolecules, rather than being present as free molecules. For this reason, effective purification of virus nucleic acids is a key step in the analysis of virus quasispecies. Isolation of DNA and RNA is a relatively simple procedure, which has been further simplified by the recent introduction of a number of commercially available kits suitable for use with a variety of samples, including body tissues, cell-culture cells, and liquid specimens.

The first part of the extraction procedure is nucleic acid release, which is usually achieved by proteolytic digestion in the presence of buffered detergents such as SDS or Triton X-100. Lysis solutions also contain EDTA, which helps maintain nucleic acid integrity by chelating divalent ions thereby inhibiting endonucleases. The next step in the extraction process is the purification of the DNA or RNA from contaminating proteins and other macromolecules. This can be achieved using a variety of methods. Both RNA and DNA can be recovered by passing the lysate through ion-exchange resins or by extraction with phenol/chloroform mixes.

An alternative method, described in *Protocol 1*, is routinely used in our laboratory for purifying genomic DNA from a variety of tissues and body fluid cells including autopsy material, semen cells (but not spermatozoa), blood cells, and cerebrospinal fluid. The resulting DNA is of sufficient purity for use in a number of PCR methods, including the amplification of full-length HIV-1 envelope genes from single-target, HIV-1 provirus DNA molecules, and the method has the added advantage that organic solvents are not used. Instead, contaminating proteins are selectively salted out in the presence of saturated sodium chloride before recovering the DNA by ethanol precipitation.

Protocol 1

DNA extraction[a]

Equipment and reagents

- Cell lysis solution 2[b]: 50 mM Tris–HCl pH 8.0, 20 mM EDTA pH 8.0, 2% (w/v) SDS
- 10 mg/ml proteinase K
- Isopropanol
- 95% and 70% ethanol
- Protein precipitating solution 3[b]: saturated NaCl
- TE buffer: 10 mM Tris–HCl, 1 mM EDTA, pH 7.4

Method

1. Add 550 μl of solution 2 to either 1–5 × 10^6 single cells or a 1–2 mm^3 tissue sample contained in a 1.5 ml microcentrifuge tube, and mix by inverting the tube five times.
2. Add 5 μl of proteinase K, mix by inverting the tube, and incubate for either 3 hours at 65 °C or overnight at 37 °C.
3. Cool the sample by placing it on ice for 10 minutes, then add 400 μl of ice-cold solution 3.
4. Mix the solutions by inverting the tube 10 times, then incubate on ice for a further 10 minutes.
5. Clarify the solution by centrifugation at 13 000 g for 10 minutes.
6. Transfer 750 μl of the supernatant to a microcentrifuge tube containing 750 μl of isopropanol, mix, then incubate on ice for 30 minutes to precipitate the DNA.
7. Centrifuge at 13 000 g for 15 minutes to pellet the DNA.
8. Use a micropipette to carefully remove and then discard the supernatant.
9. Resuspend the DNA pellet in 750 μl of 70% ethanol, incubate on ice for 5 minutes, then centrifuge at 13 000 g for 10 minutes.
10. Remove and discard the supernatant, then incubate the tube containing the pelleted DNA at room temperature to allow residual ethanol to evaporate.
11. When the DNA pellet is dry resuspend it in 50–500 μl of TE buffer.
12. Store the DNA at 4 °C for up to 1 week or −20 °C for longer term storage.

[a] This method is based on that described by Miller *et al.*, 1988 (8).

[b] These solutions are available in the DNA extraction kit supplied by Stratagene.

However, the recovery of RNA can be reduced significantly when using these methods because of the presence of RNases which are released during the cell lysis step. Therefore, lysis methods employing powerful chaotropic protein denaturants, for example guanidinium cyanate, are often used because of their inherent ability to destroy RNases, thereby preserving the RNA during the extraction procedure. One such method is described in *Protocol 2* and is particularly useful when extracting virus RNA from liquid samples ranging from 10 to 1000 μl or from small tissue samples. The solubilized RNA can be recovered a number of ways, but a simple method, developed by Boom *et al.* (9), is by adsorption to silica suspensions or silica resin. The recovery efficiency of this method is limited by the maximum amount of nucleic acid adsorbed by the silica. This method will prove beneficial in most analyses of virus quasispecies, but if RNA extraction from relatively large amounts of tissue is envisaged then an acid/phenol extraction method should be considered (10).

Protocol 2

RNA extraction

Equipment and reagents

- RNA lysis buffer: 50 mM Tris–HCl pH 6.4, 20 mM EDTA, 1.3% (v/v) Triton X-100, 5.25 M guanidinium thiocyanate
- 60 g silicon dioxide (Sigma)
- 32% (w/v) hydrochloric acid
- 5 ml glass Bijou bottles
- Ethanol
- Microcentrifuge tube homogenizers (Sigma)
- Washing buffer: 120 g guanidinium thiocyanate in 100 ml of 100 mM Tris–HCl pH 6.4
- Acetone
- TE buffer (see *Protocol 1*)

A. Preparation of silica particles

1. Suspend 60 g of silicon dioxide in demonized water in a 500 ml measuring cylinder such that the aqueous column height and width are 27 cm and 5 cm, respectively, and the total volume is 500 ml.
2. Mix the silica suspension, then allow the silica particles to sediment for 24 hours at room temperature.
3. Without disturbing the settled silica particles remove 420 ml of the supernatant by suction.
4. Add demonized water to the remaining silica suspension to a total volume of 500 ml, and mix.
5. Allow the silica to sediment for 5 hours at room temperature, then remove 440 ml by suction.
6. Add 600 μl of hydrochloric acid.
7. Transfer 4 ml aliquots of the silica suspension to 5 ml glass Bijou bottles and autoclave at 121 °C for 20 minutes.

Protocol 2 continued

B. RNA extraction

1. To a 1.5 ml sterile microcentrifuge tube, add 25–200 μl of sample, 900 μl of lysis solution, and 50 μl of the silica suspension. If extracting RNA from a solid tissue sample, add 900 μl of lysis solution to a 2 mm^3 tissue sample, homogenize the tissue using a disposable plastic homogenizer, and remove unlysed tissue samples by centrifugation at 6000 g for 5 minutes. Transfer the supernatant to a new 1.5 ml microcentrifuge then add 50 μl of the silica suspension.
2. Mix for 10 minutes, ensuring that the silica particles remain suspended.
3. Pellet the silica by centrifugation at 13000 g for 30 sec, then discard the supernatant.
4. Resuspend the silica in 1 ml of wash buffer, mix for 5 minutes, then repeat step 3.
5. Repeat step 4.
6. Resuspend the pellet in 1 ml of ethanol, mix for 5 minutes then repeat step 3.
7. Repeat step 6.
8. Resuspend the pellet in 1 ml of acetone, mix for 5 minutes then repeat step 3.
9. Remove all traces of acetone by incubating the microcentrifuge tubes, with their lids open, at 60 °C for 5–10 minutes.
10. Resuspend the silica in 50 μl of TE buffer and incubate the tubes, with their lids closed, at 60 °C for 10 minutes.
11. Pellet the silica by centrifugation at 13000 g for 3 minutes, transfer the supernatant containing the eluted RNA to a sterile 0.5 ml microcentrifuge tube and store at −70 °C.

3.1 Reverse transcription of virus RNA

Most methods of quasispecies analysis require DNA. Therefore, virus RNA genomes have to be converted into complementary or copy DNA (cDNA) in a reverse transcription (RT) reaction. A range of enzymes and reaction systems are available, but the most common methods use either the avian myoblastoma virus (AMV) or murine leukaemia virus (MLV) reverse transcriptases to extend an annealed oligonucleotide primer. The primers can either be random hexamers, which will anneal at various positions along the RNA template, or they can be an oligonucleotide, typically 15–30 bases long, whose sequence is complementary to a portion of the virus RNA located upstream of the genomic region under study. Randomly primed reverse transcription offers good flexibility such that the RT products can be used in multiple analyses of different regions of the virus. However, specifically primed RT can be more efficient, therefore the choice of method will vary between applications.

Due to the formation of relatively stable secondary structures, some virus RNAs are difficult to reverse transcribe. For this reason, we routinely use a two-step method of RT (see *Protocol 3*), where in the first step the RNA is heated to 65 °C to denature secondary structures and allow a specific oligonucleotide to anneal. The reaction is then cooled to 42 °C (the avian body temperature and therefore the optimal temperature for AMV reverse transcriptase), and the enzyme together with deoxyribonucleotides (dNTPs) and RNase inhibitor are added and primer extension takes place. If random hexamers are used to prime cDNA synthesis, it is usually necessary to allow these to anneal at room temperature before proceeding with the 42 °C incubation. The RT products can then be used directly in downstream applications such as PCR. Often the two steps of the RT reaction can be simplified and combined into a one-step reaction, but this can reduce the yield of cDNA due to inefficient primer extension. It might be necessary to heat the RNA to higher temperatures, for example 95 °C, to denature very stable secondary structures, but at these elevated temperatures RNA is thought to be very labile.

It is important to stress that RNA is very susceptible to degradation by RNases, which unfortunately are very stable enzymes and also very ubiquitous. Therefore it is essential that all equipment is sterile and that all solutions are free from contaminating DNases and RNases. RNase-free water can be prepared by pre-treatment with DEPC followed by autoclaving; however, our experience has shown that occasionally water prepared in this way can inhibit the RT, presumably because of the presence of trace amounts of DEPC. Therefore, it is preferable to buy RNase- and DNase-free water from commercial suppliers, particularly as the amount of water used is small. Commercially available RNase inhibitors are included during the RT reaction to prevent degradation of the RNA templates by RNase activity. However, it is important that these are added to the RNA after the denaturing step, otherwise they release RNases which remain active even after heat denaturation.

Protocol 3

Reverse transcription

Equipment and reagents

- 15 units μl^{-1} AMV reverse transcriptase (available from most suppliers of molecular biology enzymes)
- 5 × reverse transcriptase buffer: 250 mM Tris–HCl pH 8.5 (20 °C), 150 mM KCl, 40 mM $MgCl_2$, 5 mM dithiothreitol[a]
- 20–40 units/µl RNase inhibitor (available from most suppliers of molecular biology enzymes)
- Sterile RNase and DNase-free demonized water
- dNTP mix (equal mixture of 10 mM dATP, dCTP, dGTP, and dTTP to give a final concentration of 2.5 mM for each nucleotide)
- 5 pmol/µl oligonucleotide primer, or 6.25 $A_{260\,nm}$ units/ml of a random hexamer oligonucleotide primer (Boehringer-Mannheim (now Roche Diagnostics) or Pharmacia)

Protocol 3 continued

A. Primer annealing

1. To a sterile 0.5 ml microcentrifuge tube add 2 μl 5 × reverse transcriptase buffer, 1 μl primer, 2 μl RNase/DNase-free water and 5 μl of RNA sample.
2. Heat the sample to 65 °C for 2 minutes then cool to 42 °C for 1 minute.
3. Proceed immediately with the primer extension step B.

B. Primer extension

1. To the annealed template/primer mix add 3 μl of 5 × reverse transcriptase buffer, 6 μl of sterile distilled water, 1 μl of RNase inhibitor, 4 μl of dNTP mix, and 1 μl reverse transcriptase.
2. Incubate the samples at 42 °C for 30–60 minutes.

[a] The majority of commercially available reverse transcriptases are supplied with a purpose-formulated buffer concentrate.

Although *Protocol 3* should be adequate for most applications, amplification of difficult or long templates might require alternative approaches. The discovery that some thermostable polymerases have reverse transcriptase activity in certain buffers has meant that the whole reaction can take place at elevated temperatures. One such enzyme frequently used is Tth, this has reverse transcriptase activity in the presence of manganese ions, and DNA polymerase activity in the presence of magnesium ions. This is particularly useful for overcoming problems associated with secondary structures. Another advantage of using such enzymes is the possibility of combining the RT and PCR reactions into a single-tube method.

3.2 The polymerase chain reaction as a tool for quasispecies analysis

The polymerase chain reaction (PCR) technique has provided a method of amplifying defined regions of DNA (or cDNA obtained by reverse transcription of RNA) and is the starting point for all the analyses described in this chapter. PCR consists of repeated cycles of heat denaturation of the DNA, annealing of oligonucleotide primers to the target DNA, and extension of the annealed primers using a DNA polymerase. In theory, the process is capable of producing approximately 1 million copies of the target of DNA from as little as one molecule of template DNA after 25 cycles of amplification. However, in practice, the yield and purity of single-round PCR products after amplification of DNA or cDNA derived from clinical samples is inadequate for the analyses described in this chapter. Therefore, the sensitivity and specificity of PCR amplifications can be improved by the use of 'nested primer' reactions (11). In this technique a small aliquot of the products from an initial PCR is used as the template for a second similar reaction, using primers that are complementary to regions of the DNA internal to the primers used for the first-round reaction. This procedure can yield as much as 1 μg of product DNA from as little as one target molecule after 25 cycles for each primer pair (11, 12).

PCR product accumulation is usually verified by analysing an aliquot of the PCR reaction by submarine electrophoresis using a 2% gel (see *Protocol 5*). The DNA fragments are visualized under ultraviolet illumination following ethidium-bromide staining. To obviate the need for a separate staining step, the ethidium bromide can be included in the gel and buffer solutions at a concentration of 0.5 μg/ml of buffer. Gel electrophoresis can be carried out using either Tris–borate–EDTA (TBE) or Tris–acetate–EDTA (TAE) buffer, provided the same type and concentration of buffer is used to prepare the gel and to perform the electrophoresis. To permit size estimation a double-stranded DNA ladder of known fragment lengths is run alongside the samples.

The high sensitivity of PCR can be an inherent problem leading to the production of false-positive results. These can be caused by sample cross-contamination (which can be prevented by using filter pipette tips or positive-displacement pipettes) or, more commonly, by carryover contamination with previous PCR products. The latter problem can be overcome by using separate, designated equipment for setting up PCR's and handling PCR products and also by physically separating the areas where reactions are set up and products analysed (13). Another method of controlling PCR contamination is to use dUTP instead of dTTP in the reaction mix or in the primers and to include uracil-*N*-glycosylase (UNG) in the reaction. Previous amplification products are destroyed prior to thermal cycling in an initial incubation at 37 °C, whilst sample DNA remains unaffected. The enzyme works by removing the uracil base to leave an apyrimidic site that is sensitive to cleavage in an alkali pH, especially at elevated temperatures (14). However, the latter is incompatible with the protocol used for the UNG-mediated, asymmetrical PCR generation of single-stranded DNA (see *Protocol 17*).

Protocol 4

Generalized polymerase chain reaction (PCR) method

Equipment and reagents

- Thermal cycler (PCR) machine
- 10 × PCR buffer: 100 mM Tris–HCl pH 9.0 (25 °C), 500 mM KCl, 1.0% Triton X-100
- 5 units/μl *Taq* DNA polymerase
- 25 mM $MgCl_2$
- dNTP mix (2.5 mM of each dATP, dCTP, dGTP, and dTTP)
- Template-specific sense and antisense primers (5 pmol/μl)
- Mineral oil
- DNA or cDNA (see *Protocol 3*) template for amplification

Protocol 4 continued

Method

1. For each sample, set up a reaction mix containing 2.5 μl of 10 × PCR buffer, 2 μl of $MgCl_2$, 1 μl of each primer, 2 μl of dNTP mix, 0.1 μl of *Taq* DNA polymerase, plus sterile demonized water such that the final reaction volume after the DNA sample is added (step 3) is 25 μl.[a]
2. Overlay the reaction mix with 1–2 drops of mineral oil.
3. Add the DNA or cDNA sample (usually 1–5 μl of sample is used).
4. Transfer the tubes to the thermal cycler and carry out the PCR with a temperature profile and cycle number optimized for each primer pair.[b]
5. If necessary, analyse the PCR products using agarose gel electrophoresis (see *Protocol 5*).

[a] This is best achieved by setting up a master mix sufficient for the total number of reactions, then aliquoting this into the sample tubes, then overlaying with 1–2 drops of oil.

[b] A typical profile for amplification of a 300-base pair fragment using 20mer primers is 94 °C for 30 seconds, 55 °C for 30 seconds, 72 °C for 90 seconds.

Protocol 5

Agarose gel electrophoresis

Safety:

Warning—ethidium bromide is a very powerful carcinogen.

Wear suitable eye and skin protection when using ultraviolet light sources.

Equipment and reagents

- Electrophoretic grade agarose
- 10 × TBE buffer: 108 g Tris base, 55 g boric acid, 40 ml 0.5 M EDTA pH 8.0, in a final volume of 1 litre; or 50 × TAE buffer: 242 g Tris base, 57.1 ml glacial acetic acid, 100 ml EDTA pH 8.0, in a final volume of 1 litre
- 5 mg/ml ethidium bromide solution[a]
- Submarine gel casting tray
- Gel comb
- Submarine gel electrophoresis tank
- Power pack
- 6 × gel loading buffer: 0.25% (w/v) Bromophenol Blue, 0.25% (w/v) xylene cyanol, 30% (v/v) glycerol
- Ultraviolet transilluminator

Method

1. Prepare the gel casting tray by sealing the ends, and if necessary the sides, with autoclave tape.
2. Add 2 g of agarose to 100 ml of 1 × TBE or 1 × TAE buffer in a 250 ml conical flask.
3. Carefully boil the TAE or TBE buffer to dissolve the agarose.
4. Allow the agarose solution to cool to 60 °C, then add 10 μl of the ethidium bromide solution.

Protocol 5 continued

5. Pour the agarose solution into the gel casting tray to a depth of approximately 1.0 cm.
6. Insert the gel comb. The gel comb height should be set so as to form wells in the gel without touching the surface of the gel casting tray.
7. Allow the gel to set for 1 hour at room temperature, then remove the comb taking care not to damage the wells.
8. Place the gel in the electrophoresis tank and add sufficient TBE or TAE buffer, containing 0.1 μl of the ethidium bromide solution/ml of buffer, to fully immerse the gel. Use the same buffer type used to prepare the gel.
9. Add 2 μl of 6 × gel loading buffer to 10 μl of each DNA sample, mix, then load into the wells of the agarose gel.
10. Carry out electrophoresis at a constant voltage of 10 volts per centimetre of gel for 45–120 minutes, ensuring that the anode is located at the end furthest away from the sample wells.
11. Visualize the DNA fragments using the ultraviolet transilluminator.

3.3 Polyacrylamide gel electrophoresis

Polyacrylamide gel electrophoresis (PAGE) is capable of high-resolution separation of a number of biological molecules including DNA. Polyacrylamide gels are formed when monomers of acrylamide are polymerized in the presence of free radicals to form long chains. Inclusion of methylenebisacrylamide into the acrylamide solution results in crosslinkage of the polyacrylamide chain, and thus the generation of a porous acrylamide gel. The free radicals essential for gel polymerization are supplied by ammonium persulfate, and these are stabilized by the presence of *N,N,N′,N′*-tetramethylethylenediamine (TEMED).

The porosity of the gel, and hence the relative mobility of DNA, is determined by the percentage of acrylamide used to make the gel and also by the amount of bisacrylamide crosslinker used, although for separating DNA fragments the ratio of acrylamide to bisacrylamide is usually 29:1. Effective separation of DNA ranging from 6 to 2000 base pairs can be achieved by varying the percentage of total acrylamide (acrylamide plus bisacrylamide) used to prepare the gel. In the context of quasispecies analysis PAGE is used for the resolution of PCR products in, for example, single-stranded conformation polymorphism analysis (SSCP) and heteroduplex mapping and tracking analyses. For such applications a gel with an acrylamide concentration of 6% gives good resolution of PCR products ranging from 100 to 500 base pairs. The progression of electrophoresis can be monitored by the inclusion of marker dyes such as xylene cyanol and Bromophenol Blue.

To facilitate electrophoresis, a suitable buffer has to be included in the gel and the same buffer is used to carry out electrophoresis. Due to the relatively long run-times used in PAGE, 1 × Tris-borate-EDTA (TBE) buffer is used. Electrophoresis is usually carried out using an electrical potential of 1–10 volts per centimetre of gel (10). Higher voltages can cause unacceptable heating of the gel, resulting in DNA denaturation and altered migration rates. The temperature at which a non-denaturing gel is run is critical, particularly with analytical methods that rely on secondary structure formation, for example SSCP. For this reason, the temperature of the gel plates can be thermostatically controlled in some modern electrophoresis equipment. Another consideration in PAGE is the thickness of the gel. Gels 0.2–0.4 mm in thickness are less prone to heating and therefore give good results in methods such as SSCP analysis. However, these gels are extremely fragile and do not perform well if the DNA has to be visualized by chemical staining. By contrast gels 1–1.2 mm thick are stronger and therefore are more suitable if the DNA is to be visualized by ethidium bromide or silver staining, but the increased thickness and corresponding resistance of the gel makes them prone to overheating during electrophoresis. In addition, the amount of DNA that has to be loaded on to the gel increases with increasing gel thickness.

Modifications can be employed to improve the resolving power of PAGE for mutational analyses. The simplest is to increase the gel length or gel run, but this has a limited effect. Alternatively, additional chemicals can be added that increase the resolving power of the gel matrix. For example, the inclusion of glycerol or urea can increase the resolution of PAGE for SSCP analysis and HMA/HTA, respectively. More recently, several biochemical companies have started manufacturing and marketing ready-made polyacrylamide gel solutions, and some with reportedly better resolving power than simple acrylamide solutions. Mutation detection enhancement (MDE, Flowgen) gels (see *Protocol 7*) are one such system that we use routinely for studying HCV populations by SSCP analysis. This gel formulation has been used in a variety of applications including SSCP and HMA/HTA and has proven very effective at resolving mutations in PCR products of various lengths (15, 16).

Denaturing polyacrylamide gels are used when resolution of single-stranded DNA fragments is required, for example in DNA sequence analysis or in PCR product-length polymorphism analysis. The gel is formed in essentially the same way as a non-denaturing gel except a denaturant is included in the gel solution, the most frequently used being urea at a 6 M final concentration. In denaturing gels, the DNA fragment migration rate is proportional to the fragment size, irrespective of the base composition and sequence. Polyacrylamide gels containing urea do not dry effectively under vacuum until the urea denaturant has been removed. Fortunately, this occurs during the gel fixation step, provided the gel solution is relatively fresh and free from urea. Finally, although it is possible to silver-stain DNA resolved in denaturing gels, trace amounts of urea can adversely affect the staining process, for this reason denaturing gels are rarely stained by this method.

Protocol 6

Preparation of non-denaturing and denaturing polyacrylamide gels

Equipment and reagents

- Vertical gel electrophoresis apparatus with 40 × 20 cm glass plates
- 2 gel spacers of 0.4–1.2 mm thickness
- Non-denaturing polyacrylamide solution: 29 g of acrylamide and 1 g of bisacrylamide dissolved in 100 ml of demonized water (29:1) **or** denaturing polyacrylamide solution: 5.7 g acrylamide, 0.3 g bisacrylamide, 42g urea, 10 ml 10 × TBE dissolved in demonized water to yield a final volume of 100 ml
- Gel comb of the same thickness as the gel spacers
- Acrylease (Stratagene)
- Gel sealing tape
- 10% APS: 0.1 g ammonium persulfate in 1 ml of demonized water
- *N,N,N′,N′*-tetramethylethylenediamine (TEMED)
- 50 ml syringe

Method

1. Clean the glass plates with a detergent solution, thoroughly rinse in demonized water, then polish using ethanol applied to paper towels.
2. Spray a light coating of Acrylease to the large glass plate, allow to dry for 1 minute then remove excess Acrylease by wiping with ethanol.
3. Assemble the glass plates according to the manufacturer's instructions. In most cases this is achieved as follows: place the spacers along the side-edges of the large glass plate then place the smaller glass plate on top. Then clamp the bottom and sides of the plate/spacer/plate sandwich using bulldog clips and seal with gel sealing tape. Take particular care to ensure that the plates are fully sealed, especially at the bottom corners.
4. (a) For each 100 ml of *non-denaturing gel solution* required (which will depend upon the electrophoresis apparatus being used) add 26.6 ml of 30% acrylamide, 62.7 ml of demonized water, 10 ml of 10 × TBE, 700 μl of 10% APS, and 35 μl of TEMED.

   (b) For each 100 ml of *denaturing gel solution* required add 500 μl of 10% APS and 50 μl of TEMED to 100 ml of the denaturing polyacrylamide solution.
5. Immediately, draw the solution into a 50 ml syringe and carefully fill the space between the sealed glass plates with the gel solution, avoid introducing air bubbles.[a]
6. Lay the gel plates at an angle of approximately 10–15 degrees, insert the gel comb, clamp it in position using bulldog clips, and allow the gel to polymerize.

[a] If any air bubbles are introduced they can be expelled by gentle tapping or dislodged by carefully inserting a gel spacer thinner than those used to pour the gel.

Protocol 7

Preparation of mutation detection enhancement (MDE) gels

Equipment and reagents

- Vertical gel electrophoresis apparatus with 40 × 20 cm glass plates
- 2 gel spacers of 0.4–1.2 mm thickness
- Gel comb of the same thickness as the gel spacers
- Acrylease (Stratagene)
- 2 × MDE gel solution (Flowgen)
- 10 × TBE buffer (see *Protocol 5*)
- 10% APS (see *Protocol 6*)
- *N,N,N′,N′*-tetramethylethylenediamine (TEMED)

Method

1. Prepare the glass plates as described in *Protocol 6*.
2. For each 100 ml of gel required (which is dependent on the size of the glass plates and thickness of the spacers used) add 50 ml of 2 × MDE gel solution, 6 ml of 10 × TBE, 44 ml demonized water, 400 μl of 10% APS, and 40 μl of TEMED.
3. Pour the gel as described in *Protocol 6*.

Protocol 8

Gel electrophoresis

Equipment and reagents

- Polyacrylamide gel tank
- 1 × TBE for polyacrylamide gels, or 0.6 × TBE for MDE gels (see *Protocol 5* for 10 × TBE recipe)
- 10 μl Hamilton syringe or 10 μl gel-loading pipette tips
- Pre-cast polyacrylamide (see *Protocol 6*) or MDE (see *Protocol 7*) gel

Method

1. Place the pre-cast polyacrylamide or MDE gel into the gel electrophoresis tank.
2. Depending on whether a polyacrylamide or MDE gel is used add 1 × or 0.6 × TBE buffer to the upper and lower buffer chambers.
3. Carefully remove the comb from the gel. For sharks' tooth combs remove the comb, invert it and reinsert the comb so that the teeth just penetrate the acrylamide gel. Take care not to insert the comb too far or this will result in well distortion. Use a Pasteur pipette to rinse the wells of the gel with either 1 × or 0.6 × TBE buffer to remove unpolymerized acrylamide.
4. Load 2–10 μl of DNA sample into each well.
5. Run at a constant voltage of 20 V per cm of gel for between 12–24 hours.

Protocol 8 continued

6. Following gel electrophoresis, remove the glass plates and gel from the tank and lay them horizontally with the Acrylease-treated plate uppermost. Carefully separate the glass plates ensuring that the gel does not tear.
7. Visualize the DNA by silver staining (see *Protocol 9*) or by autoradiography (see *Protocol 11*).

3.4 Methods for labelling and staining DNA

DNA resolved in non-denaturing gels can be stained using ethidium bromide, although this method is relatively insensitive. Alternatively, the DNA band pattern can be visualized by silver staining (see *Protocol 9*) or, if a second-round radiolabelled PCR was carried out (see *Protocol 10*), by autoradiography (see *Protocol 11*). The latter method is often preferred because the non-denaturing gels used in quasispecies analyses such as SSCP and heteroduplex assays are often very thin (0.2–0.4 mm) and can easily break during the silver-staining process. However, silver staining obviates the need to use radioactive isotopes, and gel damage can be minimized by the use of thicker gels or by attaching the gel to one of the electrophoresis glass plates by the use of a chemical gel binder such as bind-silane (Sigma). Once the gel has been silver stained and the band pattern recorded, the gel is removed by scraping the plate and residual gel removed by washing the plates thoroughly. There are several silver-staining kits commercially available, although the method described in *Protocol 9* is simple to carry out and can be used to stain RNA and DNA. Irrespective of the source of reagents used for silver staining it is essential that the highest quality demonized or double-distilled water is used to prevent precipitation of the silver ions. In addition, when adding the developing solution best results are obtained if the gel is first rinsed in a small volume of developer, which is discarded before adding sufficient developer to fully submerge the gel. This removes trace amounts of unreacted silver ions and improves gel staining. When properly stained the DNA bands should appear dark-brown to black against the yellow gel background. If the background appears very dark then the purity of the water used throughout the staining process should be checked. Finally, silver staining is also a very effective protein stain, therefore always ensure that gloves are worn when handling the gel and staining solutions, and always use clean staining dishes.

Radioactively labelling the DNA during the PCR enables the DNA products to be visualized by autoradiography. The most commonly used isotope for DNA labelling is [^{35}S]dATP, which is a weak beta-particle emitter. However, safety concerns relating to the production of radioactive hydrogen sulfide gas during the amplification process has resulted in the increased use of [^{33}P]dATP. The advantage of using these isotopes is that the band patterns are well defined, unlike those produced when using the more energetic phosphorous isotope

^{32}P. However, because ^{35}S and ^{33}P are low-energy sources the gels can not be exposed to X-ray film until they have been dried under vacuum, another problem being that the highly sensitive photographic film used to visualize the DNA fragments is relatively expensive. To minimize radiolabel usage, yet at the same time ensure maximum incorporation, it is essential to carry out the PCR in the presence of reduced dNTPs (see *Protocol 10*). Although dNTP concentrations of 200 mM are typically used in PCRs, this can be reduced by as much as tenfold to ensure maximum label incorporation with minimal loss of product yield.

Protocol 9

Silver-staining polyacrylamide and MDE gels[a]

Equipment and reagents

- Non-denaturing gel fixing solution: 400 ml ethanol, 50 ml acetic acid, 550 ml of demonized water
- 10% (v/v) ethanol
- 11 mM silver nitrate
- Developing solution: 37.5 g sodium hydroxide and 10 ml of 38% methanal (formaldehyde) in a total volume of 1 litre
- 5% (v/v) acetic acid

Method

NB: All incubations are carried out at room temperature with gentle agitation.

1. Carefully transfer the gel to a tray containing sufficient gel fixing solution to fully submerge the gel and incubate for 30 minutes.
2. Replace the gel fixing solution with an equal volume of 10% ethanol and incubate for 30 minutes.
3. Replace the 10% ethanol with an equal volume of 11 mM silver nitrate and incubate for a further 30 minutes.
4. Remove the silver nitrate solution and rinse the gel twice with demonized water then twice with 100 ml of developing solution.
5. Add sufficient developer to fully submerge the gel, then incubate the gel until the nucleic acid appears dark-brown/black and the gel background appears light yellow.
6. Immediately replace the developer with an equal volume of 5% acetic acid, and incubate for 10–30 minutes.
7. Record the gel by photography or transfer the gel on to filter paper and dry it using a slab gel drier attached to a vacuum pump.

[a] Protocol kindly provided by Dr Graham Beards, Birmingham Public Health Laboratory, Birmingham B9 5SS.

Protocol 10

Production of radiolabelled DNA in a second-round PCR

Equipment and reagents

- First-round PCR products (see *Protocol 4*)
- 10 μCi/μl [^{35}S]dATP (>1000 mCi/mmol)
- 10 × PCR buffer: 100 mM Tris–HCl pH 9.0 (25 °C), 500 mM KCl, 1.0% Triton X-100
- 5 units/μl *Taq* DNA polymerase
- 25 mM $MgCl_2$
- Mineral oil
- dNTP mix (2.5 mM each of dATP, dCTP, dGTP, and dTTP)
- 5 pmol/μl template-specific sense and antisense primers
- Thermal cycler
- PCR equipment and reagents (see *Protocol 4*)

Method

1. Set up a master reaction mix containing 2.5 μl of 10 × buffer, 1.5 μl of $MgCl_2$, 17.4 μl of sterile demonized water, 0.5 μl of each primer, 0.25 μl dNTP, 0.25 μl [^{35}S]dATP, 0.1 μl of *Taq* polymerase per DNA sample.
2. For each sample, add 23 μl of the PCR mix to a microcentrifuge and overlay with two drops (25–50 μl) of mineral oil.
3. Dilute the first-round PCR products 1:20 with sterile demonized water.
4. Add 2 μl template DNA, and microcentrifuge at 13 000 g for 30 seconds.
5. Transfer the tubes to the thermal cycler and carry out the PCR (see *Protocol 4*).

Protocol 11

Autoradiography

Equipment and reagents

- Gel fixing solution: 10% (v/v) glacial acetic acid, 10% (v/v) methanol
- Staining dishes slightly larger in size than the gel
- 3M filter paper slightly larger than the gel
- Slab gel drier with vacuum pump
- Hypersensitive photographic film
- X-ray film cassette
- Photographic developing solution (e.g. Ilford PQ Universal developer diluted 1:10 in demonized water)
- Photographic fixative (e.g. Ilford Hypam Rapid Fixer diluted 1:5 in demonized water)

Method

1. Carefully transfer the glass plate on to which the polyacrylamide or MDE gel is attached into a staining dish, gel-side uppermost, containing 2 litres of gel fixing solution, taking care not to dislodge the gel. Allow the gel to fix for 20 minutes.

Protocol 11 continued

2. Transfer the gel to a sheet of filter paper by laying the sheet on top of the gel, ensuring no air bubbles are trapped, then carefully lift off the paper.
3. Place the filter paper with the gel uppermost on to the slab gel drier and dry under vacuum at 80 °C for 1–2 hours.
4. Once dry, place the filter paper gel-side uppermost into the X-ray film cassette, then under safelight conditions place a film on top of this, ensuring that the photographic emulsion side is in direct contact with the dry gel. Close the film cassette and expose the film to the gel for 18–72 hours.
5. Under safelight conditions, open the film cassette and place the film in a staining dish containing sufficient developing solution to cover the film. Incubate the film for 4 minutes at room temperature with gentle agitation.
6. Transfer the film to a staining dish containing sufficient film fixing solution to cover the film. Incubate at room temperature, with gentle agitation for 2 minutes.
7. Rinse the film in water and then dry.

4 Single-stranded conformation polymorphism analysis

Single-stranded conformation polymorphism analysis (SSCP) has been used extensively in the analysis of HCV quasispecies, most notably for the study of the effects of host immunity and interferon treatment on the composition and evolution of the virus population (17–20). In theory, SSCP can resolve single base-pair differences in DNA of up to 1 kilobase in length and identify DNA species that constitute less than 5% of the total population. The first step in SSCP analysis is PCR amplification of the region of the genome under study. It is imperative that the DNA used in SSCP analysis is derived using a PCR method that has been optimized to only produce a single band of DNA. Therefore, a nested primer PCR is most often used. To carry out SSCP analysis, double-stranded, second-round PCR products are simply heat-denatured in a gel loading buffer containing formamide, which prevents the single-stranded DNA from reannealing. The denatured PCR products are then resolved using non-denaturing polyacrylamide gels. As the PCR products migrate into the gel, secondary structures within the PCR products form in a sequence-determined manner. Therefore, DNAs of different sequences adopt different conformations and these migrate through the non-denaturing gel at different rates. Following electrophoresis the DNA band pattern in each sample can be visualized and compared (see *Figure 1*).

As with all 'population'-based analyses, it is essential that the amplified DNA is derived from a representative sample of the virus quasispecies. Therefore, it is necessary to first obtain an approximate titre of the number of cDNA molecules in the original sample. This can be estimated by nested-primer PCR amplifica-

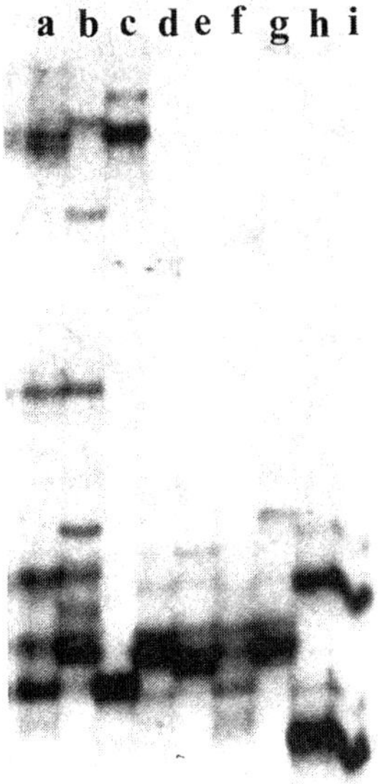

Figure 1 Single-stranded conformation polymorphism analysis of the first hypervariable region of the hepatitis C virus *E2* genes of HCV in plasma and different cell populations obtained from four different patients. Patient 1: plasma (a); B cells (b); $CD3^-$, $CD14^-$, $CD19^-$ cells (c). Patient 2: plasma (d); B cells (e). Patient 3: plasma (f); B cells (g). Patient 4: plasma (h); and $CD3^-$, $CD14^-$, $CD19^-$ cells (i). (Taken from Hatim Al-Jarrah, MSc Thesis, University of Nottingham, (1998).)

tion of aliquots of twofold dilutions of the cDNA sample followed by agarose gel electrophoresis (see *Protocol 16*). One problem associated with SSCP analysis is the run-to-run variation, which can be caused by a number of factors. The most frequent problem is variation in gel temperatures between electrophoresis runs, and this alters the amount and types of secondary structures that form, with a resulting alteration in the mobility and therefore band pattern obtained. For this reason, samples for comparison have to be resolved on the same gel.

Protocol 12

Single-stranded conformation polymorphism analysis (SSCP)

Equipment and reagents

- Formamide gel loading buffer: 95% (v/v) formamide, 10 mM EDTA, 0.1% (w/v) Bromophenol Blue, 0.1% (w/v) xylene cyanol

Method

1. Prepare a non-denaturing polyacrylamide (see *Protocol 6*) or MDE (see *Protocol 7*) gel.
2. PCR-amplify, using either the general (*Protocol 4*) or the radiolabelling PCR method (*Protocol 10*), the genomic region of interest.
3. For each sample pipette 4 μl of formamide gel loading buffer into a microcentrifuge tube.
4. Add 4 μl of the radiolabelled PCR products to the formamide gel loading buffer and mix.

Protocol 12 continued

5. Heat the tube to 95 °C for 3 minutes then snap cool by placing the tubes on either ice or a dry-ice/ethanol mix.
6. Immediately load 0.5–5 μl of the sample into the wells of the non-denaturing polyacrylamide gel.
7. Carry out electrophoresis (see *Protocol 8*), at a constant voltage of 10 volts per centimetre.
8. Visualize the DNA fragments by silver staining (see *Protocol 9*) or by autoradiography (see *Protocol 11*).

4.1 Heteroduplex and quantitative heteroduplex tracking analyses

Heteroduplex mapping was first used in the field of human genetics, and has been used to identify a number of human disease-associated genetic mutations (21). More recently, this technique has been used in the study of virus genetic heterogeneity. Initially the technique was used to differentiate HIV-1 subtypes, so-called genotyping assays (22), but more recently it has been used to study HIV quasispecies (23, 24).

The principles underlying heteroduplex mapping, tracking, and quantitative heteroduplex tracking analyses (HMA, HTA and QHTA, respectively) are the same. Basically, PCR products derived from the amplification of different variants are heat-denatured then cooled and allowed to form heteroduplexes. HTA is usually carried out by mixing labelled PCR products derived by amplification of a single (clonal) variant (probe) with an excess of unlabelled products obtained after amplification of a sample of cDNA or DNA representative of the virus population (test), which are then denatured and allowed to form heteroduplexes. Some of the labelled PCR products self-anneal forming a homoduplex that has full base complementarity throughout its length. However, because the test sample is used in excess, most of the probe DNA strands will anneal with DNA strands derived from the test sample. If the sequence is identical to the probe sequence homoduplexes will form. However, if the DNA sequence is different then double-stranded DNA with regions of base-mismatching will form—these are known as heteroduplexes. The migration rates of the heteroduplexes in non-denaturing PAGE is slower than that of the homoduplexes, and is directly proportional to the amount of genetic heterogeneity. However, it is important to appreciate that cross-hybridization between double-stranded DNA products derived from two unique variants (probe sequence and variant sequence) results in the formation of two heteroduplexes: one heteroduplex containing the sense strand from the probe hybridized to the antisense strand of the variant, and vice versa. In addition, some of the probe sense and antisense strands will reanneal to form homoduplex DNA. Therefore, if there is one distinct variant in the test sample then the HTA will yield three bands and for each additional variant there will be an additional two heteroduplex bands (see *Figure 2*). An important

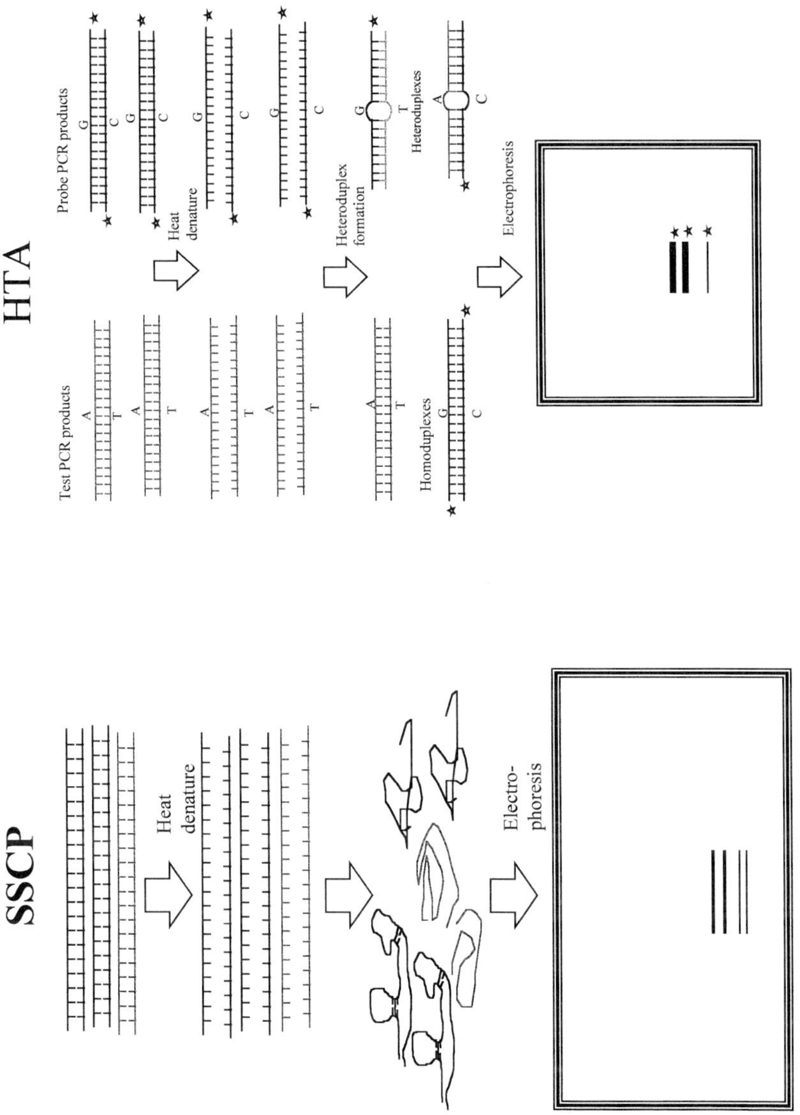

Figure 2 Diagrammatic comparison and outline of single-stranded conformation polymorphism (SSCP) and heteroduplex tracking (HTA) analyses.

refinement of the HTA technique was made by Zhu *et al.* (24), who used a single-stranded rather than a double-stranded DNA probe to generate the heteroduplexes. Therefore, each distinct variant produces only one heteroduplex, thus making it easier to quantify the number of variants and also their relative frequency in the original sample. This adaptation is known as QHTA. HMA works on the same principle, but is performed by denaturing the PCR products obtained by amplification of a sample of DNA/cDNA under study and then snap-cooling to allow both heteroduplexes and homoduplexes to form.

One advantage of HTA/QHTA over SSCP analysis is in the interpretation of the results. Unlike SSCP, the difference in the migration rates of the heteroduplexes compared to the homoduplexes is approximately proportional to the genetic distance of the two variants. For HTA, the average mobility of the two heteroduplex forms has to be calculated. Such a calculation is difficult when more than one variant is present because of the difficulty in correctly identifying which heteroduplexes were derived from which variant. Calculating relative mobilities for heteroduplexes generated in QHTA is less problematic as each variant produces either one homoduplex or heteroduplex band.

Although HTA/QHTA are powerful techniques for quasispecies analysis and subsequent prediction of virus phylogenies, there are limitations. First, the techniques only demonstrate genetic complexity and relatedness, they do not provide an insight into the nature of genetic changes. Also, the presence of DNA insertions and deletions adversely affects the mobility of heteroduplexes, such that it often gives an overestimate of the genetic distance between the test and unknown sequence (24). Obviously this effect is more noticeable as the percentage of inserted or deleted nucleotides increases. Finally, the relative position of the mismatches within the heteroduplex is an important determinant of mobility.

Protocol 13

Heteroduplex tracking analysis (HTA)

Equipment and reagents

- Labelled probe PCR products derived from amplification of a single virus variant cDNA/DNA molecule[a]
- Test PCR products derived by amplification of cDNA/DNA representative of the virus quasispecies under study
- 10 × annealing mix: 1 M NaCl, 100 mM Tris–HCl pH 7.8, 20 mM EDTA
- 0.2 M EDTA
- 6 × gel loading buffer (see *Protocol 5*)
- Pre-cast, non-denaturing polyacrylamide (see *Protocol 6*) or MDE (see *Protocol 7*) gel.
- Electrophoresis equipment and reagents (see *Protocol 8*)
- Equipment and reagents for autoradiography (see *Protocol 11*)

Method

1. Add 1 μl of 0.2 M EDTA to each PCR product tube to inactivate the *Taq* polymerase.

Protocol 13 continued

2. For each sample add 1 μl of 10 × annealing buffer and 1 μl of a 1:10 dilution of the EDTA-treated probe PCR product to a microcentrifuge tube.
3. Add 8 μl of the each test PCR product to the microcentrifuge tube.
4. Heat the tubes to 95 °C for 2 minutes, then cool rapidly in ice.
5. Add 2 μl of the gel loading buffer to each sample.
6. Load 2–10 μl of the DNA sample into the wells of a pre-cast, non-denaturing polyacrylamide (see *Protocol 6*) or MDE (see *Protocol 7*) gel.
7. Carry out the electrophoresis for between 4–18 hours at 10 volts per centimetre, depending on the length and thickness of the gel. (see *Protocol 8*).
8. Visualize the DNA products using autoradiography (see *Protocol 11*).

[a] This can be derived by amplification of plasmid-containing cloned DNA encompassing the genomic regions of interest or it can be PCR products derived from amplification of a single cDNA/DNA molecule.

4.2 Length polymorphism analysis (LPA)

Certain virus genomes contain regions that are highly variable in length. Again, this property is typified by the HIV-1 envelope gene whose hypervariable regions demonstrate an unusually high level of length variation, consisting of insertions or deletions of usually three base pairs (11, 25, 26), thus ensuring maintenance of the open reading frame. A typical procedure for LPA is described in *Protocol 14*. Briefly, LPA consists of the resolution of radiolabelled PCR products, which have been rendered single-stranded by heat denaturation, using denaturing PAGE. This technique is able to discriminate between variants differing by only 1% in length, representing as little as 10% of the total population. The resolving power of the technique decreases as the length of the PCR product under study increases; the typical size for PCR products for use in LPA is between 100 and 300 base pairs. The optimal time for carrying out the electrophoresis, and the amount of DNA to load on to the denaturing gel has to be determined empirically for each application. However, it should be noted that doublet bands are sometimes observed if the electrophoresis has been carried out for too long—this is caused by subtle differences in the migration rates of the sense and antisense strands due to differences in the base composition of each strand. For this reason it is advisable to run denatured PCR products obtained after amplification of a single variant alongside the samples under test. Estimation of the frequency of individual length variants can be made using a densitometer, and the size of these can be estimated by comparing the variant band migration rates with those of a radiolabelled DNA ladder which has been run alongside the sample (see *Figure 3*).

Protocol 14

Length polymorphism analysis

Reagents

- Formamide gel loading buffer: 98% demonized formamide, 10 mM EDTA pH 8.0, 0.25% xylene cyanol, 0.25% Bromophenol Blue
- Denaturing polyacrylamide gel (see *Protocol 6*)
- Second-round PCR products (see *Protocol 10*)
- Electrophoresis equipment and reagents (see *Protocol 8*)
- Equipment and reagents for autoradiography (see *Protocol 11*)

Method

1. PCR-amplify the virus cDNA or provirus DNA using a nested-primer radiolabelling PCR method (see *Protocol 10*).
2. Prepare a denaturing polyacrylamide gel (see *Protocol 6*).
3. For each sample pipette 4 μl of formamide gel loading buffer into a microcentrifuge tube.
4. Add 4 μl of second-round PCR products to the formamide gel loading buffer.
5. Incubate the sample at 95 °C for 3 minutes to denature the PCR products then snap-cool the sample by placing the tubes on either ice or a mixture of dry-ice and ethanol.
6. Immediately carry out electrophoresis (see *Protocol 8*), using 0.5–5 μl of each sample.
7. Visualize the DNA by autoradiography (see *Protocol 11*).

4.3 Point mutation assays

Point mutation analyses are frequently used for detecting drug-resistant genotypes of viruses such as HIV-1. Several assay formats have been described including discriminatory PCR (27), probe extension assays (28, 29), and line probe assays (30), all of which require prior knowledge of the type and location of the drug-resistance mutation. Of these, the simplest to perform is probably discriminatory PCR. In this assay, two discriminatory primers, usually sense primers with 3′ termini which are complementary to either the wild-type or resistant DNA sequence, are used in separate PCRs with a common antisense primer (see *Figure 4*). If the population consists of either wild-type or resistant viruses then PCR products will only accumulate in the reaction containing the corresponding discriminatory primer. If the population consists of both wild-type and mutant viruses then specific products will accumulate in both reaction tubes (27, 31). The general PCR method described in *Protocol 4* can be used with discriminatory primers, and usually the discriminatory primers are used in the second round of a nested primer PCR.

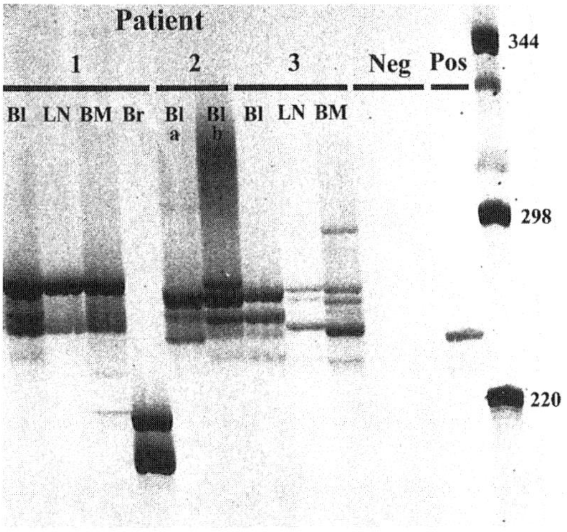

Figure 3 Length polymorphism analysis of the first and second hypervariable regions of the HIV-type 1 envelope protein in various tissues and body fluids obtained from three different patients. Patients 2 and 3 are epidemiologically related transmission events (baby/mother, respectively). Bl, blood; LN, lymph node; BM, bone marrow; Br, brain. Lower case letters indicate two sequential samples from the same patient. Neg/Pos, negative and positive PCR controls, respectively. The size of the DNA ladder is indicated on the right-hand side (base pairs).

Protocol 15

Point mutation assay[a]

Equipment and reagents

- Microtitre plate or strips
- 1 mg/ml stock solution of streptavidin in distilled water
- 100 mM Tris-HCl pH 7.5
- Phosphate-buffered saline (PBS) pH 7.2: 8 g NaCl, 0.2 g KCl, 1.44 g Na_2HPO_4, 0.24 g KH_2PO_4 in 88 ml, adjust pH by adding 0.1 M NaOH or 0.1 M HCl to pH 7.2, adjust volume to 1000 ml using demonized water
- Phosphate-buffered saline/BSA (PBSB): PBS containing 1% bovine serum albumin
- PCR equipment and reagents (see *Protocol* 4)
- TTB: 100 ml Tris-HCl pH 7.5 containing 0.5% Tween-20
- 0.2 M NaOH
- PMA buffer: 40 mM Tris-HCl pH 7.5, 20 mM $MgCl_2$, 50 mM NaCl, 0.1% Tween-20)
- 10 pmol/ml oligonucleotide probe in PMA buffer
- 0.1 M dithiothreitol (DTT)
- Extension mix: 936 μl 10 × PCR buffer (see *Protocol* 4), 468 μl 0.1 M dithiothreitol, 6084 μl of distilled water, 104 μl 10% Tween-20, 8 μl *Taq* polymerase (5 U/μl)
- 20 μM of each ddNTP–FITC
- Horseradish peroxidase conjugated anti-FITC (diluted 1:1000 in PBS, 0.1% Tween-20, 1% bovine serum albumin)
- 1 mg/ml 3′,3′,5′,5′-tetramethylbenzidine (TMB) (0.1 mg of TMB dissolved in 100 μl of dimethylsulfoxide)
- 0.1 M sodium acetate pH 6.0
- Hydrogen peroxide 30% v/v
- 1 M H_2SO_4

Protocol 15 continued

Method

1. Dilute the stock streptavidin 1/50 in 100 mM Tris–HCl pH 7.5.
2. Add 75 μl to the wells of a microtitre plate or strip, sufficient for the number of assays to be run.
3. Incubate at room temperature for 18 hours.
4. Discard the diluted streptavidin solution and add 100 μl of PBSB to each well.
5. Incubate at room temperature for 1–2 hours.
6. PCR-amplify the virus subgenomic region under study (see *Protocol* 4).
7. Wash a streptavidin-coated plate twice with 100 ml TTB.
8. Dilute the PCR product 1/10 in TTB and add 25 μl to each of four microtitre plate wells.
9. Incubate the plate at 37 °C for 1 hour with shaking.
10. Discard the diluted PCR product and wash the plate five times with TTB.
11. Add 100 μl of 0.2 M NaOH to each well and incubate at room temperature for 15 minutes.
12. Wash each well five times with TTB.
13. Add 75 μl of the probe diluted in PMA buffer and incubate the plate at 50 °C for 1 hour.
14. Remove unbound probe by washing five times with TTB.
15. Dilute each ddNTP–FITC (88 μl of each ddNTP–FITC in 1728 μl of labelling mix), add 75 μl to each corresponding microtitre well, and incubate at 50 °C for 30 min.
16. Wash five times with TTB.
17. Add 75 μl of diluted conjugate to each well and incubate at 37 °C for 1 hour with shaking.
18. Wash five times with TTB.
19. Prepare the TMB substrate (available from immunochemical suppliers) by adding 100 μl of TMB solution to 9.9 ml of sodium acetate, then add hydrogen peroxide to give a final concentration of 0.01%.
20. Add 100 μl of the TMB substrate and incubate at room temperature for 15–20 min.
21. Stop the reaction by adding 100 μl 1 M H_2SO_4.
22. Measure the absorbance at 450 nm.

[a] Kindly provided by Drs Anna Ballard and Liz Boxall, Birmingham Public Health Laboratories, Birmingham Heartlands Hospital, Birmingham, B9 5SS.

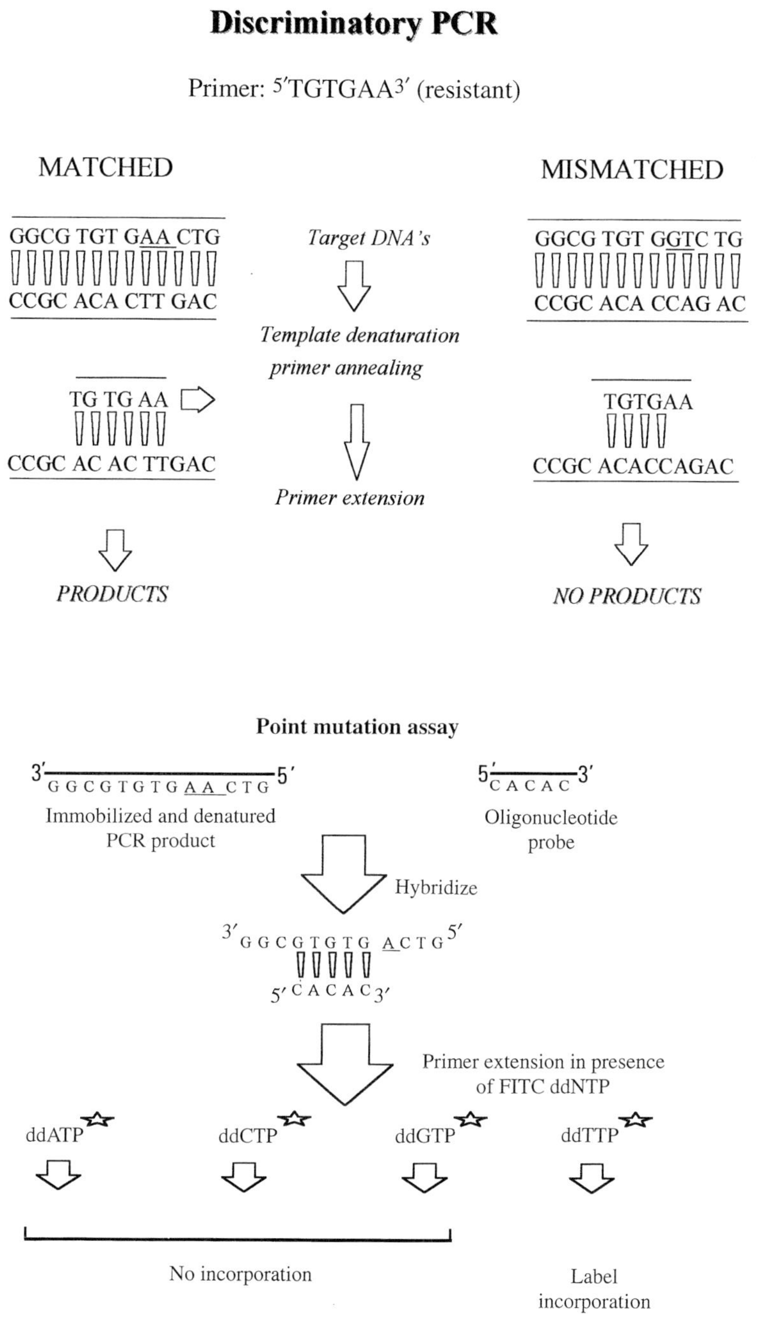

Figure 4 Diagram showing the principles of discriminatory PCR and point mutation assays for the detection of drug-resistance mutations.

A similar method for detecting point mutations is the probe extension assay (*Protocol 15*). In this assay, a probe is hybridized to immobilized and denatured PCR products. An extension solution containing a suitable DNA polymerase and either labelled ddATP, ddCTP, ddGTP, or ddTTP (ddNTPs) is added. For each test sample four separate hybridizations and extensions are set up, each one containing a different ddNTP. The ddNTPs incorporated during the extension step will be determined by the DNA sequences of individual PCR strands immediately downstream of the hybridized probe (see *Figure 4*). Earlier assays utilized radioactive ddNTPs, but alternative enzyme immunoassay formats are generally favoured in most diagnostic applications.

5 Sequence analysis

Although sequence analysis of virus quasispecies provides information for a relatively small sample of variants constituting a quasispecies, the quality of information obtained is unsurpassed. Most sequencing methods use PCR-generated templates that can be sequenced directly or cloned into vectors, from which the sequence of the insert DNA may be analysed and the gene products expressed and studied (3). However, *Taq* polymerase, the enzyme most frequently used to perform PCR, introduces errors during replication at a frequency of approximately 1 base/9000 per replication (32), and recombination between products from different target molecules can occur (33). Therefore, it is difficult to assess whether data obtained from cloned material is representative of the original sample. Point mutations introduced during thermal cycling can be reduced by using enzymes with a higher fidelity than *Taq* polymerase, for example *Pfu* or *Pwo* polymerases, but the problem of recombination remains. In addition, the population heterogeneity determined from analyses of cloned PCR products can be severely underestimated. This problem arises when the number of clones analysed exceeds the number of variants in the original DNA/cDNA sample. Therefore, to negate this problem it is advisable to determine the virus titre in the DNA sample before amplification, then to use an aliquot of DNA/cDNA containing a representative number of virus variants in the PCR which is used for subsequent cloning and sequencing.

Direct sequencing of PCR products overcomes the problem of errors introduced by *Taq* polymerase, as only the most frequent (consensus) sequence present will be observed (34). This approach has been used to assess the relative frequency of drug-resistance mutations present in HIV-1 populations (35). However, directly sequencing the products from amplifications of a mixture of very variable molecules may not yield an easily determinable sequence. One approach developed to overcome this problem is to dilute the DNA template to the single molecule level and then carry out multiple PCR amplifications at this end-point dilution (see *Protocol 16*). A certain frequency of the reactions would be positive and, depending on the frequency, most would represent products derived from a single template DNA (or cDNA) molecule. This approach has been used very successfully to discriminate between different variants of HIV-1 (see *Figure 5*) (7, 11, 12, 26, 31).

Point mutations introduced during the amplification process do not present a problem in subsequent sequencing reactions as they will be present on only a minority of the amplimers, and PCR recombination can not occur as only one virus variant is present. Amplification of two or more genetically distinct template molecules, or errors introduced by *Taq* polymerase during the first cycle of the PCR, would appear as a band in two or more lanes at the same place in the sequence, and therefore any sequences containing such artefacts can be discarded. However, it is important to verify that doublet bands are not being caused by enzyme stops of sequencing compressions.

Sequence determination from some denatured PCR products is difficult, but several modifications to improve the methods have been adopted; e.g. snap-cooling of the PCR product after denaturation and carrying out the sequencing reaction in the presence of 10% dimethylsulfoxide (DMSO) to prevent strand reannealing (36), or the inclusion of DNA binding proteins to prevent the formation of secondary structures (37). An alternative to this is to produce single-stranded DNA for sequencing, either during the PCR by limiting the concentration of one of the primers (asymmetric PCR) (34) or by strand separation of the products following PCR by using streptavidin-biotin affinity separation (38). We have developed an asymmetric PCR method (see *Protocol 17*) in which the sense primer has uracil residues in place of thymidine residues. Following 20–30 cycles of PCR the primer is digested by the addition of uracil-*N*-glyco-

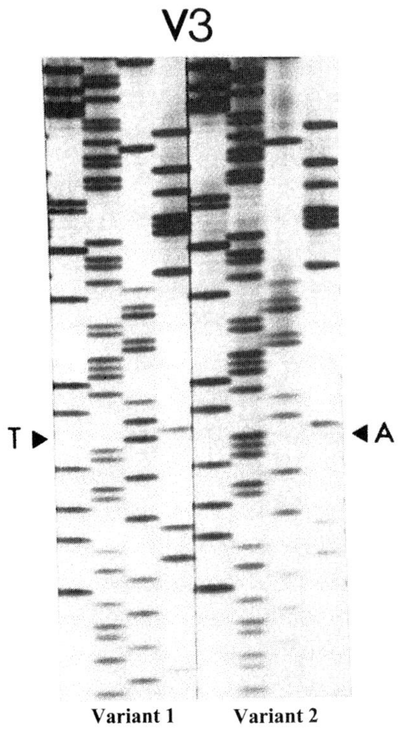

Figure 5 Sequencing gel showing differentiation between two variants of HIV-1 using a limiting dilution PCR and direct sequencing approach.

sylase, with subsequent PCR cycles resulting in the production of single-stranded DNA (39). DNA sequencing can then be carried out using a primer located downstream of the PCR primer digested during the asymmetric PCR.

Protocol 16

Isolation and PCR amplification of single virus DNA/cDNA molecules

Equipment and reagents

- DNA dilution buffer (human DNA in sterile demonized water 100 ng/μl)
- Equipment and reagents for nested primer PCR (see *Protocol 4*)
- Equipment and reagents for agarose gel electrophoresis (see *Protocol 5*)

Method

1. Prepare a twofold dilution series of the sample DNA/cDNA in DNA dilution buffer to give a final volume of 7.5 μl per dilution point.
2. For each DNA dilution set up (in triplicate) a nested primer PCR (see *Protocol 4*) using 2 μl of DNA per reaction.
3. Analyse the products of the second-round PCR using agarose gel electrophoresis (see *Protocol 5*).
4. Determine the end-point dilution, i.e. the dilution at which approximately one of the three triplicate PCRs is positive.
5. Having determined the end-point dilution for the sample DNA/cDNA, set up multiple nested primer PCRs using DNA/cDNA diluted to the end-point dilution.
6. Identify positive PCRs using agarose gel electrophoresis.
7. If the number of positive reactions is significantly different than 30%, adjust the dilution of the DNA/cDNA sample and repeat the nested primer PCR using DNA/cDNA diluted to the new end-point.

Before PCR products can be used in DNA sequencing reactions, unincorporated nucleotides, and often oligonucleotide primers, have to be removed (see *Protocol 18*). A good method for removing unincorporated nucleotides is to treat the PCR products with shrimp alkaline phosphatase (SAP) followed by heat inactivation of the enzyme. If the PCR primers are used at relatively low concentrations in the PCR (0.2–1.0 μM) then these do not usually interfere with subsequent sequencing protocols. If necessary, unincorporated primers can be removed by treating the PCR products with exonuclease I prior to the SAP treatment. However, exonuclease I digests single-stranded DNA and therefore can not be used following asymmetric PCR amplification. Finally, low amounts of PCR products can adversely affect sequence quality when using manual sequencing methods. This problem can be circumvented by using a sequence labelling mix in which one of the three unlabelled nucleotides, depending on the sequence immediately downstream of the sequencing primer, is omitted (40).

Protocol 17

Asymmetric PCR amplification

Equipment and reagents

- 10 × PCR buffer (see *Protocol 4*)
- PCR equipment and reagents (see *Protocol 4*)
- 5 units/μl *Taq* DNA polymerase
- 25 mM $MgCl_2$
- dNTP mix (see *Protocol 4*)
- 5 pmol/μl template-specific sense and antisense primers (with dU replacing dT in one of the primers, usually the sense primer)
- Sterile DNase-/RNase-free demonized water
- DNase-/RNase-free mineral oil
- 1 unit/μl uracil-*N*-glycosylase (UNG; Amersham International)
- UNG dilution buffer (Amersham International)

Method

1. For each sample, prepare a second-round PCR mix (reaction volume of 25 μl) (see *Protocol 4*).
2. Overlay the reaction mix with 1–2 drops of mineral oil.
3. Add 1 μl of each first-round PCR to the corresponding second-round PCR tube. Centrifuge the microcentrifuge tubes at 13 000 g for 30 seconds.
4. Transfer the tubes to the thermal cycler and carry out the PCR for 30 cycles (see *Protocol 4*).
5. Dilute the UNG 1:4 in UNG dilution buffer then add 1 μl (0.25 units) to each PCR tube.
6. Incubate at 37 °C for 30 minutes.
7. Continue thermal cycling for a further 15 cycles with the same temperature profile used for the initial 30 cycles.

Finally, recently developed automated sequencing methods and cycle sequencing methods have made the process less laborious. The former method requires access to specialized analytical equipment but does save time as data collection is automated. When using a fluorescently labelled di-deoxy terminator chemistry to directly sequence PCR products it is important to choose a system that produces even peak heights, otherwise sequencing reactions containing mixed populations or errors introduced during PCR amplification can be missed. PCR products generated for manual sequencing methods are very suitable templates for use in automated reactions, and to save costs the amount of reagents used in the sequencing reactions can be reduced (see *Protocol 20*). This method is suitable for subsequent analysis using Applied Biosystems sequence analysers.

Protocol 18

Removal of unincorporated nucleotides and primers from PCR products

Equipment and reagents

- 1 unit/μl shrimp alkaline phosphatase (SAP; United States Bioscience)
- 1 unit/μl exonuclease I (United States Bioscience)

Method

1. For double-stranded PCR products carry out steps 2–5 described below. For products derived from asymmetric PCRs carry out steps 4 and 5 otherwise the single-stranded PCR product will also be digested.
2. Remove unincorporated primers by adding 1 μl of exonuclease I to each PCR reaction then incubate for 15 minutes at 37 °C.
3. Heat-inactivate the exonuclease I by incubating the sample at 75 °C for 5 minutes.
4. Add 1 μl of SAP to each sample and incubate for 30 minutes at 37 °C.
5. Heat-inactivate the SAP by heating the sample to 75 °C for 15 minutes.

Protocol 19

Manual dideoxynucleotide sequencing of PCR products

Equipment and reagents

- 13 units/μl Sequenase Version 2.0 (United States Bioscience)
- 5 × Sequenase buffer
- Sequenase dilution buffer
- 5 pmol/μl sequencing primer
- 1–5 μg/μl single-stranded DNA binding protein (SSBP)
- 0.5 μg/μl proteinase K (Sigma)
- dNTP/ddNTP nucleotide termination mixes for use with Sequenase Version 2
- 10 μCi/μl [^{35}S]dATP (>1000 mCi/mmol)
- Labelling dNTP mix: 16 μM of each dGTP, dCTP, dTTP
- 100 mM dithiothreitol (DTT)
- Formamide gel loading buffer (see *Protocol 12*)
- Equipment and reagents for a denaturing polyacrylamide gel (see *Protocol 6*)

A. Reaction preparation

1. Prepare the PCR products for sequencing (see *Protocol 18*).
2. For each sequencing template label four tubes G, A, T, C (termination tubes).

Protocol 19 continued

3. Add 2.5 μl of the ddNTP mix to the corresponding labelled tube. For example, add 2.5 μl of mix ddA to the tube labelled A, add 2.5 μl of mix ddG to the tube labelled G and so on.
4. Make up a primer annealing mix containing 2 μl of 5 × Sequenase buffer, 0.5 μl primer, 0.25 μl SSBP, 2.25 μl of sterile demonized water per sequencing template.
5. Prepare the labelling sequencing mix consisting of 1 μl DTT, 2 μl labelling dNTP mix, 1.75 μl Sequenase dilution buffer, 0.25 μl Sequenase Version 2, and 0.25 μl of [^{35}S]dATP per template.[a] Store the labelling mix on ice.
6. Immediately proceed to Part B.

B. Annealing step

1. To 5 μl of each sequencing template add 5 μl of the annealing mix.
2. This step is dependent upon whether the sequencing templates are double-stranded or single-stranded (asymmetric) PCR products.
   (a) For double-stranded templates heat the sample to 95 °C for 2 minutes then snap-cool on either wet ice or a mixture of dry ice and ethanol.
   (b) For single-stranded templates heat the sample to 60 °C for 5 minutes then cool slowly to room temperature.
3. Immediately proceed to Part C.

C. Labelling step

1. Add 5.25 μl of the labelling mix to each template and incubate for 2–5 minutes at room temperature.
2. Transfer 3.5 μl of the labelled template to each of the corresponding G, A, T, and C termination mix tubes. Avoid cross-contaminating the termination mixes.
3. Incubate the termination tubes at room temperature for 5 minutes.
4. Add 1 μl of proteinase K solution to each termination tube and incubate for either 10 minutes at 60 °C, 60 minutes at 37 °C, or 12–18 hours at room temperature.
5. Add 4 μl of the formamide gel loading buffer. Either store at −20 °C for up to 2 weeks or proceed with step 6.
6. Heat the samples to 85 °C.
7. Immediately proceed with denaturing polyacrylamide gel electrophoresis (see *Protocol 8*).

[a]One nucleotide can be omitted from the labelling mix depending on the sequence immediately downstream of the sequencing primer (40).

Protocol 20

Fluorescently labelled, dye-terminator automated sequencing method

Equipment and reagents

- Prism Big dye terminator sequencing kit (Applied Biosystems)
- 5 pmol/μl sequencing primer
- 3 M sodium acetate solution pH 5.2
- 95% ethanol
- 70% ethanol
- Thermal cycler

Method

1. Prepare the PCR products for sequencing (see *Protocol 18*)
2. For each sequencing template, add 4 μl of the sequencing reaction mix to a 200–500 μl microcentrifuge tube and place the tubes on ice.
3. Make up a primer mix consisting of 0.5 μl of primer and 2 μl of water per template.
4. Add, sequentially, 2.5 μl of the primer mix and 4.5 μl of the sequencing template to the microcentrifuge tubes containing the sequencing reaction mix.
5. Transfer the tubes to a thermal cycler and carry out the reaction using a temperature and cycling profile of 95 °C for 10 seconds, 50 °C for 5 seconds, and 72 °C for 4 minutes.

5.1 Sequence manipulation and phylogenetic analyses

A very important component of DNA sequence-based quasispecies analysis is data manipulation and interpretation. This in itself is a very complex issue and cannot be adequately treated in this chapter. However, such analyses do provide important information relating to the nature and rates of sequence change and therefore of the underlying selective pressures exerted on the virus population. Before any phylogenetic analyses can be carried out the DNA sequences have to be aligned. For readers who are unfamiliar with DNA sequence manipulation a tutorial webpage highlighting the different methods is available at http://biobase.dk/Embnetut/Gcg/index.html.

The most frequently used alignment programs are 'pileup', which is available in the Wisconsin GCG package, and Clustalw. The latter is available online (http://ch.nus.sg/bio/clustalw/clustalw.html), is relatively easy to use, and has a number of output formats available for subsequent analyses. Phylogenetic analyses are used to determine the relationship between variants within a virus quasispecies and to study evolutionary changes—a good introductory text to the principles of phylogenetic inference has been written by Li and Grauer (41). Again there are various computational packages available for phylogenetic and evolutionary analyses, including PAUP (details of the availability and cost are available online at http://www.webcom.com/~sinauer/new.shtml paup) and

PHYLIP (available free to registered users from http://evolution.genetics.washington.edu/phylip.html).

References

1. Steinhauer, D. A. and Holland, J. J. (1986). *Annual Reviews in Microbiology*, **41**, 409–33.
2. Smith D. B., McAllister J., Casino C., and Simmonds P. (1997). *Journal of General Virology*, **78**, 1511–19.
3. Meyerhans, A., Cheynier, R., Albert, J., Seth, M., Kwok, S., Sninsky, J., Morfeldt-Manson, L., Asjo, B., and Wain-Hobson, S. (1989). *Cell*, **58**, 901–910.
4. Major, M. E., Nightingale, S., and Desselberger, U. (1993). *Journal of General Virology*, **74**, 2531–7.
5. Speck, R. F., Wehrly, K., Platt, E. J., Atchison, R. E., Charo, I. F., Kabat, D., Chesboro, B., and Goldsmith, M. A. (1997). *Journal of Virology*, **71**, 7136–9.
6. Schuitemacker, H., Koot, M., Kootstra, N. A., Derckson, M. W., De Goede, R. E.Y., van Steenwijk, R. P., Lange, J. M. A., Shatternkerk, J., Miedema, F., and Tersmette, M. (1992). *Journal of Virology*, **66**, 1354–60.
7. Simmonds, P., Zhang, L. Q., McOmish, F., Balfe, P., Ludlam, C. A., and Leigh Brown, A. J. (1991). *Journal of Virology*, **65**, 6266–76.
8. Miller, S. A., Dykes, D. D., and Polesky, H. F. (1988). *Nucleic Acids Research*, **16**, 1215.
9. Boom, R., Sol, C. J. A., and van de Noordaa, J. (1990). *Journal of Clinical Microbiology*, **28**, 495–503.
10. Sambrook, J., Fritsch, E. F., and Maniatis, T. (1989). *Molecular cloning—a laboratory manual.* (2nd edn). Cold Spring Harbor Laboratory Press, New York.
11. Simmonds, P., Balfe, P., Ludlam, C. A., Bishop, J. O., and Leigh Brown, A. J. (1990). *Journal of Virology*, **64**, 5840–50.
12. Balfe, P., Simmonds, P., Ludlam, C. A., Bishop, J. O., and Leigh Brown, A. J. (1990). *Journal of Virology*, **64**, 6221–33.
13. Kwok, S. and Higuchi, R. (1989). *Nature*, **339**, 237–8.
14. Longo, M. C., Berninger, M. S., and Hartley, J. L. (1990). *Gene*, **39**, 125–8.
15. Karrington, M., Miller, T., White, M., Gerrard, B., Stewart, C., Dean, M., and Mann, D. (1992). *Human Immunology*, **33**, 208–12.
16. Soto, D. and Skumar, S. (1992). *PCR Methods and Applications*, **2**, 96–8.
17. Enomoto, N., Kurosaki, M., Tanaka, Y., Marumo, F., and Sato, C. (1994). *Journal of General Virology*, **75**, 1361–9.
18. Bukh, J., Miller, R. H., and Purcell, R. H. (1995). *Seminars in Liver Disease*, **15**, 41–63.
19. Maggi, F., Fornai, C., Vatteroni, M. L., Giorgi, M., Morrica, A., Pistello, M., Cammarota, G., Marchi S, Ciccorossi, P., Bionda, A., and Bendinelli, M. (1997). *Journal of General Virology*, **78**, 1521–5.
20. Peters, T., Schlayer, H. J., Hiller, B., Rosler, B., Blum, H., and Rasenack, J. (1997). *Journal of Virological Methods*, **64**, 95–102.
21. Keen, T. J., Inglehearn, C. F., Lester, D. H., Bashir, R., Jay, M., Bird, A. C., Jay, B., and Bhattacharya, S. S. (1991). *Genomics*, **11**, 199–205.
22. Delwart, E. L., Shpaer, E. G., Louwagie, J., McCutchan, F. E., Grez, M., Rubsamen-Waigmann, H., and Mullins, J. I. (1993). *Science*, **262**, 1257–61.
23. Delwart, E. L., Sheppard, H. W., Walker, B. D., Goudsmit, J., and Mullins, J. I. (1994). *Journal of Virology*, **68**, 6672–83.
24. Zhu, T., Wang, N., Carr, A., Nam, D. S., Moor-Jankowski, R., Cooper, D. A., and Ho, D. D. (1996). *Journal of Virology*, **70**, 3098–107.
25. Moddrow, S., Hahn, B. H., Shaw, G. M., Gallo, R. C., Wong-Staal, F., and Wolf, H. (1987). *Journal of Virology*, **61**, 570–8.
26. Ball, J. K., Holmes, E. C., Whitwell, H., and Desselberger, U. (1994). *Journal of General Virology*, **75**, 867–79.

27. Larder, B. A., Kellam, P., and Kemp, S. D. (1991). Zidovudine resistance predicted by detection of mutations in DNA from infected lymphocytes. *AIDS*, **5**, 137–44.
28. Kaye, S., Loveday, C., and Tedder, R. S. (1992). *Journal of Medical Virology*, **37**, 241–6.
29. Tedder, R. S., Kaye, S., Loveday, C., Weller, I. V. D., Jeffries, D., Norman, J., Weber, J., Bourelly, M., Foxall, R., Babiker, A., Darbyshire, J. H. (1998). Comparison of culture- and non-culture-based methods for quantification of viral load and resistance to antiretroviral drugs in patients given zidovudine monotherapy. *Journal of Clinical Microbiology*, **36**, 1056–63.
30. Stuyver, L. Wyseur, A., Rombout, A., Louwagie, J., Scarcez, T., Verhofstede, C., Rimland, D., and Schinazi, R. F. (1997). *Antimicrobial Agents and Chemotherapy*, **4**, 284–91.
31. Sheehy, N., Desselberger, U., Whitwell, H., and Ball, J. K. (1996). *Journal of General Virology*, **77**, 1071–81.
32. Tindall, K. R. and Kunkell, T. A. (1988). *Biochemistry*, **28**, 6008–13.
33. Meyerhans, A., Vartanian, J.-P., and Wain Hobson, S. (1990). *Nucleic Acids Research*, **18**, 1687–91.
34. McCabe, P. C. (1989). *PCR protocols: a guide to methods and applications* (ed. M. A. Innis, D. H. Gelfand, J. J. Sninsky, and T. J. White), pp. 76–84. Academic Press, New York.
35. Albert, J., Wahlberg, J., Lundeberg, J., Cox, S., Sandstrom, E., Wahren, B., and Uhlen, M. (1992). *Journal of Virology*, **66**, 5627–30.
36. Winship, P. R. (1989). *Nucleic Acids Research*, **17**, 1266.
37. Ball, J. K. and Desselberger, U. (1992). *Analytical Biochemistry*, **207**, 349–51.
38. Hultman, T., Stahl, S., Hornes, E., and Uhlen, M. (1989). *Nucleic Acids Research*, **17**, 4937–45.
39. Ball, J. K. and Curran, R. (1997). *Analytical Biochemistry*, **253**, 264–7.
40. Tsang, T. C. and Bentley, D. R. (1989). *Nucleic Acids Research*, **16**, 6238.
41. Li, W.-H. and Grauer, D. (1991). *Fundamentals of molecular evolution*. Sinauer Associates, Sunderland, MA.

Chapter 7
In-vitro replication of RNA viruses

R. Banerjee, M. Igo, R. Izumi, U. Datta, and A. Dasgupta
Department of Microbiology and Immunology, UCLA School of Medicine, 10833 Le Conte Avenue, Los Angeles, CA 90095, U.S.A.

1 Introduction

RNA viruses are very diverse and utilize many distinct strategies for replicating their genomes. The genome organization of these viruses vary widely, thus the translation and replication strategies employed must also be varied. However, despite all the complexities of RNA viruses and their differing replication strategies, there are many similarities at the molecular level (1). Some of the common *in vitro* techniques employed to study these RNA replication strategies as well as methodologies to study *in vitro* RNA–protein interactions will be discussed in this chapter.

RNA viruses consist of the positive-stranded and the negative-stranded viruses. In the positive-stranded RNA viruses such as poliovirus, introduction of the virus genome into a susceptible host is sufficient for replication of the RNA genome. This is due to the fact that the introduced or transfected RNA acts both as mRNA and as the template for replication. Replication of the poliovirus RNA genome proceeds through the formation of a complete negative-strand RNA, which serves as the template for the synthesis of the progeny virion RNA molecules (2, 3). In sharp contrast, the genomes of negative-stranded viruses cannot serve as mRNA. In VSV (vesicular stomatitis virus), a negative-stranded virus, the virus-encoded RNA is encapsidated with the virus-encoded nucleocapsid protein; the resulting nucleoprotein complex (RNP) then serves as the template for transcription and replication to form anti-genomic RNP, which in turn serves as template exclusively for replication (4, 5). Amongst the positive-stranded viruses, the best-studied examples are poliovirus and Brome mosaic virus. The poliovirus system is well characterized, which is facilitated by the development of an *in vitro* transcription and replication system that yields viable infectious plaques (6). This system is described in greater detail in this chapter.

The plant virus Brome mosaic virus (BMV) is a member of the Alphavirus family. The members of this group share a conserved core set of RNA replication genes and fairly conserved features of RNA replication. Past studies of BMV indicate the involvement of host proteins in virus replication and they are thought to be integral components of the virus replication machinery. One detailed study (7) reported replication of the BMV genome by using yeast cells transformed with individual BMV genes. This study concluded that some of the host factors essential for the virus to propagate can be substituted in yeast cells, and, additionally, they can function either in *cis* or *trans* as demand arises.

2 *De-novo* synthesis of poliovirus

Although poliovirus is one of the simplest known animal viruses with a single-stranded RNA genome surrounded by a capsid, questions regarding the mechanisms of polyprotein synthesis initiation, genome replication, the function of some individual virus mature proteins, as well as the role of some host cellular proteins in virus replication remain to be solved. An ideal experimental system would be the reconstitution of viable virus in a cell-free system using *de-novo* synthesis. This objective was originally achieved in 1991 (6). This section describes the steps we follow for the *de-novo* synthesis of poliovirus in our laboratory (8–10). These procedures are now followed by many different groups studying poliovirus replication. Either virion RNA, isolated on caesium chloride gradients, or infectious T7-transcribed genomic RNA is added to uninfected HeLa S10 cytoplasmic extracts, which are able to sustain the translation of the virus RNA, replication, and assembly into infectious virus particles so that viable plaques can be obtained. The cytoplasmic S10 extract is able to support the functionally active 3D polymerase expression and is able to faithfully execute all steps of virus replication (8). The methodology used for the preparation of HeLa S10 extract is described in this section, although a detailed description can be obtained from the references at the end of this chapter (6, 9).

2.1 Isolation of poliovirus genomic RNA

Protocol 1

Isolation of poliovirus genomic RNA

Equipment and reagents

NB: Prepare all reagents in DEPC-treated water

- DEPC-treated water (Add 0.1% (v/v) diethylpyrocarbonate (DEPC) to sterile distilled water and incubate at room temperature overnight. Inactivate DEPC by autoclaving.)
- Ultracentrifuge (e.g. Sorvall, Beckman, etc.) with Sorvall SS-34 rotor (or equivalent) and a swing-out bucket rotor
- Phosphate-buffered saline (PBS) pH 7.1
- 10 × RSB: 100 mM Tris–HCl pH 7.5, 100 mM NaCl, 15 mM $MgCl_2$
- 10% SDS
- Lysis buffer: 10 ml 10 × RSB, 1 ml NP-40, 89 ml DEPC-treated water
- 10 × TNE buffer: 100 mM Tris–HCl pH 7.5, 100 mM NaCl, 100 mM EDTA
- 1 × SDS buffer: 10 ml 10 × TNE buffer, 5 ml 10% SDS, 85 ml DEPC-treated water

Protocol 1 continued

- 15% and 30% linear sucrose gradient in 1 × SDS buffer
- Phenol/chloroform
- Ethanol
- 3 M sodium acetate pH 5.15

Method

1. Infect 5 litres of a HeLa cell suspension with poliovirus at m.o.i. of 25.
2. At 7 hours after infection, wash the cells with PBS, spin at 1000 g, 10 min at 4 °C. Store the pellet at −70 °C.
3. Thaw the pellet. Lyse infected cells in 1 × RSB containing 1% NP-40 for 30 min with intermittent mild shaking on ice. Use 8–10 ml of the lysis buffer per litre of original cell suspension. Do not vortex the cells.
4. After lysis, spin in a Sorvall SS-34 rotor or equivalent for 5 min at 300 g at 2–4 °C (pre-cool the rotor). Collect the turbid supernatant. Discard the pellet.
5. Adjust the supernatant to 0.5% SDS using 10% SDS. Spin the supernatant at 45 000 r.p.m. for 2 hours at room temperature.[a]
6. Discard the supernatant (cytoplasm). Rinse the virus pellet very gently with 1 ml of 1 × SDS buffer and remove the attached cellular DNA (white stringy material).
7. Add the minimum amount of 1 × SDS buffer to cover the pellet and put the tubes containing the virus pellet on a shaker in slow motion overnight at room temperature.
8. After the virus pellet has dissolved, spin very briefly in a microcentrifuge to remove remaining cellular debris. Collect the clear supernatant.
9. Make a 34 ml linear sucrose gradient with 15% and 30% sucrose in 1 × SDS buffer containing 0.5% SDS. Layer the supernatant carefully on top of the gradient. Spin for 3 hours at 24 000 r.p.m. in a swing-out bucket rotor at room temperature. Turn the brake off so the gradient is not disturbed.
10. Collect 1.5 ml factions from the gradient starting with the bottom of the tube, by piercing with a needle.
11. Read the OD_{260} using a 1:10 to 1:50 dilution of each fraction in 1 × SDS buffer and 15% sucrose. Expect to see two peaks—the virus peak is the first from the bottom of the tube. Pool the active fractions.
12. Dilute the pooled fractions at least 1:1 with 1 × SDS buffer and spin down the virions at 45 000 r.p.m. for at least 3 hours (overnight is best) at room temperature.
13. Discard the supernatant and redissolve the virion pellet in a small volume of 1 × SDS buffer (i.e. less than 500 μl).
14. Purify the virus RNA by phenol:chloroform extraction (1:1) (twice) and precipitate using 2.5 volumes of ethanol and 1/10 volume 3 M sodium acetate pH 5.5.
15. After precipitating, pellet the RNA and wash with 80% ethanol, then dissolve the RNA in DEPC-treated water. Adjust the OD_{260} such that the final RNA concentration is 1 mg/ml. Store aliquots at −70 °C.

[a] If it is necessary to increase the volume for the spin, add 1 × SDS buffer to the top of the tube.

2.2 HeLa S10-extract preparation

It is essential that the cells used are healthy and growing at optimal rates for the preparation of S10 extracts.

Protocol 2

HeLa S10-extract preparation

Equipment and reagents

- Ultracentrifuge and appropriate rotors (Beckman JS-4.0 and Beckman JA-20, or equivalents)
- Dounce homogenizer with a type B pestle
- Sterile (autoclaved) 30 ml Corex™ tubes (Sorvall)
- Isotonic buffer: 35 mM Hepes pH 7.4, 146 mM NaCl, 11 mM glucose
- 10 × buffer: 0.2 mM Hepes pH 7.4, 1.2 M KCH_3CO_2, 40 mM Mg $(CH_3CO_2)_2$, 50 mM DTT
- Hypotonic buffer: 20 mM Hepes pH 7.4, 10 mM KCl, 1.5 mM $Mg(CH_3CO_2)_2$, 1 mM DTT
- 200 mM $CaCl_2$
- 4 M KCl
- Micrococcal nuclease (Sigma) or S7 nuclease (Boehringer Mannheim)
- 200 mM EGTA
- 50 ml conical tubes

Method

1. Harvest 2.5 litres (4×10^5 cells per ml) of uninfected HeLa cells by centrifugation at 3000 g and wash three times with isotonic buffer. Pool all the cells together in a 50 ml tube and note the packed cell volume (PCV).
2. Resuspend the pellet in 1.5 volumes of hypotonic buffer and incubate on ice for 10 min.
3. Break the cells in a Dounce homogenizer with 25 strokes of a type B pestle.
4. Add 0.1 volume of 10 × buffer and centrifuge at 500 g for 10 min at 4°C in a Beckman JS-4.0 rotor (or equivalent) to pellet the nuclei.
5. Pour the supernatant into a 30 ml Corex tube and centrifuge at 12 000 g for 15 min at 4 °C in a Beckman JA-20 rotor (or equivalent).
6. Pour the S10 supernatant into a 50 ml conical tube and adjust to 1 mM $CaCl_2$ using 200 mM stock solution. Treat with 75 U of S7 nuclease at 20 °C for 15 min. Inactivate nuclease by adjusting the S10 supernatant to 2 mM EGTA.
7. Centrifuge the S10 extract again at 12 000 g for 15 min at 4°C in a JA-20 rotor (or equivalent). Store the extract in 400 μl aliquots at −70 °C.

2.3 Translation initiation-factor preparation

Protocol 3

HeLa initiation-factors (IF) preparation

Equipment and reagents

- Ultracentrifuge and appropriate rotors (Beckman 70-Ti, JA-20, 60-Ti rotors, or equivalent)
- Hypotonic buffer: 20 mM Hepes pH 7.4, 10 mM KCl, 1.5 mM $Mg(CH_3CO_2)_2$, 1 mM DTT)
- Dounce homogenizer with a type B pestle
- 4 M KCl
- Dialysis buffer: 5 mM Tris–HCl pH 7.5, 100 mM KCl, 0.05 mM EDTA, 1 mM DTT, 5% glycerol

Method

1. Harvest and homogenize uninfected HeLa cells as in *Protocol 2*.
2. After homogenization, remove the nuclei without the addition of any salts (no 10 × buffer) by centrifugation at 500 g for 10 min at 4 °C.
3. Centrifuge the post-nuclear supernatant for 15 min at 12 000 g at 4 °C in a JA-20 rotor (or equivalent).
4. Centrifuge the S10 supernatant for 1 hour at 270 000 g in a Beckman 70 Ti rotor (or equivalent) to pellet the ribosomes. If necessary, pellet the ribosomes at 170 000 g in a Beckman 60-Ti rotor (or equivalent) for 2 hours.
5. Discard the supernatant and resuspend the ribosomal pellet in hypotonic buffer at 240 A_{260} U/ml (this requires a very small volume, usually less than 2 ml). Adjust the ribosomes to 0.5 M KCl using 4 M KCl and stir gently for 15 min on ice. Spin the salt-washed ribosomes again at 270 000 g for 1 hour in a Beckman 70-Ti rotor at 4 °C.
6. Dialyse the supernatant for 2 hours in dialysis buffer at 4 °C. Centrifuge the dialysed sample at 5000 g for 5 min at 4 °C to remove denatured proteins that may have precipitated. Store the HeLa cell IFs in 50 μl aliquots at −70 °C.

Note: If the final spin at 270 000 g requires diluting the sample to obtain the correct volume for spinning, then dilute the sample with hypotonic buffer with 0.5 M KCl and after the spin concentrate the supernatant to a concentration of 240 A_{260} U/ ml before dialysis.

2.4 Coupled transcription and translation

Protocol 4

Coupled transcription and translation

Equipment and reagents

- 10 × nucleotide reaction mix: 10 mM rATP, 2.5 mM rGTP, 2.5 mM rCTP, 2.5 mM rUTP, 600 mM KCH_3CO_2, 300 mM creatine phosphate, 4 mg/ml creatine kinase, 155 mM Hepes–KOH, pH 7.4
- DEPC-treated water (see *Protocol 1*)
- RNase A and T1
- DMEM (Dulbecco's modified essential medium) containing 10% fetal bovine serum
- HeLa cell monolayers (90–100% confluent)

Method

1. For a 50 μl reaction, mix 25 μl of HeLa S10 extract (see *Protocol 2*), 10 μl HeLa IFs (see *Protocol 3*), 5 μl of 10 × nucleotide reaction mix, and 25 μg of virus RNA (see *Protocol 1*) per ml. Bring the volume to 50 μl with DEPC-treated water.
2. Incubate the reaction for 6–12 hours at 34 °C.
3. Treat the reaction with 0.8 μg of RNase A and 76 U of RNase T1 for 20 min at 25 °C.
4. Dilute the sample with DMEM appropriately and assay for p.f.u. on HeLa cell monolayers.

2.5 Poliovirus plaque assay

Protocol 5

Poliovirus plaque assay

Equipment and reagents

- 60 mm culture plates
- Serum-free DMEM
- DMEM with 10% fetal bovine serum
- Overlay solution: 51% of 1.8% Noble agar, 47% of 2 × DMEM (no serum), 2% of fetal bovine serum. (Make fresh before use. Melt the agar in a microwave oven. Incubate at 50 °C to prevent solidification. Warm DMEM and serum to 37 °C. Once ready for overlay, mix the three solutions together and overlay the cells.)
- Crystal violet: 0.1% (w/v) crystal violet, 20% (v/v) ethanol

Method

1. Grow HeLa cells in 60 mm culture plates to about 90% confluency in DMEM with 10% fetal bovine serum. Aspirate off the medium and wash with serum-free DMEM once. Aspirate.
2. Dilute the virus in serum-free DMEM.

Protocol 5 continued

3. Add 500 μl of virus dilution to each plate.
4. Place the plates in 5% CO_2 incubator at 37 °C. Rock the plates to cover with liquid every 5 min for a total incubation time of 30 min. Remove the virus inoculum by aspiration.
5. Add 5 ml of overlay solution.
6. Allow to harden at room temperature for about 15 min.
7. Place in a CO_2 incubator at 37 °C for 48 hours.
8. Carefully remove the agar overlay. Add 500 μl of 0.1% crystal violet solution. Stain for 30 seconds. Aspirate the stain. Allow to air-dry.
9. Count the plaques against the purple-stained, intact cell monolayer.

3 Expression of virus proteins with enzymatic and RNA binding activities

Numerous assays of interest require virus protein at concentrations and purities that is often difficult to achieve by purification of the virus particles themselves. In such instances it becomes necessary to overexpress virus gene products. To achieve this, the gene of interest is ligated to an engineered plasmid or other expression vector, which in turn is introduced into a living cell. The vectors normally used for these purpose have a selectable marker, transcription promoter elements, translation control sequences such as an ATG start codon, and a polylinker sequence for insertion of gene of the protein of interest. Moreover, the level of expression of the foreign gene product can often be controlled using inducible promoters.

If the protein of interest is not post-translationally modified or there are no substantial or confirmed reports of post-translational modification, it can be conveniently expressed in an *Escherichia coli* overexpression system. Due to the ease of manipulation and quick growth in inexpensive media, many groups favour *E. coli* expression systems. Several of these systems are available, and the majority take advantage of affinity chromatography between an immobilized ligand and an affinity tag on the protein of interest. If the expressed protein is toxic to the cells then using a system with a lower baseline expression can help to overcome this problem.

One expression system used with great success in our studies utilizes the pET vector, where the fusion protein carries six histidine residues in tandem either at the C or N terminus (11–13) (*Figure 1*). This is a good choice over many other systems available since the small and uncharged (at physiological pH) histidine tag, even if not cleaved, does not seem to disturb or alter the activity of most expressed proteins. A stretch of six histidines (compared to three or four) has been shown to bind more reliably and efficiently to the affinity resin under strong denaturing conditions using agents such as 6 M urea or guanidine

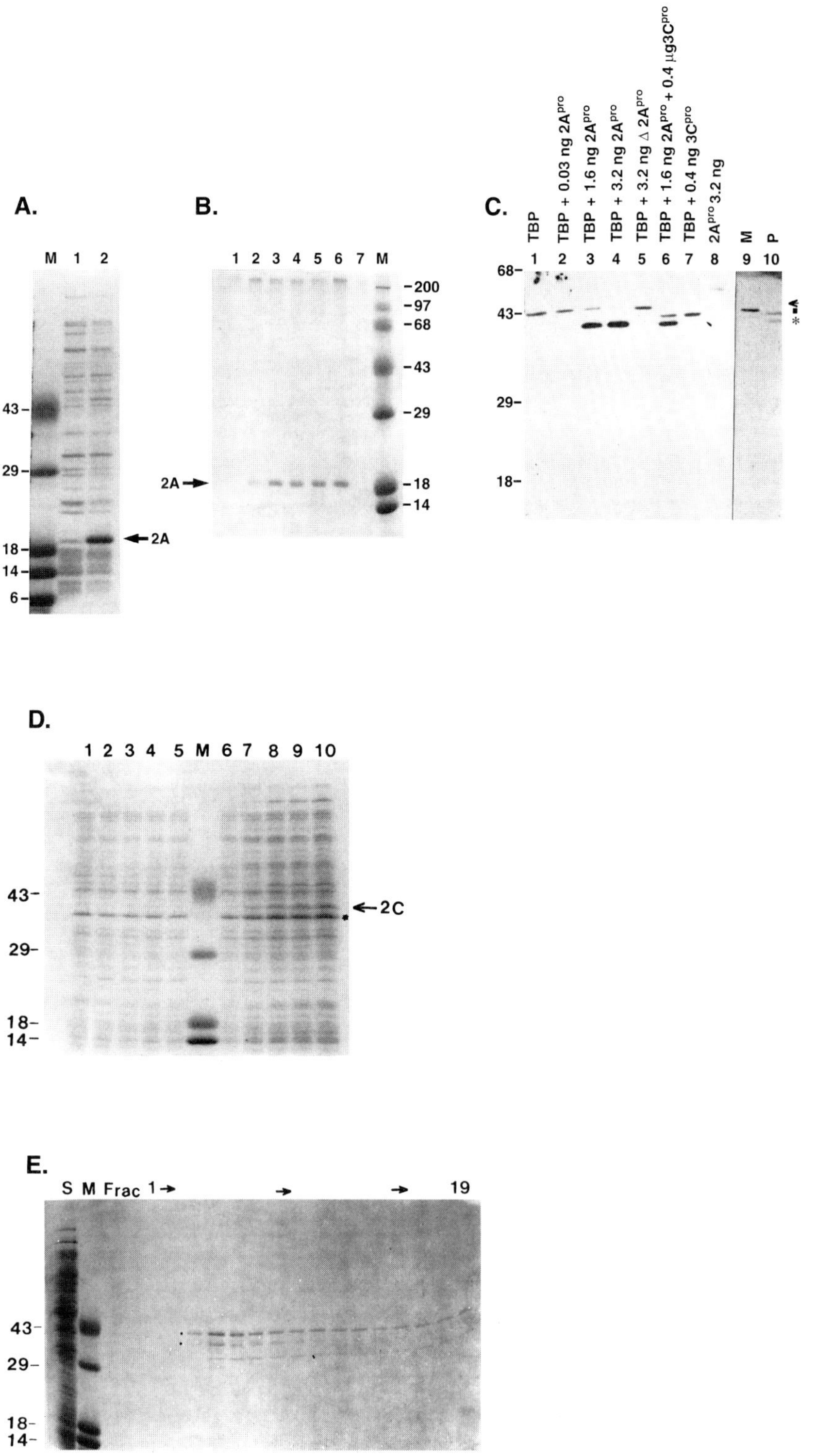
A.
M 1 2
43
29
18
14
6
2A
B.
1 2 3 4 5 6 7 M
200
97
68
43
29
18
14
2A
C.
TBP
TBP + 0.03 ng 2Apro
TBP + 1.6 ng 2Apro
TBP + 3.2 ng 2Apro
TBP + 3.2 ng Δ 2Apro
TBP + 1.6 ng 2Apro + 0.4 μg3Cpro
TBP + 0.4 ng 3Cpro
2Apro 3.2 ng
M
P
1 2 3 4 5 6 7 8 9 10
68
43
29
18
*
D.
1 2 3 4 5 M 6 7 8 9 10
43
29
18
14
2C
E.
S M Frac 1
19
43
29
18
14

Figure 1 Expression and purification of virus protease, 2A, and a virus non-structural protein, 2C. *E. coli* BL21(DE3) cells were transformed with recombinant plasmids encoding either histidine-tagged poliovirus 2A protease, or polypeptide 2C, or control pET15b plasmid and grown in LB medium containing antibiotic at 37 °C and induced with IPTG. (A) The pellet fraction isolated from cells expressing virus protease 2A, ($2A^{pro}$) were Coomassie-stained. Lane 1 represents the protein profile from control cells and lane 2 represents the protein profile from cells harbouring recombinant plasmid pET-2A expressing $2A^{pro}$. Lane M is the pre-stained, molecular-weight marker proteins, with the sizes of the proteins indicated. (B) The pellet fraction obtained after cell lysis was used for isolating inclusion bodies, solubilized in lysis buffer, and loaded on to a Sepharose column with immobilized Ni^{2+}. Histidine-tagged $2A^{pro}$ was eluted in buffer containing 100 mM imidazole (elution buffer) and fractions collected. Proteins were resolved on SDS-PAGE gel and Coomassie-stained. The migration of $2A^{pro}$ is marked by an arrow. (C) Protease activity assay of the purified polypeptide using eukaryotic transcription factor, TBP as substrate (34). A monoclonal antibody against TBP was used for detecting the cleaved product by Western blot analysis. Purified $2A^{pro}$ is used at three different concentrations (lanes 2, 3, and 4) to detect protease activity. A truncated virus 2A protease ($\Delta 2A^{pro}$) purified under identical conditions, with the active protease domain deleted is used as control (lane 5). An additional virus-encoded protease, 3C (purified using a similar strategy) is used to distinguish between the cleaved products of TBP (lanes 6 and 7). Lane 8 contains purified $2A^{pro}$. Mock- and poliovirus-infected HeLa cell extracts (lanes 9 and 10, respectively) were also analysed. (D) Coomassie stain of whole-cell extract expressing virus 2C polypeptide. Lanes 1–5 show the protein profiles of a whole-cell extract obtained from pET 15b-transformed cells at 0, 1, 2, 3, and 4 hours post-induction, respectively, while lanes 6–10 show the protein profiles of cells obtained from pET15b-2C transformed cells at the same time points. Lane M contains molecular-weight marker proteins and the numbers on the left indicate the molecular mass of the corresponding marker proteins in kilodaltons. Expression of His–2C and a truncated 2C in lanes 6–10 as visualized by Coomassie blue staining are marked by an arrow and an asterisk, respectively. (E) The pellet fraction obtained following lysis was solubilized in buffer containing 6 M urea and loaded on to a column packed with TALON resin (immobilized Co^{2+}). Histidine-tagged 2C was eluted using imidazole (elution buffer) and fractions collected. An aliquot of fractions (Frac) was analysed by SDS-PAGE and polypeptide-stained using Coomassie Blue. Lane S shows the protein profile of the solubilized pellet fraction that was loaded on to the column. Lane M, molecular-weight marker proteins, with the numbers indicating the molecular masses in kilodaltons.

hydrochloride. Denaturing conditions are often used to solubilize the expressed protein when it localizes to the inclusion bodies or the bacterial pellet. Functional native protein can be recovered after slow renaturation. At times, a thrombin or enterokinase cleavage site is available between the tandem histidine residues and the translational start site of the protein of interest. This feature allows for the cleavage of the histidine tag when necessary, and represents a short stretch of amino acid residues. Many resins are currently available, each tailored to purify 6 × histidine-tagged recombinant proteins. One resin we use for isolation of our proteins is available as TALON (Clontech). This resin, with immobilized Co^{2+}, is compatible with many common chemicals and reagents and allows proteins to be purified under denaturing conditions.

The gene encoding the protein of interest is cloned into the multiple cloning site of the expression plasmid by standard techniques. The recombinant plasmid is transformed into a suitable *E. coli* host strain, usually BL21(DE3) cells (12).

This strain contains a chromosomal copy of the gene for T7 RNA polymerase and the lysogen DE3—a lambda derivative that expresses the polymerase under the control of the *lac* promoter. The expression of the histidine-tagged protein following transcription of the target DNA in the plasmid is induced by the addition of isopropyl β-D-thiogalactopyranoside (IPTG). IPTG is a non-metabolizable analogue of lactose, the natural inducer of *lac* promoter. *Protocol 6* only suggests starting conditions and it may need to be optimized for individual proteins.

Protocol 6

Overexpression of proteins in *E. coli*

Equipment and reagents

- LB (Luria-Bertani) medium and plates: 1% bactotryptone, 0.5% yeast extract, 0.5% NaCl, 0.1% glucose, adjust to pH 7.2. Add 50 μg/ml ampicillin (or other appropriate antibiotic for selection of the recombinant plasmid)
- *E. coli* BL21(DE3) cells or other suitable cell line for expression
- Autoclaved flasks and 37 °C warm room or incubator for growing cells
- Reagents and equipment for subcloning DNA fragments, transformation of *E. coli* cells, screening recombinants, and protein analysis by SDS-PAGE quantitation
- Sonicator with microtip
- Sorvall centrifuge (RC5B) with rotors and bottles or other comparable centrifuge
- 100 mM IPTG stock solution
- 0.45 μm membrane filter
- Sonication buffer: 20 mM Tris–HCl, 500 mM NaCl, 2 mM imidazole, 0.1% NP-40, final pH 8.0
- Lysis buffer: 20 mM Tris–HCl, 100 mM NaCl, 2 mM imidazole, 0.1% NP-40, 6 M urea, final pH 8.0
- Wash buffer: 20 mM Tris–HCl, 100 mM NaCl, 10 mM imidazole, 0.1% NP-40, 6 M urea, final pH 8.0
- Elution buffer: 20 mM Tris–HCl, 100 mM NaCl, 100 mM imidazole, 0.1% NP-40, 6M urea, final pH 8.0
- Dialysis buffer: 50 mM Tris–HCl pH 7.4, 100 mM KCl, 1 mM DTT, 20% glycerol, 0.1% NP-40, and urea at 4, 2, and 0 M final concentration
- Dialysis bags (appropriate molecular weight cut-off) and clips for securing
- PEG 8000 (polyethylene glycol)
- TALON resin (Clantech) or a suitable metal immobilized resin column for affinity purification
- Coomassie Blue

Method

1. Grow a 5.0 ml culture of BL21(DE3) cells, or another *E. coli* expression cell line harbouring the recombinant plasmid, overnight at 37 °C, using appropriate antibiotic selection media.
2. Next day, inoculate a portion of the growing cells into a larger volume of LB media containing appropriate antibiotic, and grow to an OD_{600} of approximately 0.4–0.6.
3. Induce protein expression by adding 0.4–1.0 mM (final concentration) IPTG or a suitable inducer to the growth media. Continue the incubation for an additional 3–4 hours.[a]

Protocol 6 continued

4. Harvest the cells by centrifugation at 5000 g for 10 min at 4 °C and aspirate the medium.[b]
5. Lyse cells by freeze–thaw and sonication. Carry out the sonication by placing the tube in an salt/ice bath.[c]
6. Resuspend the inclusion-body pellet homogeneously in cold lysis buffer and incubate at 4 °C for 2–3 hours.[d]
7. Remove the insoluble membranous material by centrifugation at 10 000 g for 20 min.
8. Isolate the solubilized protein from this final spin supernatant using metal-affinity chromatography. Filter the supernatant through a 0.45 μm membrane before loading it on to the column to prevent the column from clogging.
9. Pack the resin is into a suitable sized column taking care no air bubbles are trapped, then equilibrate the column with lysis buffer. Apply the supernatant containing the solubilized protein obtained from step 8 to the column and allow it to flow under gravity, then collect the eluant and reapply it to the resin once more.
10. Wash the resin 3–4 times with lysis buffer, followed by wash buffer 3–4 times.[e]
11. Elute the protein by adding elution buffer to the column, collect the eluate in small fractions. Run a small aliquot of each sample on an SDS-polyacrylamide gel to resolve the proteins and stain with Coomassie blue.
12. Pool the fractions containing the largest proportions of the protein of interest and transfer to a dialysis bag with an appropriate molecular-weight cut-off. Dialyse against changes of buffer containing gradual reductions in urea concentration (4 M, 2 M, and twice in 0 M) at 4 °C for a minimum of 3 hours each for each dialysis.
13. After the last dialysis in 0 M urea, transfer the samples into a fresh dialysis tube and proteins concentrated by dialysis at 4 °C against solid PEG 8000 or sucrose. Replace the PEG or sucrose at regular intervals until the desired volume is reached. The dialysis bag can be tightened as the extract is gradually concentrated. Alternatively, use a microconcentrator (Centricon, Amicon) to concentrate the samples. Remove the samples once the desired concentration is reached and store in aliquots at −70 °C.

[a] The time of induction and IPTG concentration is variable for different proteins and needs to be determined empirically.

[b] The cells obtained can be stored as a pellet, but we recommend making a uniform suspension in sonication buffer before storing at −70 °C.

[c] A change of viscosity is visually apparent when cells are lysed after a few short pulses. It is best to avoid long sonication to prevent overheating. NB: Sonicator power settings are variable and dependent on the sonicator being used and the size of the tube holding the cells. If unsure, use a different power setting during each burst.

[d] It is important for the efficient recovery of expressed proteins that inclusion bodies be uniformly resuspended and not left as clumps. A single short pulse of sonication may help in resuspension.

[e] The volume of buffer used is dependent on the column bed volume, and it is best to use 5 times the bed volume for each wash.

4 RNA binding assays

Once a virus protein has been purified from an overexpression system or infected cell lysate it is important to be able to use the purified protein in functional assays. The genomes of RNA viruses generally encode proteins that interact with the virus RNA sequence to form a nucleoprotein complex (14–17). This complex formation is a prerequisite for genome replication. Some of these peptides or their precursor polypeptides have the capability to act in *cis*, while others act in *trans*. The role of some of these proteins have been determined by their ability to interact with RNA and form a functional nucleoprotein complex. This complex, once formed, may in turn either attract or approximate host factors or accessory virus functional proteins. Thus several virus and host cellular proteins are required for the complete replication of the virus genome. Assessing the ability and affinity of a protein to bind specifically to RNA sequences can be analysed using the following protocols.

4.1 *In vitro* transcription for labelled probe preparation

For simplicity this procedure has been subdivided into four stages (*Protocols 7–10*). The starting point is to linearize a recombinant plasmid containing the probe sequence as a DNA fragment (18, 19).

Protocol 7

Restriction digestion

Equipment and reagents

- Water bath or incubator maintained at 37 °C
- TE buffer: 10 mM Tris–HCl pH 8.0, 1 mM EDTA
- DNA sample in water or TE buffer
- Restriction endonuclease, appropriate for the restriction sites required
- 10 × compatible restriction endonuclease buffer
- 0.8% mini agarose gel[a] containing 0.5–1.0 μg/ml ethidium bromide, and other electrophoresis equipment and reagents
- 6 × DNA loading dye: 0.25% (w/v) Bromophenol Blue, 0.25% (w/v) xylene cyanol, 30% (v/v) glycerol in water
- 0.5 M EDTA (optional)
- 1 mg/ ml acetylated BSA (optional), to be added at 1 μl per 10 μl final volume

Method

1. Set up the following in a clean microcentrifuge tube:

DNA to be digested[b]	0.5–1.0 μg in water or TE buffer
10 × restriction buffer	2.0 μl
Restriction enzyme	5–10 units

   Add water to a final 20 μl volume and mix gently.
2. Incubate at 37 °C, or the appropriate temperature recommended for the enzyme, for 3–4 hours.

Protocol 7 continued

3. Check if the DNA has been completely linearized by removing a small portion (1–2 μl) and checking on a agarose gel containing 0.5–1.0 μg/ ml ethidium bromide and comparing this with an aliquot of undigested DNA.[a]
4. Add 6 × DNA loading dye to the tube, at termination of the reaction, and load on to an agarose gel and start the electrophoresis. The gel should be run far enough so as to separate the linearized plasmid from the supercoiled species. Linearized DNA migrates slower than the undigested supercoiled plasmid.

[a] The agarose gel percentage is determined by the size of the DNA. Generally, for all practical purposes a 0.8% mini agarose gel should suffice.

[b] **NB**: The amount of DNA to be cleaved and/or the reaction volume can be altered, but the proportion and composition of the components should be constant. The total volume of enzyme being added should not exceed 10% of the reaction volume to avoid high glycerol concentration, since glycerol may inhibit enzyme function. It is important that no trace level of RNase be present in the starting DNA material. However, if its presence is suspected it is best to purify the starting DNA material by phenol:chloroform extraction and ethanol-precipitation.

Protocol 8

Agarose gel purification

Equipment and reagents

- Microcentrifuge and sterile tubes
- Agarose gel containing digested DNA (see *Protocol 7*)
- DEPC-treated water (see *Protocol 1*)
- Scalpel or razor blades
- Spin-X tubes (Corning Costar)
- 3 M sodium acetate pH 5.5
- Ethanol
- SpeedVac

Method

1. Excise the linearized DNA band from the agarose gel using a clean scalpel or blade.
2. Chop the excised DNA gel slice into small pieces using DEPC-treated water to aid in mincing the gel.
3. Transfer the DNA-embedded agarose gel pieces into a Spin-X tube.
4. Incubate the tube containing gel pieces at −70 °C for 30 min.
5. Leave the tube at room temperature for an additional 20–30 min.
6. Spin the tube in a microcentrifuge at room temperature for 10 min using top speed. Discard the upper part containing the fine gel pieces above the filter and transfer the eluate into a fresh microcentrifuge tube. Add a one-tenth volume of 3 M sodium acetate (pH 5.5) followed by 2 volumes of 100% ethanol and precipitate the linearized DNA at −70 °C for 30 min.

Protocol 8 continued

7. Recentrifuge at 4 °C for 15 min, aspirate the supernatant and wash the DNA pellet once using 70% ethanol, followed by a second quick wash with 100% ethanol. Finally dry the residual ethanol using speed vacuum and dissolve the DNA pellet in a suitable volume of DEPC-treated water (10–20 μl). Store at 4 °C until ready to make labelled probe.

In vitro transcription reactions are widely used to synthesize RNA from recombinant plasmids. The gel-purified, linearized DNA from *Protocols 7* and *8* is now ready for use in generating labelled RNA probes to be used in RNA binding analysis. Usually a single radiolabelled rNTP is used for labelling the RNA sequence, and the most commonly used nucleotide is UTP. It is usually not advisable to use ^{32}P-labelled ATP for RNA probe preparation since, for unknown reasons, less label is generally incorporated. If a label other than UTP is available and needs to be used, the composition of the cocktail should be altered in a similar way. Unincorporated nucleotide can be partially removed from the labelled probe by ethanol precipitation. DNA template can be removed, if necessary, by digestion with DNase I (RNase-free) following the transcription reaction. Since it is usually necessary to obtain the full-length labelled RNA transcript free from degraded or truncated species, the purification of the RNA probe is described in greater detail in the following section.

Protocol 9

Transcription reaction

Equipment and reagents

- 5 × transcription reaction buffer: 200 mM Tris-HCl pH 7.5, 30 mM $MgCl_2$, 10 mM spermidine, 50 mM NaCl
- 100 mM DTT
- 40 U/μl RNAsin ribonuclease inhibitor (Promega or other supplier)
- 20 mM each of rATP, rGTP, and rCTP (ribonucleotide cocktail)
- 1 mM UTP (usually diluted from a 10 mM stock using DEPC-treated water)
- 10 μCi/ μl [α-^{32}P]UTP (3000 Ci/mmol; e.g. Amersham, Pharmacia, Biotech)
- Linearized gel-purified DNA template (see *Protocol 8*) at 0.5–1.0 μg per μl
- 20 U/μl SP6, T7, or T3 RNA polymerase (Promega or other supplier)
- Stop solution: 8 M urea, 0.025% (w/v) Bromophenol Blue, 0.025% (w/v) xylene cyanol

Method

1. To a labelled, nuclease-free, microcentrifuge tube add the components in the order shown:

5 × Transcription buffer	4 μl
100 mM DTT	1 μl

Protocol 9 continued

RNasin	1 μl
20 mM ribonucleotide cocktail	1 μl
1 mM UTP	1 μl
[α-^{32}P]UTP	5 μl
Linearized, purified DNA template	1 μl
RNA polymerase	1 μl
DEPC-treated water to 20 μl final volume	

2. Spin the tube briefly in a microcentrifuge.
3. Incubate the transcription reaction mixture for an hour at 37 °C water bath.
4. Add an equal volume of stop solution.

The RNA probe generated using *Protocols 7–9* is now ready for purification. This step is highly recommended since truncated and degraded RNA species can be separated allowing isolation of only the full-length transcript. Using gel-purified, full-length RNA transcript also gives the most reliable results and often avoids inconsistent data.

Protocol 10

Purification and elution of the labelled probe

Equipment and reagents

- Scintillation counter (and equipment and scintillant)
- Sequencing or large protein gel apparatus with high-voltage power pack
- X-ray film, plastic wrap, and clean fresh blade or scalpel
- Radiation protection shield
- 8% acrylamide:8 M urea solution for sequencing gel[a]
- 10% ammonium persulfate (APS)
- TEMED (N,N,N′,N′-tetramethylethylenediamine)
- DEPC-treated water (see *Protocol 1*)
- Elution buffer: 0.1% (w/v) SDS, 0.5 M NH_4OAc, 10 mM $Mg(OAc)_2$, 1 mM EDTA
- 1 × TBE sequencing gel running solution
- Stop solution: 8 M urea, 0.025% (w/v) Bromophenol Blue, 0.025% (w/v) xylene cyanol
- 1 ml syringe with a fine-gauge needle
- Potassium acetate solution (add 11.5 ml of glacial acetic acid to 60 ml of 5 M KOAc solution and make up the volume to 100 ml with DEPC-treated water)
- 0.5 M sodium phosphate solution
- Whatman DE filter paper

Method

NB: Follow local rules governing the use of radioisotopes.

1. Cast a sequencing-type thin gel according to the manufacturer's instruction for the equipment.[b] To 35 ml of acrylamide:urea solution add 0.35 ml of 10% APS and 25 μl

Protocol 10 continued

TEMED. Pre-run the sequencing gel at 2000 V for 30 min in 1 × TBE buffer before loading the RNA samples.

2. Gently flush the wells using a 1 ml syringe with a fine-gauge needle immediately prior to loading, since urea tends to accumulate in the wells and interferes with running of the samples On termination of the transcription reaction (see *Protocol 9*), add an equal volume of stop solution to the tube, and load the sample into pre-marked wells of the gel.[c].
3. After loading the samples, start the gel with the power setting fixed at a constant 50 W for an appropriate time, dependent on the RNA probe length. For an RNA of 100–125 bases, run until the Bromophenol Blue dye position is approximately 2–5 cm from the bottom of the gel.
4. At the end of the run carefully separate the glass plates. Gently cover the gel on the glass plate with plastic wrap.
5. In a dark room, firmly lay X-ray film over the plastic-covered gel. Notch one side of the film for alignment after developing.
6. Expose the gel for 15–20 min and develop the film. The exposure time is variable and left to individual users, but is primarily dependent on the label incorporation into RNA.
7. Realign the developed film with the gel on the plate with the aid of the orientation notch. On the glass plate, use a marker pen to identify and mark the correct full-length RNA transcript or the band of interest.
8. Remove the plastic wrap and use a clean blade to excise the probe band from the gel. Be sure to do this step with a protective shield between yourself and the gel.
9. Mince the gel into fine pieces using elution buffer, and transfer gel pieces into a pre-labelled microcentrifuge tube. Add 500–750 μl of elution buffer to the tube and incubate for 10–12 hours at room temperature.
10. Centrifuge the contents of the tube briefly (2–3 min) in a microcentrifuge at 4 °C and transfer the clear aqueous layer into a new tube.
11. Add 1/10 reaction volume of potassium acetate solution and incubate on ice for 5–10 min. Centrifuge once again for 5 min at 4 °C in a microcentrifuge.
12. Aspirate the supernatant into a fresh clean tube and ethanol-precipitate the eluted RNA probe at −70 °C for a few hours. Leave the tube overnight for efficient precipitation of the RNA.
13. Centrifuge the tube containing RNA for 20–30 min at 4 °C in a microcentrifuge, remove the supernatant, and wash the pellet once with 70% ethanol and then with 100% ethanol, then dry in a vacuum. **Caution**: Although it may be difficult to visualize a distinct pellet for RNA, do not under any circumstance closely examine it, because of the radiation hazard.
14. Resuspend the dried pellet in 20–50 μl of DEPC-treated water. Store the probe RNA samples in small aliquots at −80 °C. Avoid multiple freeze–thaw cycles as this accelerates probe degradation.

Protocol 10 continued

15. Quantify the RNA label as follows: spot 1 μl on a Whatman DE filter paper. After filter binding, wash sequentially using 0.5 M sodium phosphate solution (7 ml), water (10 ml), and ethanol (1–2 ml). Air-dry the paper and monitor counts incorporated using a scintillation counter. Use 20 000–200 000 c.p.m. per reaction tube for assay depending on the method used.

[a] Acrylamide–urea gels are preferred since they provide the best resolution.

[b] If a sequencing gel apparatus is unavailable, then a standard protein gel assembly will suffice, but separation of bands may not be ideal.

[c] It is best to load the same sample in several wells side by side.

4.2 Labelled nucleoprotein complex formation

An essential aspect of investigating virus replication strategies is to analyse complex formation *in vitro*. Stable complex formation between the virus RNA and proteins of virus or cellular origin is absolutely necessary before analysis by either gel shift assay or protein–RNA-crosslinking. Care should be taken that all tubes and solutions used for these procedures are nuclease-free, otherwise complex formation may be significantly affected.

Protocol 11

Labelled nucleoprotein complex formation

Equipment and reagents

- 10 × binding buffer: 25 mM Hepes pH 7.9, 250 mM KCl, 20 mM $MgCl_2$, 50% glycerol
- 500 mM DTT (freshly prepared)
- 50 mM ATP
- Yeast tRNA (Boehringer Mannheim) (Commercially available tRNA must be phenol:chloroform extracted, ethanol-precipitated, and finally suspended in DEPC-treated water at a concentration of 10 μg per μl.)
- RNasin RNase inhibitor (e.g. Promega)
- Uniformly labelled and purified [^{32}P]RNA probe as described in *Protocol 10*. (Dilute the probe using DEPC-treated water to approximately 20 000 c.p.m./μl.)
- Purified virus protein of interest or cellular protein extract
- Water baths maintained at 45 °C and 30 °C
- RNA loading dye: 50% (v/v) glycerol, 0.5% (w/v) xylene cyanol, 0.5% (w/v) Bromophenol Blue

Method

1. To a sterile, labelled microcentrifuge tube add the components in the following order:

10 × binding buffer	2.5 μl
500 mM DTT	1.0 μl
50 mM ATP	1.0 μl

Protocol 11 continued

Yeast tRNA	1.5 μl
RNasin (40 U/μl)	1.0 μl
Purified virus protein or cell extract	1–5 μl

DEPC-treated water to final volume of 25 μl (leave room for probe addition)

2. Incubate the labelled probe RNA obtained from *Protocol 10* in a 45 °C water bath for 30 min prior to adding the binding reaction.
3. Pre-incubate the RNA binding reaction tube (minus probe RNA) in the 30 °C water bath for 5 min.
4. Add the labelled probe to the RNA binding components' tube and continue the incubation for an additional 10–15 min to allow the protein to bind to the RNA probe.
5. Stop the reaction by adding 5 μl of 6 × RNA loading dye to the 25 μl sample and mixing gently pipetting up and down. Load into the sample wells of a non-denaturing gel as described below for the gel shift assay (see *Protocol 12*), or proceed to RNA-protein UV-crosslink analysis (see *Protocol 14*).

4.2.1 Gel shift assay

This assay is also known as gel retardation or electrophoretic mobility shift assay (EMSA) (*Figure 2*). It is one of the most common methods used to detect nucleic acid–protein interactions. The underlying principle is that the binding of a protein to a uniformly labelled RNA reduces the mobility of the RNA in non-denaturing polyacrylamide gel, and thus the complex can be separated from the unbound probe electrophoretically. Some critical factors affecting gel shifts include the signal strength, the signal-to-noise ratio, the protein concentration, and the reproducibility. The protein concentration optimum for the required signal may vary between proteins and amongst different extracts prepared of the same protein. As a general rule, the protein concentration should be checked each time a new cell extract or a fresh preparation of the purified protein is made. The approximate concentration of the purified protein can be determined by conventional protein assays and should be kept constant in all subsequent experiments. For this reason it is essential to titrate the protein concentration for the assay and then select a midpoint for further analysis.

4.2.2 UV-crosslink analysis

In this method, a purified protein of virus or cellular origin, or a cellular extract, is incubated with a uniformly labelled RNA probe (see *Figure 3*). The protein–RNA nucleotidyl complex(es) formed are then subjected to ultraviolet (UV) light, causing covalent bonds to form between the RNA and the protein(s). Since UV light is absorbed by plastics, the best method is to shine the light directly down on top of the reaction contained in a microtitre plate. Wear a protective UV

Protocol 12
Gel shift assay

Equipment and reagents

- Labelled nucleoprotein complex (see *Protocol 11*)
- Clean pair of glass plates with clips and side-spacers, or a suitable vertical-gel apparatus
- Rain X or Sigmacote (Sigma)
- Syringe with fine-gauge needle
- Glass container (slightly larger than the gel)
- Whatman chromatography paper
- X-ray film
- 40% acrylamide:bisacrylamide solution (29:1; e.g. Bio-Rad)
- 20 × TBE: 121 g Tris base, 61 g boric acid, 40 ml of 0.5 M EDTA (pH 8.0), make up to one litre with water
- Tank buffer: 0.25 × TBE
- TEMED
- 10% ammonium persulfate solution (APS) (prepared fresh)
- Fixing solution: 10% methanol, 10% acetic acid in water

Method

1. Clean the glass plates with detergent and water followed by ethanol. Use Rain X or Sigmacote (Sigma) on one plate of the apparatus to allow the gel stick to one plate. Use tape to mark the treated plate to consistently treat one side of one plate and thus avoid using two coated plates.
2. Assemble the two plates with side and bottom spacers in between, and seal the plates carefully with tape.
3. Place the gel assembly evenly on a bench surface and prepare the gel mix as follows:

Acrylamide:bisacrylamide solution	4.65 ml
20 × TBE	437 μl
10% APS	0.7 ml
TEMED	7 μl
Distilled water to a final volume	35 ml

4. Load the gel mixture between the plates taking care that no bubbles are formed, and gently insert the comb. Clip the comb securely to the top edge until the gel polymerizes, which normally takes approximately 20–45 min.
5. Dilute the tank buffer.
6. Remove the tape from the gel apparatus. Secure the plates on the gel box and add running buffer (0.25 × TBE). Flush the wells with 0.25 × TBE solution with the aid of a syringe fitted with a fine-gauge needle so that residual polyacrylamide is washed off. Also remove air bubbles trapped beneath the gel with the aid of a bent needle.
7. Pre-run the gel for at least 30 min before loading the samples in the cold room. Allow the current to drop steadily from 25 mA to approximately 2–5 mA.

Protocol 12 continued

8. At termination of reaction incubation, load 50% of each of the samples containing the ^{32}P-labelled probe, protein, and RNA dye (obtained from earlier step) into the sample wells and run the gel in the cold room at a constant 120–150 V. Determine the gel running time empirically, based on the length of the probe.[a]
9. On completion, remove the gel from the glass plates and place in a glass container.[b] Fix the gel using a liberal volume of fixing solution for 10 min at room temperature with gentle agitation.[c]
10. Carefully remove the fixing solution. With the gel in the glass container place a slightly large sized piece of Whatman paper on top of the gel. Gently lift the paper with the gel attached to it and cover the gel with a plastic wrap.
11. Dry the gel and expose it to X-ray film.

[a] In our experience, for samples with a 100-nucleotide probe length, the Bromophenol dye should be approximately 2.0–2.5 cm from the bottom on a gel of a total height of 15 cm. This gives good resolution of the retarded complex from the free probe.

[b] A laboratory spatula is a handy tool to pry open the glass plates.

[c] For further analysis on the retarded ribonucleoprotein complex go to *Protocol 13* without fixing.

Protocol 13

Analysis of the ribonucleoprotein complex

Equipment and reagents

- Equipment and reagents as *Protocol 12*
- 0.5 × TBE buffer
- Clean sharp blade
- 2 × SDS-PAGE sample buffer
- Second-dimension SDS-PAGE gel (cast using a broad-well comb)
- Silver stain

Method

1. Perform gel shift assay as described in *Protocol 12*. Separate the two glass plates and with the gel attached to one of the plates cover the exposed part of the gel with plastic wrap using all necessary precautions for radioactivity.
2. Expose the gel to an X-ray film in the dark at room temperature for 30–60 min and develop the film. Be sure to notch one edge of the gel for alignment.
3. Align the developed film with the wet gel. Excise the retarded complex containing nucleoproteins with a clean sharp blade and remove the plastic wrap.
4. Transfer the gel slice to a microcentrifuge tube and incubate in 1 ml of 0.5 × TBE buffer for an hour at 37 °C.
5. Aspirate the TBE solution and equilibrate the gel slice with 2 × SDS-PAGE sample buffer at 37 °C for 30 min followed by 10 min incubation at 65 °C.

Protocol 13 continued

6. Gently place the treated gel slice in the sample well of a second-dimension SDS-PAGE gel,[a] and electrophorese at the appropriate voltage and current routinely used.
7. After completion of the run, silver-stain the SDS gel to visualize the proteins.[b] Alternatively, transfer the proteins to a nitrocellulose membrane (*Protocol 17*) and analyse by Western blot as described in *Protocol 19*.

[a] When casting the gel, use a broad-well comb so it is easier to slide the gel inside the wells.

[b] Silver staining has the advantage of being more sensitive than Coomassie Blue staining and can detect as little as 20 ng of protein. Many kits are now commercially available and are simple to use by following the manufacturer's directions.

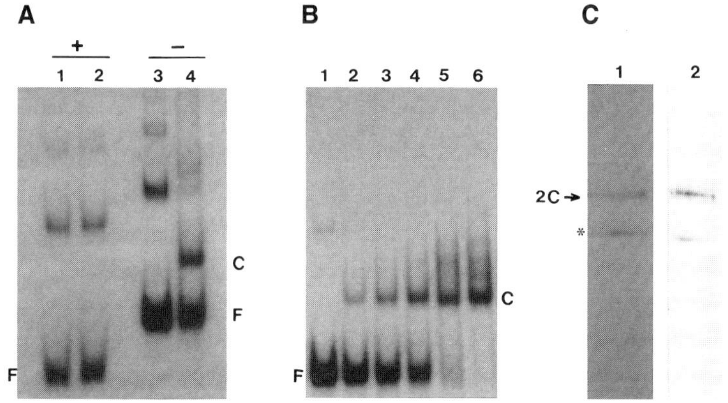

Figure 2 Binding of 2C to the 3′-cloverleaf structure of poliovirus negative-strand RNA. Uniformly ^{32}P-labelled, gel-purified, full-length RNA probe corresponding to the 5′ first 100 nucleotides of the positive strand or the 3′ terminal 100 nucleotides of the negative strand were prepared according to *Protocol 9*. (Panel A) The [^{32}P]RNA probe was added to the binding reaction containing 2C protein to form the nucleoprotein complex, and the complex analysed by gel shift assay. Lanes 1 and 2 show the binding reaction with positive-strand RNA probe and lanes 3 and 4 represent binding reactions using negative-strand RNA probes. Lanes 1 and 3 are control reactions without added virus 2C protein and lanes 2 and 4 contains 60 ng of purified 2C protein. The free probe (F) and the gel retarded complex (C) are shown. (Panel B) Dose-dependent binding of 2C protein to a negative-strand RNA probe. The gel-purified, negative-strand RNA probe was added to the binding reaction with increasing amounts of purified 2C, and the resulting complex analysed by gel shift. (Panel C) Analysis of the retarded nucleoprotein complex by silver stain and Western blot. The nucleoprotein complex (C) observed in the gel shift assay was isolated as described in *Protocol 13*, and the gel-eluted protein was resolved by SDS–PAGE and silver stained (lane 1) or transferred on to a nitrocellulose membrane and developed using anti-2C antibody in a Western blot analysis. The migration of 2C and that of truncated 2C is marked by an arrow and an asterisk, respectively.

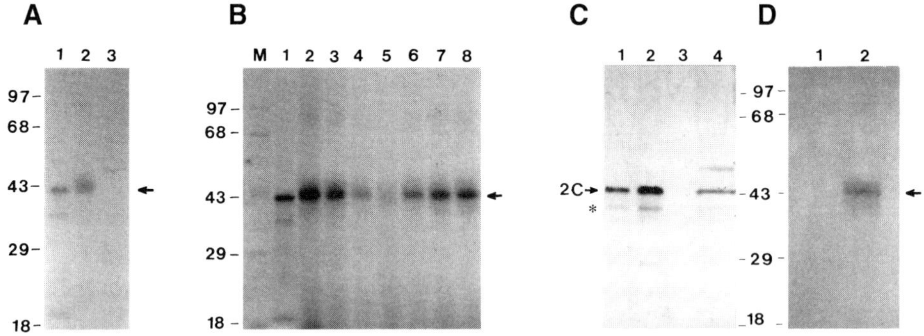

Figure 3 UV-crosslinking analysis of poliovirus 2C protein interaction with the positive- and negative-strand RNA probes. (A) UV-crosslinking and ^{32}P-label transfer to purified poliovirus 2C after the binding reaction performed as described in *Protocol 14*. Samples were resolved on an SDS-PAGE gel and proteins analysed. Lane 1, [^{35}S]methionine-labelled 2C protein translated *in vitro* using rabbit reticulocyte lysate. Lanes 2 and 3, purified virus 2C protein UV-crosslinked to either the negative-strand probe (lane 2) or the positive-strand probe (lane 3). (B) Specificity of UV-crosslinked 2C with the ^{32}P-labelled negative-strand RNA. Increasing quantities of cold, specific, homologous RNA (negative strand) or non-specific, heterologous RNA were used. Cold, competing RNAs were added to the binding reaction mixture containing 2C prior to the addition of labelled RNA. Labelled RNA was added 5–10 min later and reactions were UV-crosslinked, RNase digested and analysed by 14% SDS-PAGE. Lane M, ^{14}C-labelled protein molecular-weight markers. Lane 1, [^{35}S]methionine-labelled 2C protein translated *in vitro* using rabbit reticulocyte lysate. Lane 2, UV-crosslinked 2C in the absence of any competitor. Lanes 3–5, reaction mixture containing 50, 250, and 500 ng of cold, competitor homologous RNA, respectively. Lanes 6–8, reaction mixture containing 50, 250, and 500 ng of cold, competitor heterologous RNA. Arrows in the panel indicate the position of 2C migration and the numbers on the side correspond to the position and approximate molecular masses in kDa of the ^{14}C-labelled marker proteins. (C) Western blot analysis of affinity-purified 2C protein crosslinked to a negative-strand RNA probe. Lanes 1 and 2, 60 and 120 ng of affinity-purified virus 2C protein (direct loading). Lanes 3 and 4, UV-crosslinked reactions using a negative-strand RNA probe without and with 2C protein, respectively. Proteins were visualized by developing the membranes with alkaline phosphatase-conjugated secondary antibodies. (D) Autoradiogram of the UV-crosslinked 2C protein transferred on to a nitrocellulose membrane. Lanes and 2 contain duplicate samples from crosslinked reactions with (lane 2) and without (lane 1) 2C protein.

face-shield in addition to taking routine radioactive protection. A mixture of RNases is added to the reaction tubes to digest the unprotected excess RNA, and the reaction mixture is then subjected to SDS-PAGE. Since the protein is now labelled, the molecular weight of the protein from a crude extract can be determined. The migration of the RNA–protein complex on a protein denaturing gel is compared to a standard molecular weight marker or to [^{35}S]methionine-labelled, *in vitro* transcribed and translated protein. It is not unusual for UV-crosslinked proteins to migrate slightly slower than unlinked proteins. A distinct advantage of UV-crosslinking is the detection and identification of protein by size from a crude cellular extract that binds specifically with high affinity, in contrast to proteins that bind weakly.

Protocol 14

UV-crosslink analysis

Equipment and reagents

- Vacuum gel dryer
- Hand-held UV light
- ELISA microtitre plate
- Labelled nucleoprotein complex (see *Protocol 11*)
- RNase A (20 μg/ml) and T1 (10 000 u/ml) mixture
- SDS protein gel components and apparatus to run gels
- *In vitro* translated [^{35}S]methionine-labelled protein
- [^{14}C]methylated protein, molecular weight markers (Gibco) and pre-stained rainbow markers (Amersham)
- Fixing solution: 10% acetic acid, 10% methanol in water
- Fluorographic agent: 1 M sodium salicylate solution or Amplify®, Amersham
- 2 × SDS sample buffer (*Protocol 13*)
- Water bath maintained at 37 °C
- Heating block

Method

1. Follow *Protocol 11* for nucleoprotein complex formation. Use labelled RNA probe at 100 000 c.p.m./μl.
2. Following incubation and complex formation, transfer the reaction contents of the microcentrifuge tube to a pre-marked ELISA multiwell plate.
3. Place a hand-held UV light on the plate, making sure that light shines on the wells containing the samples. Continue treatment under UV light for 20–30 min,[a] preferably in a dark room.
4. On termination, transfer the contents from the plate to a fresh microcentrifuge tube, add 5 μl of the cocktail of RNase A and T1, and incubate at 37 °C for 45–60 min.[b]
5. Add an equal volume of 2 × SDS sample buffer. Denature the protein samples at 95 °C for 5 min in a heating block. Cool on ice and load into the wells of a denaturing gel and resolve.
6. Simultaneously load standard molecular-weight marker proteins and a lane containing the [^{35}S]methionine-labelled sample to compare with the UV-crosslinked protein(s) migration. Electrophorese at constant current until the dye runs off the gel and the proteins are resolved.[c]
7. Disassemble the apparatus and remove the gel. Fix by incubating in fixing solution for 15 min at room temperature with gentle agitation.[d]
8. Place the treated gel on a piece of Whatman paper and dry on a gel dryer. Expose the dried gel to X-ray film until the band of interest is visible—this may vary from 10–24 hours.

Protocol 14 continued

[a] The optimum time varies for different RNA–protein interactions and subsequent complex formation.

[b] This treatment digests the excess and unbound probe, except for the RNA fragment protected by the specifically bound crosslinked protein.

[c] We find the multicoloured, protein molecular-weight markers to be very useful for determining appropriate gel running times.

[d] If signal intensity is low and ^{14}C-labelled marker proteins are used it is best to incubate the gel with an fluorographic agent for 10–15 min or as recommended by the manufacturer. This treatment results in impregnation of the gel with the enhancer, which gives a better signal resolution.

Under certain conditions, it becomes important to analyse the stability of the probe. This is particularly the case when the probe is mutagenized to observe the effect of RNA mutation on protein binding. An apparently negative result of RNA–protein complex formation due to a mutation in the probe could be due to the instability of the probe under the conditions of the binding assay. This test should be performed in parallel with UV-crosslinking analysis.

Protocol 15

Probe stability analysis

Equipment and reagents

- High-voltage power pack
- Sequencing gel or large protein gel apparatus
- 8% acrylamide:8 M urea solution for sequencing gel
- Binding reaction with nucleoprotein complex (see *Protocol 11*)
- 3 M sodium acetate pH 5.5
- Water-saturated phenol:chloroform (1:1 (v/v))
- RNA loading dye: 8 M urea, 0.025% (w/v) Bromophenol Blue, 0.025% (w/v) xylene cyanol
- DEPC-treated water
- X-ray film
- 100% and 70% ethanol

Method

1. Follow steps 1–4 as in Protocol 11 for labelled nucleoprotein complex formation.
2. Following the incubation at step 4 (*Protocol 11*), the binding reaction is made up to 0.45 ml with DEPC-treated water. Add 50 μl of 3 M sodium acetate (pH 5.5).
3. Add water-saturated phenol:chloroform mixture (1:1 (v/v)), secure the cap tightly and vortex the tube for a minute followed by a spin at room temperature for 3 min. Transfer the upper aqueous layer (approximately 500 μl) containing RNA to another pre-labelled, nuclease-free microcentrifuge tube.
4. Repeat the phenol:chloroform extraction once more. Discard the tube with the residual organic layer to the liquid radioactive waste.

Protocol 15 continued

5. Precipitate the labelled RNA probe from the aqueous phase by adding ethanol (2 volumes) and incubating at −70 °C for 2–3 hours. This step can be extended overnight without loss of RNA.
6. Centrifuge the tubes in a microcentrifuge at 4 °C, wash the RNA pellet once with 70% ethanol, then once with 100% ethanol.
7. Dry the RNA pellet using a vacuum pump.
8. Resuspend the dried RNA pellet in approximately 10 μl DEPC-treated water, and then add an equal volume of RNA loading dye.
9. Pre-run the gel and flush the wells as described in *Protocol 10*. Load 50% of the sample on to the acrylamide:urea gel and resolve the RNA once again.
10. On completion, fix and dry the gel and expose to X-ray film briefly. Ensure the probe is not degraded.

An important aspect of this type of study involving nucleic acid–protein interactions is the specificity of binding. One of the ways to address this is by cold competition, wherein increasing levels of unlabelled ('cold') homologous competitor RNA is added to a fixed amount of protein and probe RNA in the binding reaction. It is important that a heterologous RNA of comparable size should also be included in parallel reactions. When unlabelled, specific RNA is added in excess to a binding reaction containing an identical binding site to the RNA transcript probe, it competes for the probe RNA. The effect of this is a gradual decrease in signal intensity on addition of increasing amounts of the cold, competitor RNA. However, upon addition of non-specific, heterologous RNA to the reaction tube, where there is no specific binding site for the protein, the RNA–protein interaction signal does not alter. *Protocol 16* for *in vitro* transcription can be used to generate unlabelled RNA for such experiments.

Protocol 16

Specificity of UV-crosslinking analysis

Equipment and reagents

- All transcription reagents as shown in *Protocol 9* except radiolabelled UTP
- Gel-purified, linearized DNA for specific and non-specific cold RNA preparation (use at 0.5- 1.0 μg/μl)
- Water-saturated phenol:chloroform mixture (*Protocol 15*)
- 100% and 70% ethanol
- Agarose gel apparatus and gel running buffer
- 10 mM stock of each ribonucleotide
- RQ DNase (RNase-free; Promega)
- 3 M sodium acetate pH 5.5
- DEPC-treated water

Protocol 16 continued

Method

1. In a sterile, nuclease-free microcentrifuge tube add the components in the order below:

5 × transcription buffer	20.0 μl
100 mm DTT	10.0 μl
RNasin (40 U/μl)	2.5 μl
Linearized DNA	2.0 μl
10 mM ATP	5.0 μl
10 mM CTP	5.0 μl
10 mM GTP	5.0 μl
10 mM UTP	5.0 μl
RNA polymerase (20 U/μl)	2.5 μl

DEPC-treated water to 100 μl final volume.

2. Following addition of all the components, spin the tube briefly (30 sec) in a microcentrifuge.
3. Incubate the reaction mixture at 37 °C for 60–90 min, then add RQ DNase (RNase-free) and continue the incubation for an additional 15–30 min at 37 °C.
4. Dilute the reaction tube components to 0.45 ml using DEPC-treated water and add 50 μl of 3 M sodium acetate (pH 5.5).
5. Follow steps 3–7 in *Protocol 15*.
6. Resuspend the dried RNA pellet in DEPC-treated water. Quantify the RNA and dilute accordingly as required for competition assays. Run a small amount of the RNA on a denaturing or native agarose gel and check the integrity of the RNA obtained (this is important).

Protocol 17

Transfer of proteins to nitrocellulose

Equipment and reagents

- SDS-PAGE gel apparatus and power pack
- Nitrocellulose membrane (Schleicher and Schuell) cut to gel size
- Filter paper or blotting sheets cut to gel size
- Flat-ended forceps
- Corning glass trays
- Blotting apparatus (tank blotter or semidry apparatus)
- Transfer buffer: 25 mM Tris base, 190 mM glycine, 20% methanol (do not adjust pH)
- Magnetic stirrer in the cold room
- Pre-stained protein molecular-weight markers for reference

Protocol 17 continued

Method

1. Run the sample of interest on an appropriate SDS-PAGE gel using pre-stained protein molecular-weight markers as a reference.
2. After removing the gel from the assembly, cut a piece of nitrocellulose membrane large enough to cover the separating gel after the stacking gel has been removed. Wet the membrane for 2–5 min with distilled water, followed by a second wash in transfer buffer (using a separate dish) for the same length of time.[a] Cut four sheets of filter paper to the same size as the nitrocellulose and soak in transfer buffer.[b]
3. Soak the tank sponges in transfer buffer and lay on the tank blot-holder. Carefully place one piece of filter paper on top of the sponge, and make sure no air bubbles are trapped between them. Place another layer of filter paper, and again remove any bubbles.[c]
4. Carefully place the equilibrated gel on the top filter paper. Notch one corner for orientation purposes. Gently remove bubbles from beneath the gel.
5. Place the pre-soaked nitrocellulose on top of the gel. Remove bubbles and place the other two pieces of filter paper over the nitrocellulose, followed by the second sponge. Make sure there are no bubbles between the layers, as this will reduce transfer in those areas. Mark the notch end of the gel on the membrane with a ballpoint pen.
6. Put the sandwich assembly in the apparatus and fill the tank chamber with transfer buffer. Place the entire chamber on top of a stir plate, and recirculate the buffer. Electrophorese at 4 °C for 2 hours to overnight.[d]
7. Remove the sandwich assembly from the tank and open it carefully. Discard the filter paper, and, with the aid of forceps, remove the nitrocellulose. Check the transfer efficiency by observing the coloured molecular markers on the gel. If required, store the nitrocellulose for future staining after allowing to air-dry. Store between folds of filter paper at 4 °C.

[a] This prepares the nitrocellulose for efficient transfer and retention of the proteins.

[b] It is best to use a flat-ended forceps to manipulate the nitrocellulose and best to handle the edge only. Always wear gloves, otherwise residues left on the membrane may alter the results.

[c] Rolling a plastic 5 or 10 ml pipette over the nitrocellulose is helpful.

[d] Be careful of the orientation of the gel with respect to the nitrocellulose and the electrodes in the apparatus. The protein will migrate towards the negative electrode. If your gel or nitrocellulose is placed incorrectly, the proteins will migrate into the buffer and be lost.

4.3 Northwestern analysis

Another useful technique for detecting the interaction between RNA and protein is the Northwestern blot, in which proteins (of virus or cellular origin) are immobilized on a membrane and probed using a specific labelled RNA. Non-specific binding of the probe can be minimized by pre-hybridizing the membrane in the presence of non-specific (heterologous) RNA prior to probe hybridization.

This technique has its limitations however, since, unlike gel retardation assays, binding specificity cannot be addressed using cold, competitor RNA.

Protocol 18

Detection of RNA–protein interactions

Equipment and reagents

- Labelled purified RNA of interest, prepared according to *Protocols* 9 and 10
- 50 × Denhardt's solution (Sigma)
- SSB solution: 10 mM Tris pH 7.0, 50 mM NaCl, 1 mM EDTA, 1 × Denhardt's solution in DEPC-treated water (freshly made)
- Proteins transferred on to nitrocellulose membrane from *Protocol* 17
- *E. coli* tRNA
- Hand-held Geiger counter (optional)
- Water bath maintained at 70 °C
- X-ray film

Method

1. Add 10 ml SSB with 25 μg/ml *E. coli* tRNA to the nitrocellulose and incubate at room temperature for 1–2 hours (pre-hybridization step).
2. Denature the RNA probe at 70 °C for 5 min.[a]
3. Dilute the (denatured) probe in 10 ml of fresh SSB to approximately 200 000 c.p.m. and incubate with the nitrocellulose for 1 hour with gentle shaking.
4. Remove the probe solution and discard. Wash the nitrocellulose 3–5 times using 30–50 ml of a fresh SSB solution. Monitor, if desired, the specific bound probe on the nitrocellulose between washes with a hand-held Geiger counter. Remember to dispose of all radioactive solutions appropriately.
5. Cover the nitrocellulose with plastic wrap and expose to X-ray film for between 4 hours to overnight depending on the interaction.

[a] This step should be excluded if RNA secondary structure is suspected to be important in the interaction with the protein.

4.4 Western analysis

Western analysis, often referred to as immunoblotting, is used to detect or identify proteins recognized by specific polyclonal or monoclonal antibodies. The proteins are first separated by SDS-PAGE electrophoresis and then electrophoretically transferred to a membrane using either a tank or a semidry transfer apparatus. The choice of membrane includes nitrocellulose or polyvinylidene fluoride (PVDF), and the transfer of proteins can be monitored by staining the membrane with Ponceau S. Following the immobilization of the proteins on to the membrane, the membrane free sites are blocked by a blocking agent. The primary antibody then recognizes and binds to its specific protein antigen on the membrane. A secondary antibody directed against the IgG of the primary antibody is added and this binds to the bound primary antibody. Many of the commonly used secondary antibodies are conjugated to alkaline phosphatase or peroxidase enzymes, to allow for chemiluminescence or colorimetric detection of the antibody–antigen complex. A detailed methodology for both these approaches is described below.

Protocol 19

Colorimetric detection

Equipment and reagents

- Protein samples transferred on to nitrocellulose (see *Protocol 17*)
- Blocking solution: 50 mM Tris–HCl pH 7.6, 150 mM NaCl, 0.05% (v/v) Tween-20, 3% (v/v) non-fat milk
- Primary antibody on unlabelled antibody specific to the target protein
- Secondary antibody Anti IgG coupled to an enzyme such as alkaline phosphatase or horseradish peroxidase
- Phosphate-buffered saline solution (PBS) pH 7.4
- Wash solution: 50 mM Tris–HCl pH 7.6, 150 mM NaCl, 0.05% (v/v) Tween-20
- 100 mM Tris–HCl pH 9.5
- Nitroblue tetrazolium and 5-bromo-4-chloro-3-indolyl phosphate (NBT–BCIP) solution
- Ponceau S (optional) (Sigma)

Method

1. Stain the nitrocellulose membrane with immobilized proteins with Ponceau S solution at room temperature, and destain using distilled water, if desired.[a]
2. Incubate the membrane with blocking solution for 30 min at room temperature with gentle shaking.
3. Aspirate the blocking solution and incubate with the primary antibody diluted in wash solution.[b]
4. After antibody incubation, aspirate the solution and wash the membrane 3–4 times in wash solution for 5 min each wash.[c]
5. Incubate the membrane with the secondary antibody for 1 hour at room temperature. Dilute the secondary antibody fresh, normally 1:1000-fold in blocking solution, but follow the manufacturer's recommended dilution.
6. Repeat step 3, aspirate the solution and wash the membrane 3–4 times with wash solution.
7. Wash the membrane once with PBS for 5 min, followed by an additional wash with 100 mM Tris–HCl pH 9.5 for 5 min at room temperature.
8. Develop the membrane using the NBT–BCIP solution, which is normally diluted 1:1 in 100 mM Tris–HCl pH 9.5.[d]

Protocol 19 continued

9. When the colour of required intensity is reached, wash the membrane with water to arrest the reaction. After it is dried, store the membrane between the folds of filter or blotting paper.

[a] This will not inhibit antibody binding in future steps, but efficient transfer can be determined before proceeding for detection. Also, you can use either a small plastic tray or a heat-sealable plastic bag for antibody detection. Volumes described here are for small tray staining. Adjust the volume appropriately for heat-sealable bags.

[b] The dilution and incubation time may vary for different antibodies and both need to be optimized.

[c] This antibody solution can be reused several times as the amount of antibody that actually binds to the filter does not significantly deplete the solution.

[d] BCIP substrate hydrolysis produces an indigo precipitate after oxidation with NBT.

Protocol 20

Chemiluminescent detection

Equipment and reagents

- Blocking reagent (Boehringer Mannheim)
- Tris-buffered saline (TBS): 100 mM Tris–HCl pH 7.5, 0.8% (w/v) NaCl
- TBS-T: TBS containing 0.1% (v/v) Tween-20
- Primary antibody (antibody against the protein of interest)
- POD-conjugated secondary antibody
- Chemiluminescent reagent (e.g. Pierce)
- Small plastic tray (a pipette-tip box lid works well) or heat-sealable plastic bags
- Timer, film cassette, X-ray film, plastic wrap, marker pen, and flat forceps

Method

1. Take 10–15 ml of the 1% blocking reagent, made to the manufacturer's directions, and wash the nitrocellulose for between 1 hour and overnight.
2. Pour off the blocking reagent and discard. Dilute the primary antibody in 0.5% of the blocking reagent according to the manufacturer's suggestion (usually 1:200–1:1000). Wash in primary antibody solution for 1–2 hours.[a]
3. Wash the nitrocellulose membrane three times for 15 min (each time) in TBS-T at room temperature.
4. Dilute the secondary antibody in 0.5% of the blocking reagent (usually 1:1000–1:10 000 dilution). Incubate the nitrocellulose membrane for 1 hour. Discard the solution, since the POD-conjugated antibodies cannot be reused.
5. Wash the nitrocellulose membrane as in step 3.
6. Prepare the following items during the final wash. Pour an appropriate volume of the chemiluminescent reagents into 15 ml tubes. Follow the manufacturer's directions—some are ready to use, others must be diluted. Gather together the film,

Protocol 20 continued

cassette, flat forceps, a piece of plastic wrap 2–3 times the size of the nitrocellulose, a timer, and a marker pen. Take everything into the darkroom.

7. Pour off the final wash. Add the chemiluminescent reagent on the membrane, and wait 1–2 min. Wrap the nitrocellulose carefully using the plastic wrap. Avoid bubbles on the front side of the blot, and be sure the 'film-facing' side of the nitrocellulose is covered only once by the plastic wrap.
8. Transfer and place the nitrocellulose in the cassette, with a film. Wait 2 min and develop the film, adding a new film quickly and restarting the timer. Use a range of times to obtain the best results, and note that exposure times may differ between experiments.[b, c] Use a marker pen to mark each film with the time of exposure as it leaves the developer—this minimizes confusion.

[a] The primary antibody/blocking reagent can be stored at −20 °C and used for multiple blots, as the concentration of antibody is relatively high relative to the amount of antigen bound to the nitrocellulose.

[b] Typical exposure times are between 2 seconds and 45 min for chemiluminescence.

[c] If no protein is detected but you are sure there was a sufficient amount present, wash with TBS (not TBS-T) since Tween can disrupt the binding of some antibodies.

5 Membrane binding of virus proteins

The poliovirus-encoded, membrane-associated precursor polypeptide 3AB has been shown to play an important role in virus genome replication. Specifically, it has been implicated in virus RNA synthesis, and perhaps may be used for VPg delivery to the 5′ ends of plus- and minus-strand RNA during replication (20, 21). Expression of the polypeptide in eukaryotic cells and examination of its specific localization encompasses two of the commonly used techniques, known as indirect immunofluorescence and an *in vitro* membrane binding assay, for which the methodology is described in this section.

5.1 Indirect immunofluorescence

Immunofluorescence is an immunochemical method used to investigate the subcellular localization of a target protein, either expressed constitutively or following transfection. Cells transfected with the engineered DNA of interest, upon treatment will emit a bright fluorescence colour that can be easily visualized under a fluorescence microscope (*Figure 4*).

Transfection is a broad term used to describe the introduction of exogenous DNA efficiently into eukaryotic cells by physical or chemical means. This process is normally used to analyse *in vivo* gene expression. The DNA coding for the protein of interest is cloned into an appropriate vector prior to its introduction. It is important that a cell line appropriate for the specific gene of interest be used in this type of study. Also, the transfection conditions should be optimized for each cell line being studied. Some cells are comparatively easier

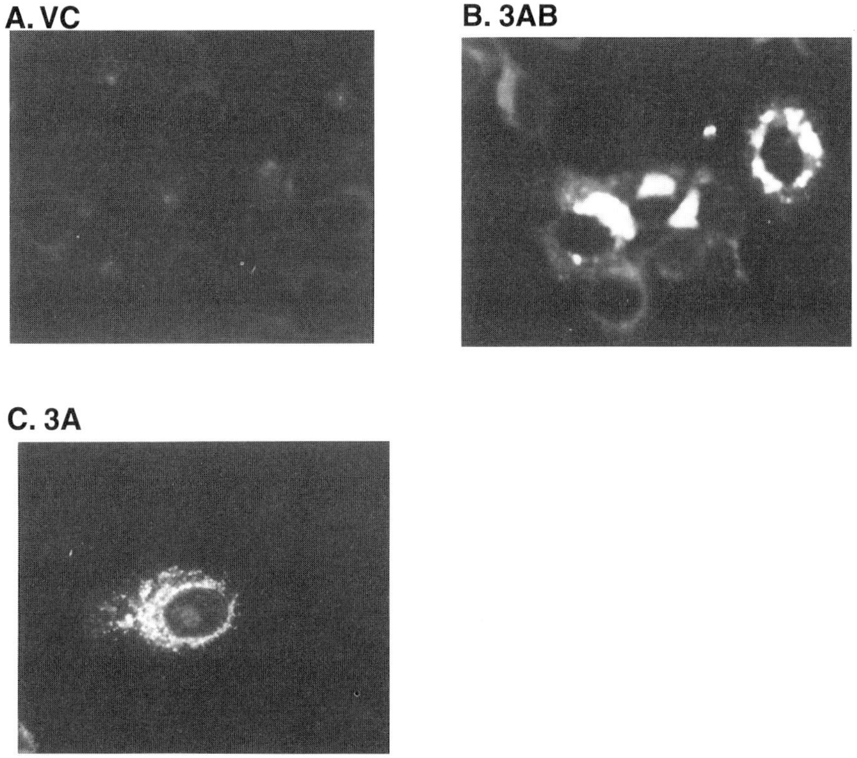

Figure 4 Analysis of poliovirus 3AB and 3A polypeptides in HeLa cells by indirect immunofluorescence. VC: vaccinia virus-infected control (18).

to transfect than others for undefined reasons. Currently, there are four techniques widely used for introducing DNA into cells: the calcium phosphate method (22, 23), DEAE-dextran transfection (24), electroporation (25), and liposome-mediated transfection (26, 27). The choice of method is left to individual users since each method has its own advantage and disadvantage. Cost-effectiveness is another factor to consider. An effective method should incorporate sufficient transfer of the recombinant DNA, with low toxicity and reasonable expression for its *in vivo* effect studies. It is most important that the isolated DNA used for transfection (using any of the techniques mentioned above) should be of good quality with no protein, RNA, or other contaminants. An absorbance ratio A_{260}/A_{280} of approximately 1.8–2.0 is a quick check and a reasonable indicator of the DNA quality, in addition to checking on a agarose gel. This is an important contributing factor for success or failure of transfection. The last method of transfection using liposomes is routinely used in our studies.

The mechanism by which cationic and neutral lipids mediate transfection of DNA or RNA into the cell is unclear. It is thought that the negatively charged phosphate groups on the DNA backbone interact electrostatically and bind to the positively charged surface of liposomes. The resulting charge then, presumably associates and mediates binding to negatively charged sialic acid residues on the cell surface and mediates delivery of the DNA to the cells interior, finally resulting in expression of the exogenous DNA target sequence. In transient transfection, transcription of the transfected gene and subsequent expression can be analysed between 1 and 4 days after transfection medium is applied to cells; in contrast to stable transfection where the introduced genes are integrated into genomic DNA, thereby resulting in the generation of a cell line. In the latter case, several weeks may be required for selection and characterization of stable cell lines before they can be used for the required studies. This time frame is in addition to the time required for manipulating the recombinant clones of interest. Such cell lines constitutively express the protein of interest. However, advanced methods are currently available that can effectively switch on and off the expression of recombinant proteins (28, 29). Transient transfection can be performed in conjunction with recombinant vaccinia virus infection. On productive infection, the virus expresses recombinant RNA polymerase, which drives the transcription of the introduced gene and the resulting expression of the protein.

Protocol 21

Lipofection

Equipment and reagents

- Liposome reagent (e.g. Lipofectin®, Gibco)[a]
- Glass coverslips
- Cell-specific medium and supplements
- DNA to be transfected

Method:

1. Grow cells on coverslips using appropriate growth medium supplemented with the required cell-type serum to approximately 60–80% confluency. If possible, maintain the same cell density for all transfections for the best results.
2. Transfect a fixed quantity of the recombinant plasmid DNA construct into cells in the growth phase on the coverslips, using 30 μg of Lipofectin® for a defined time.
3. Determine the optimum DNA concentration and exposure time empirically for each cell type.[b]

[a] Other commercially available liposome-mediated transfection reagents can replace lipofectin, but their efficiency needs to be established at the beginning of the experiment.

[b] The optimum amount of DNA will vary widely depending on the type of DNA and the cell line being studied. Different cell types will need more or less DNA for same surface areas. Also, it is best to follow the manufacturer's suggestion to obtain a reasonable transfection efficiency.

The specificity and affinity of an antibody targeted towards the protein antigen inside the cell is then exploited. A typical analysis involves four basic steps:

(a) fixing and permeabilizing cells;

(b) blocking sites for non-specific interaction;

(c) specific interaction of the antibody and antigen followed by FITC-conjugated secondary antibody;

(d) final analysis of the sample by fluorescence microscope after the sample has been mounted.

The following protocol is recommended for examining the internal distribution of the antigen using fluorescence analysis and the infection of cells for RNA polymerase expression as described earlier.

Protocol 22

Fluorescence analysis

Equipment and reagents

- Tissue culture slides with wells
- 2×10^5 cells/well
- Complete culture medium
- Complete culture medium with appropriate serum (5–10% FCS)
- Virus cocktail[a]
- PBS^{++} (PBS containing 0–1% w/v $CaII_2$, $MgCl_2$)
- PBS-T: PBS^{++} containing 0.02% Triton X-100
- Washing buffer: PBS^{++} solution containing 0.05% Tween-20
- Immunofluorescence blocking buffer (IBB): PBS^{++}, 0.2% gelatin, 0.05% Tween-20
- Fluorescein isothiocyanate (FITC)-conjugated secondary antibody
- Mounting medium (Sigma)
- 10% buffered formalin phosphate (Fisher)
- Primary antibody specific for the target protein
- Humidified chamber
- Coverslips

A. Slide infection

1. Plate 2×10^5 cells into tissue-culture slide wells and allow to grow overnight.
2. Prepare the virus cocktail (vaccinia virus) estimating there will be 4×10^5 cells in the wells. Use an m.o.i of 1.
3. Infect cells by washing the wells with PBS^{++} and then evenly distribute the virus cocktail over the cells.
4. Incubate for 1 hour with occasional gentle shaking.
5. Wash the cells with warm complete medium once.
6. Add 2.0 ml of complete medium and incubate for the desired time.

Protocol 22 continued

B. Fluorescence analysis

1. Wash the cells once with ice-cold PBS^{++}, and fix the cells for 15 min at room temperature in 10% buffered formalin phosphate. Do not exceed the time as this is a critical factor for the signal intensity.
2. Wash a second time with PBS^{++} and permeabilize the cells using PBS-T for 5 min at room temperature.[b]
3. Block the slide in immunofluorescence blocking buffer (IBB) for 2–3 hours at room temperature. Alternatively, carry out this step at 4 °C overnight or for 1 hour at 37 °C. Exercise care from this step and do not allow the slides to dry.
4. Incubate the slide with the primary antibody diluted in IBB for 1.5 hours at 37 °C in a humidified chamber. Use several sequential dilutions of the antibody if using for the first time or if the titre is unknown. Optimize the dilution of the primary antibody to give a clean signal.
5. Gently wash the slide five times using washing buffer. Each wash is done for 5 min at room temperature. Care must be taken to avoid exposing the slides to light, as such exposure may fade the fluorescence irreversibly.
6. Incubate the slides with the compatible secondary antibody (anti-primary FITC, 1:1000 diluted in IBB). Incubate for 1.5 hours at 37 °C in a humidified chamber.[c]
7. Wash three times with washing buffer for 5 min (each wash) at room temperature with gentle shaking.
8. Wash twice with PBS^{++} for 5 min as step 7.
9. Finally wash once briefly with double-distilled water and apply mounting buffer. Overlay with a coverslip avoiding air bubbles. Store the slides at 4 °C and in darkness. Store the slides wrapped with Parafilm® if they are to be stored for longer than 1–2 days.

[a] This contains the recombinant vaccinia virus expressing the RNA polymerase diluted in cell medium. The polymerase drives the expression of the recombinant plasmid DNA *in vivo*. The propagation of recombinant vaccinia virus is beyond the scope of the present chapter and readers are recommended to refer to relevant literature (18, 30, 31).

[b] This treatment solubilizes the membrane phospholipids' component. Exceeding this time may lead to major alterations in cell structure and distortion of images.

[c] FITC (fluorescein isothiocyanate) is a popular fluorophore. However, others such as Texas Red and rhodamine can be substituted and are available from many suppliers. It is best to follow the manufacturer's recommended dilution.

5.2 *In vitro* membrane binding assay

In vitro translation involves using a cell-free system to translate protein from the mRNA(s). The mRNA is obtained by *in vitro* transcription of the cloned gene by added RNA polymerase. Endogenous RNA translation is eliminated to prevent background labelling, and the system is adapted for faithful translation of exogenously added mRNA. The protein can be labelled using translational grade [^{35}S]methionine and detected by autoradiography. The concentration of methionine used can be adjusted depending on the number of methionine residues in the protein being translated. The exposure time of the SDS-PAGE gel can be altered to get clean resolution of the protein of interest. The choice of radiolabel is not limited to methionine and other amino acids could be substituted in the reaction. However, methionine is the most commonly used amino acid. It is important that calcium salts should not be added since they can reactivate the endogenous nucleases in the lysate. *Protocol 23* describes the methodology we use, and which gives consistently good results.

Protocol 23

In vitro membrane binding assay

Equipment and reagents

- Rabbit reticulocyte lysate, nuclease-free (e.g. Promega)
- *In vitro* transcribed mRNA substrate (of the protein to be translated) in DEPC-treated water suspended at approximately 0.5–1.0 μg/μl (see *Protocols 7–9, 16*)
- 40 U/μl RNasin, ribonuclease inhibitor (e.g. Promega)
- 1 mM complete amino acids mixture (minus methionine)
- 10 μCi/μl [^{35}S]methionine (1000 Ci /mmol; e.g. Amersham)
- Canine pancreatic microsomal membranes (Promega) (Add 1.8 μl per 50 μl translation reaction.)

Method

1. Obtain mRNA by *in vitro* transcription of linearized DNA (see *Protocols 7–9, 16*).
2. Perform a translation reaction in the presence of 3.5 equivalents of canine microsomal membranes. Also include a control translation reaction with all the components except microsomal membrane. Add the following components in a microcentrifuge tube to a standard translation system:

Rabbit reticulocyte lysate (nuclease-treated)	35 μl
RNasin	1 μl
1 mM amino acid mixture (without methionine)	1 μl
[^{35}S]methionine	2 μl
RNA substrate	1 μl

   Adjust the final reaction volume to 50 μl, using DEPC-treated water.
3. Incubate the reaction at 37 °C for 1 hour.
4. On termination, microcentrifuge the lysate containing the labelled translational products for 15 min at 4 °C to separate the particulate and soluble fractions.
5. Examine the fractions for the presence of the labelled proteins following immunoprecipitation using either polyclonal or purified IgG directed against the *in vitro* translated virus protein.

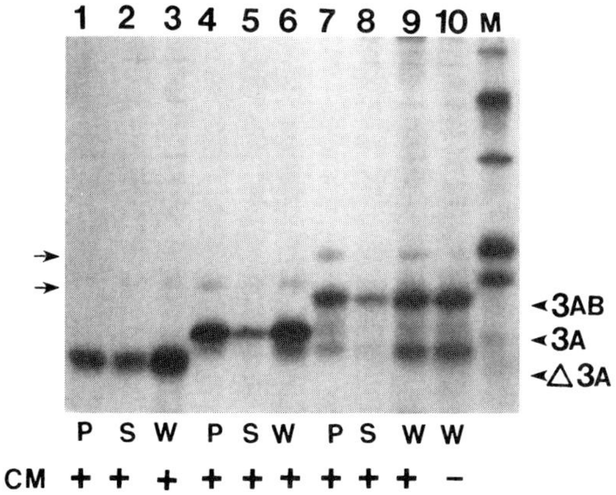

Figure 5 Distribution of poliovirus 3AB, 3A, and Δ3A between particulate (P) and soluble (S) fractions during *in vitro* translation. *In vitro* transcribed 3AB, 3A, and 3A mRNA were translated in the presence of [^{35}S]methionine in a rabbit reticulocyte lysate. Translation reactions were performed in the absence and presence of canine microsomal membrane as described. Particulate and soluble fractions were separated and analysed by SDS-PAGE following immunoprecipitation using a specific antibody. W, whole (unfractionated) reaction mixture. Lanes 1 to 3, Δ3A-RNA; lanes 4–6, 3A-RNA; lanes 7–10, 3AB-RNA. Arrows indicate modified 3AB and 3A (particularly in lanes 4, 6, 7, and 9).

The results from this assay can be used as additional evidence to confirm the *in vivo* data obtained from immunofluorescence analysis (32). Binding of virus proteins to the membranes preferentially localizes to the pellet fraction (*Figure 7.5*). The purpose of this is that poliovirus RNA replication occurs in the cellular microenvironment of the membrane and many proteins are known to bind cellular structures. This assay is based on the fact that when translation of proteins is performed in the presence of microsomal membranes, new proteins tend to associate with the pellet fraction if they have a predominantly membrane binding affinity. The total lysate is separated into particulate and soluble fractions, and the membrane-bound labelled protein can be effectively immunoprecipitated. Immunoprecipitation reaction is done with specific antibodies targeted towards the protein to be examined. Translation is normally performed in the presence of [^{35}S]methionine or a labelled amino acid, so the immunoprecipitated protein can be detected after resolution on a denaturing SDS protein gel followed by autoradiography of the dried gel.

Acknowledgements

This work was supported by National Institutes of Health grants AI 18272 and AI 27451 to A.D. R.I. was supported by a Howard Hughes graduate fellowship. We would like to thank all members of the Dasgupta laboratory (past and present) for their support and encouragement.

References

1. Fields, B. N., Knipe, D. M., and Howley, P. M. (Eds) (1996). *Fields Virology* (3rd edn). Lippincott-Raven, Philadelphia, USA.
2. Dasgupta, A. (1983). *Virology*, **128**, 245–51.
3. Dasgupta, A., Zabel, P., and Baltimore, D. (1980). *Cell*, **19**, 423–9.
4. Whelan, S. P. J, Ball, A. L., Barr, J. N., and Wertz, G. T. W. (1995). *Proc. Natl Acad. Sci. USA*, **92**, 8388–92.
5. Pattnaik, A. K., Ball, A.L., LeGrone, W., and Wertz, G. T. W. (1992). *Cell*, **69**, 1011–20.
6. Molla, A., Paul, A. V., and Wimmer, E. (1991). *Science*, **254**, 1647–50.
7. Janda, M. and Ahlquist, P. (1993). *Cell*, **72**, 961–70.
8. Barton, D. J., Morasco, B. J., and Flanegan, J. B. (1996). In *Methods in enzymology*, Vol. 275 (ed. Kuo, L. C., Olsen, D. B. and Carroll, S. S.), pp. 35–57. Academic Press, San Diego, CA.
9. Barton, D. J. and Flanegan, J. B. (1993). *J. Virol.*, **67**, 822–31.
10. Barton, D. J., Black, P., and Flanegan, J. B. (1995). *J. Virol.*, 5516–27.
11. Rosenberg, A. H., Lade, B. N., Chui, D., Lin, S., Dunn, J. J., and Studier, F. W. (1987). *Gene*, **56**, 125–35.
12. Studier, F. W. and Moffat, B. A. (1986). *J. Mol. Biol.*, **189**, 113.
13. Studier, F. W., Rosenberg, A. H., Dunn, J. J., and Dubendorff, J. W. (1990). In *Methods In enzymology*, Vol. 185 (ed. Goeddel, D. U.), pp. 60–89. Academic Press, San Diego, CA.
14. Andino, R., Rieckhof, G. E., and Baltimore, D. (1990). *Cell*, **63**, 369–80.
15. Andino, R., Rieckhof, G. E., Achacoso, P. L., and Baltimore, D. (1993). *EMBO J.*, **12**, 3587–98.
16. Banerjee, R., Echeverri,.E., and Dasgupta, A. (1997). *J. Virol.*, **71**, 9570–8.
17. Banerjee, R. and Dasgupta, A. [Manuscript in preparation.]
18. Ausubel, F. M., Brent, R., Kingston, R. E., Moore, D. D., Siedman, J. G., Smith, J. A., and Struhl, K. (eds) (1999). *Current protocols in molecular biology* pp. 3.0.1–3.9.8. Wiley, New York.
19. Sambrook, J., Fritsch, E. F., and Maniatis, T. (1989). *Molecular cloning—a laboratory manual* (2nd edn), pp. 10.27–10.37. Cold Spring Harbor Press, New York.
20. Graham, F. L. and van der Eb, J. (1973). Virology, **52**, 456.
21. Chen, C. and Okayama, H. (1987). Mol. Cell. Biol., **7**, 2745.
22. McCutchan, J. H. and Pagano, J. S. (1968). *J. Natl Cancer. Inst.*, **41**, 351.
23. Potter, H., Wier, L., and Leder, P (1984). *Proc. Natl Acad. Sci. USA*, **81**, 7161.
24. Wong, T. K., Nicolau, C., and Hofschneider, P. H. (1980). *Science*, **215**, 166.
25. Felgner, P. L., Gadek, T. R., Holm, M., Roman, R., Chan, H. W., Wenz, M., Northop, J. P., Ringold, G. M., and Danielsen, M. (1987). *Proc. Natl Acad. Sci. USA*, **84**, 7413–17.
26. Gossen, M. and Bujard, H. (1992). *Proc. Natl Acad. Sci. USA*, **89**, 5547–51.
27. Gossen, M., Freundlieb, S., Bender, G., Muller, G., Hillen,W., and Bujard, H. (1995). *Science*, **268**, 1766–9.
28. Semler, B. L., Anderson, C. W., Hanecak, R., Dorner, L. F., and Wimmer, E. (1982). *Cell*, **28**, 405–12.
29. Semler, B. L., Kuhn, R. J., and Wimmer, E. (1988). In *RNA genetics*, Vol. 1 (ed. E. Domingo, J. J. Holland, and P. Ahlquist), pp. 23–48. CRC Press, Boca Raton, FL.
30. Moss, B., Elroy-Stein, O., Mizukami, T., Alexander, W. A., and Fuerst, L. T. R. (1990). *Nature* (London), **348**, 91–2.
31. Hruby, D. E. and Grosenbach, D. (1995). In *Methods in molecular genetics*, Vol. 7 (ed. Adolph, K. W.), pp. 45–64. Academic Press,
32. Datta, U. and Dasgupta. A. (1994). *J. Virol.*, **68**, 4468–77.

Chapter 8

Packaging segmented and non-segmented RNA virus genomes

J. Barr* and J.W. McCauley†

*Institute of Virology, University of Glasgow, Church Street, Glasgow G11 5JR, U.K.

†Compton Laboratory, Institute of Animal Health, Compton, Newbury, Berkshire RG20 7NN, U.K.

1 Introduction

1.1 Assembly and packaging defined

The process of virus assembly describes the events by which the mature virion is formed from its constituent components within an infected cell. Because viruses come in such a wide variety of structural arrangements and consequently have adopted incredibly diverse assembly strategies, this definition is by necessity very broad. The range of processes included in this term 'assembly' are extremely varied and range from simple associations of a genome with no more than a few types of protein (Tobacco Mosaic Virus, TMV, or ϕX174) to the formation of a virion comparable in size to a small bacterium and containing over 100 different proteins (poxviruses). A detailed description of assembly is beyond the scope of this chapter, or indeed this book, and so the reader wishing to know more about the various assembly strategies adopted by viruses would be best advised to look to one of the many more specific texts.

In this chapter we are concerned with the process of 'genome packaging', this term describing the stage of the assembly process in which the genome itself is incorporated into the virus particle. Genome packaging thus only represents one aspect of the whole assembly process and may be both preceded and followed by additional processes that lead to the formation of the mature virion.

RNA viruses all have genomes that are wrapped in a proteinaceous capsid structure to give what is termed either the nucleocapsid or the ribonucleoprotein (RNP) complex. However, for many viruses, notably the enveloped viruses, this process of incorporation of the genome into the nucleocapsid structure is distinct from our definition of packaging, and it is usually given its own name which is encapsidation. As an example, for the negative-strand viruses, the nucleocapsid structure is the functional template for RNA synthesis and does not constitute an infectious virus particle, many more assembly steps must occur for the virus to form the mature virion. For these negative-stranded viruses, packaging, as we have defined it above, is the process by which the genome, in the form of the nucleocapsid, is inserted into the mature virus particle.

1.2 Genome selection

A fundamental problem that faces a virus as it prepares to assemble its assorted components into virion structures, is how to select the genome out of the complex mixture of RNAs that are present inside an infected cell, which will be both viral and cellular in origin. This problem has clearly been overcome, and a measure of how well this has been achieved can be assessed for any type of virus by calculating what proportion of the released viruses are capable of initiating a productive infection. This measure is known as the particle:p.f.u. ratio, and for some viruses this figure approaches 1. A virus with a low particle:p.f.u. ratio must possess a packaging procedure capable of ensuring that the RNA packaged into the virion is a functional genome. However, there are other viruses with particle:p.f.u. ratios well over 1000, and for these viruses genome packaging may be a less selective process which unavoidably assembles genomes that are defective in some way.

Commonly, the genome packaging selection is driven by the specificity of a virus-encoded protein for the genome RNA through a protein–nucleic acid interaction. This recognition may be through a specific sequence or a through the presentation of a particular secondary structure. If further stages of assembly are required then these steps usually rely upon recognition of this complex by other virus-encoded components. For example, in the case of the non-segmented negative-strand RNA viruses, genome selection occurs at the stage of genome encapsidation when nucleocapsid proteins specifically enwrap the viral genome. Once formed, the nucleocapsid structure is then recognized by the viral matrix protein, which in turn is able to associate with the viral G protein located at the plasma membrane. In this instance then, virus assembly consists of an initial protein–nucleic acid interaction followed by two layers of protein–protein interactions.

In this chapter, we have chosen examples from two RNA viruses, vesicular stomatitis virus (VSV) and influenza virus, each of which has adopted different mechanisms of assembly, and consequently, packaging. Whilst this chapter is by no means a comprehensive exploration of RNA virus genome packaging, these examples will introduce and illustrate the practical considerations needed for studying this fascinating stage of the virus life cycle.

2 Genome packaging in non-segmented, negative-strand RNA viruses

For negative-stranded RNA viruses in general, only those nucleocapsids that contain negative-stranded genomes are capable of initiating a productive infection. What drives this situation is the fact that the viral RNA-dependent RNA polymerase (RdRp) will only recognize a genome which is encapsidated with N protein. Consider a virion containing an encapsidated negative strand. On entering a cell, it will be recognized by the associated RdRp and will act as a template for anti-genome synthesis. This incoming genome will also support transcription to yield N protein mRNA, which will allow the nascent anti-genome molecules to be encapsidated. These encapsidated anti-genomes will themselves be recognized by the viral RdRp, and thus the genomes and anti-genomes can be amplified by further rounds of RNA synthesis resulting in a productive infection. On the other hand, if a positive-stranded anti-genome were to enter the cell, the outcome would be much different. Although the encapsidated anti-genome could act as a template for replication, it could not support N protein mRNA synthesis, so the resulting nascent genomes would remain naked and consequently would not be recognized by the RdRp, and the infection would proceed no further. Because of this situation, it would seem to be advantageous for the virus, in the interests of economy, to be able to exclusively package negative-strand genomes. The packaging process adopted by the prototypic non-segmented negative-stranded virus, VSV, is currently the best characterized of all the viruses from this group. VSV is able to selectively package only negative-stranded genomes, and, very recently, inroads have been made into understanding the molecular basis for this specificity. However, not all negative-strand viruses exhibit the same level of specificity as does VSV. Rabies virus, classified along with VSV in the Rhabdoviridae family, does not appear to use a mechanism that actively selects negative-stranded nucleocapsids for packaging (1). However, despite this lack of selection, the majority of rabies virus particles *do* contain negative-strand genomes. This is achieved due to the accumulation of a similar majority of negative-stranded genomes compared with anti-genomes within the infected cell, and so the proportion of genomes compared with anti-genomes in released particles reflects their relative abundance within infected cells. This example shows how two viruses from the same family and with many molecular features in common, have devised totally different approaches to the packaging problem.

2.1 Investigation of genome packaging: analysis of nucleocapsid RNA polarity before and after packaging into virions

When initiating an investigation into the genome packaging process of a negative-strand virus, a good starting point is to investigate whether a genome selection procedure is in operation which actively selects either positive- or negative-polarity genome length, nucleocapsid RNAs from the cellular pool. This question may be approached by an analysis of the polarity of nucleocapsid RNAs isolated from both within the cytoplasm of the infected cell where they are a potential substrate for packaging, and also from released virus particles into which the nucleocapsids have been packaged. If the relative abundance of positive- and negative-polarity RNAs are found to be exactly the same in nucleocapsids harvested from both the cytoplasmic and released virion sources, then it would seem that nucleocapsids are not being selected for packaging on the basis of their polarity.

A simple and direct approach to this question is to analyse the relevant RNAs by Northern blot analysis. This procedure involves immobilizing the viral RNA by transfer to a solid support, and then determining the polarity of the RNA by hybridization to strand-specific radiolabelled probes. This procedure involves many steps, including:

- initial growth of virus;
- harvesting of viral RNA;
- separation of RNAs using denaturing agarose–formaldehyde gel electrophoresis;
- radiolabelled RNA probe preparation; and finally
- the Northern blot analysis itself.

These methods are routine laboratory procedures, and as such, alternative protocols will be described in standard laboratory texts. However, for the sake of completeness, all the necessary steps required for this procedure are included here.

It is important to note that this protocol will only allow qualitative detection of either polarity of an RNA strand. The signal emitted from different hybridized probes cannot be used directly as a measure of the abundance of their corresponding target RNAs because two different probes will have different physical properties, which will make the comparison inaccurate. To assess the relative abundance of the positive- and negative-sense viral RNAs, the relevant probes must first be used in Northern blot analysis using known quantities of target-immobilized RNAs, thus determining a relationship between the probe signal and the corresponding RNA abundance. Then by comparing the intensity of the probe signal when hybridized to the viral RNA to this standard, the abundance of each viral RNA can be determined in real terms.

Finally, it should be noted that VSV is a serious animal pathogen and permission must be granted by the appropriate regulatory body before the virus can be handled in the laboratory.

Protocol 1

Growth of VSV in BHK 21 cells

Equipment and reagents

- 60 mm dishes of 80–90% confluent BHK 21 cells
- Phosphate-buffered saline (PBS) pH 7.2 containing 0.025 M $MgCl_2$
- Stock of VSV
- Glasgow minimal Eagle's medium (GMEM; Gibco-BRL), 5% fetal calf serum (Gibco-BRL)

Protocol 1 continued

Method

1. Draw off the medium from 60 mm dishes of 80–90% confluent BHK 21 cells.
2. Wash once with 3 ml of PBS.
3. Add VSV in 200 μl of PBS, at a p.f.u./cell ratio of between 0.1 and 1.
4. Incubate for 60 minutes at 37 °C, with occasional agitation.
5. Draw off the inoculum, wash twice with PBS, then overlay with 2 ml of GMEM, 5% FCS.
6. Incubate at 37 °C for 12 hours.
7. Draw off the supernatant, and then spin it at 1500 g for 5 minutes. Discard the pellet which contains unattached cells and cell debris. Keep the supernatant at 4 °C throughout.

There are several procedures that can be used to concentrate the virus particles present in the clarified supernatant prepared according to *Protocol 1*. By far the most straightforward approach is to simply pellet the virus using high-speed centrifugation; however, this process inevitably co-pellets other large molecular weight components from the supernatant.

Protocol 2

Harvesting and purifying virus from supernatant fluids

Equipment and reagents

- TE buffer: 10 mM Tris-HCl pH 7.4, 1 mM EDTA
- Beckman TLA45 rotor (or equivalent)

Method

1. Spin the clarified supernatant at 125 000 g in a Beckman TLA45 Rotor for 90 minutes at 4 °C to pellet the virus. Alternative: 21 000 r.p.m. (52 000 g) in a Sorval type 30 fixed angle rotor (or equivalent) for 90 minutes at 4 °C to pellet the virus.
2. Carefully remove the supernatant by vacuum aspiration.
3. Resuspend the virus pellet in TE buffer.

An additional step, which yields a rather more pure preparation, is to subject the concentrated virus generated using *Protocol 2* to sucrose gradient centrifugation. This approach will allow separation of the standard virion particles from the contaminating defective interfering (DI) particles that tend to arise when an infection is initiated with high m.o.i.

Protocol 3

Sucrose gradient centrifugation

Equipment and reagents

- Ultracentrifuge
- Beckman TST41 and TLA45 rotors (or equivalents)
- Gradient-former
- TE buffer: 10 mM Tris–HCl pH 7.4, 1 mM EDTA
- 20% (w/v) and 60% (w/v) RNase-free sucrose in TE buffer
- Phenol
- Chloroform

Method

1. Overlay a 20% to 60% sucrose (in TE) gradient with the resuspended virus harvested from infected cells made in *Protocol 1*.
2. Spin the gradient in a Beckman SW41 rotor (or equivalent) at 125 000 g for 90 min.
3. Remove the virus fraction from the gradient, dilute in 5 volumes of TE and pellet by centrifugation for 1 hour at 45 000 r.p.m. in a Beckman TLA45 rotor (or equivalent).
4. Resuspend the virus pellet in 400 μl TE. Harvest the viral RNA by performing sequential phenol and chloroform extractions (1 phenol then 1 $CHCl_3$, equal volumes). Recover the viral RNA by ethanol precipitation and resuspend the pellet in TE. Store at −70 °C.

In addition to harvesting viral RNA from nucleocapsids which have been packaged into released particles, viral nucleocapsids can be harvested from the cytoplasm of infected cells before they are assembled into virions. *Protocol 4* involves treating the VSV-infected cells with a mild detergent, which disrupts the plasma membrane but leaves the nuclear membrane intact. The nuclear fraction is then pelleted by centrifugation and discarded. Nucleocapsids are precipitated from the remaining cytoplasmic extract by immunoprecipitation using an antibody that recognizes the VSV nucleocapsid protein.

Protocol 5 describes a method for separating RNAs with respect to their size under denaturing conditions. The quantities stated in the procedure described below are calculated for a standard midi-sized gel apparatus designed to run gels having a volume of approximately 250 ml. This protocol involves the use of formaldehyde and, because of the toxicity of this substance, many of the steps described below should be performed in a fume hood.

The method of choice for the Northern blot procedure (*Protocol 7*) is to make an RNA probe by run-off transcription using T7 RNA polymerase. The sequence that is to be expressed as the radiolabelled probe needs to be inserted immediately downstream of the T7 RNA polymerase promoter on a suitable transcription plasmid such as pGEM1 (Promega). For the purposes of the experiment

Protocol 4

Harvesting nucleocapsids from cytoplasmic extracts

Equipment and reagents

- BHK monolayers (see *Protocol 1*)
- 60 mm dishes of 80–90% confluent BHK 21 cells (see *Protocol 1*)
- Rose lysis buffer (RLB): 1% (v/v) NP-40, 0.4% deoxycholate, 66 mM EDTA, 10 mM Tris–HCl pH 7.4
- Rose wash buffer (RWB): 1% (v/v) NP-40, 0.5% (w/v) deoxycholate, 0.1% (w/v) SDS, 150 mM NaCl, 10 mM Tris–HCl pH 7.4
- NTE: 0.1 M NaCl, 10 mM Tris–HCl pH 7.4, 1 mM EDTA
- Anti-VSV nucleocapsid (N) protein antibody
- Staphylococcus protein-A (Sigma)
- Sterile rubber policeman or cell scraper

Method

1. Infect BHK monolayers as described in *Protocol 1*.
2. Allow the infection to proceed for 6 to 8 hours.[a]
3. Remove the medium from the monolayers by vacuum aspiration, and transfer the 60 mm dishes to a tray of ice to chill the cells.
4. Scrape the cells into 400 μl of ice-cold RLB, transfer to a microcentrifuge tube, and incubate on ice for 7 minutes with occasional and gentle vortexing.
5. Pellet the cell nuclei by centrifugation at 12 000 *g* for 2 minutes in a benchtop centrifuge. Transfer the supernatant to a new microcentrifuge tube.
6. Add 100 μl of NTE, 0.1% (w/v) SDS to the cytoplasmic extract, vortex, and add a suitable quantity of anti-VSV N protein antibody. Incubate for between 4 and 16 hours at 4 °C with gentle agitation.
7. Add 100 μl of staphylococcus protein-A preparation and agitate for 1 hour at 4 °C, with gentle agitation.
8. Pellet the staph-A complex by centrifugation at 12 000 *g* for 30 seconds. Wash the pellet with three rounds of centrifugation and resuspension using RWB.
9. Resuspend the washed pellet in 400 μl of NTE + 0.1% SDS and boil for 2 minutes to dissociate the nucleocapsids from the staph-A protein.
10. Pellet the staph-A by centrifugation at 12 000 *g* for 1 minute, and retain the supernatant containing the VSV nucleocapsids.
11. Harvest the RNA from the nucleocapsids by extraction with 400 μl phenol followed by 400 μl chloroform, and concentrate the RNA by ethanol precipitation.

[a] Virus particles are released as soon as 2 hours after the onset of infection. The 6–8-hour incubation period is a compromise between allowing sufficient amplification of virus and loss of virus through the release of assembled particles.

Protocol 5

Denaturing agarose–formaldehyde gel electrophoresis

Equipment and reagents

- Midi-sized gel apparatus (Gibco-BRL)
- Agarose
- 1.0 M sodium phosphate pH 7.0
- 37% formaldehyde stock solution
- Formaldehyde gel running buffer: 215 ml formaldehyde (37% commercial stock), 24 ml of 1 M sodium phosphate pH 7.0, H_2O up to 1200 ml
- 99% formamide
- Peristaltic pump for buffer recirculation
- 4 × sample buffer: 7.15 ml formaldehyde (37%), 80 mM sodium phosphate pH 7.0, H_2O to 10 ml. Bromophenol Blue to colour
- Loading buffer: 0.5% (w/v) SDS, 25% (v/v) glycerol, 25 mM EDTA

Method

1. Dissolve 2 g agarose in 160 ml distilled water by heating in a microwave oven.
2. Allow the agarose to cool to 50 °C, then add 4 ml of 1.0 M sodium phosphate, pH 7.0.
3. Add 36 ml of 37% formaldehyde stock solution to the agarose and mix by stirring. Pour the mixture into the gel apparatus.
4. Allow the gel to harden at room temperature for 1 hour, then cover with running buffer in the gel running apparatus.
5. Prepare the sample to be applied to the gel by mixing with an equal volume of 4 × sample buffer and 2 volumes of formamide (99%).
6. Heat the sample at 65 °C for 10 minutes, chill quickly on ice.
7. Add one volume of loading buffer, mix and load the sample into the gel.
8. Run the gel for 17 hours at 30 V (or as appropriate) with recirculation of the running buffer.

Protocol 6

Preparation of radiolabelled probes

Equipment and reagents

- DNA template and appropriate restriction enzyme
- Phenol
- Chloroform
- 10 mM each of ATP, CTP, GTP
- 1 M DTT
- T7 RNA polymerase
- Radiolabelled nucleotide, typically [^{35}S]UTP (NEN) (1250 Ci/mmol)
- 10 × T7 RNA polymerase transcription buffer (normally supplied with the enzyme): 400 mM Tris–HCl pH 8.0, 80 mM $MgCl_2$, 500 mM NaCl, 20 mM spermidine–HCl
- RNase-free DNase (RQ DNase; Promega)
- RNase inhibitor (RNasin; Promega)
- G-50 spin column (Pharmacia; LKB)

Protocol 6 continued

Method

1. Linearize the DNA template using a restriction enzyme that cuts at the distal end of the DNA template relative to the promoter.
2. Extract the linearized template first with phenol, then with chloroform, and recover the plasmid DNA by ethanol precipitation and centrifugation.
3. Wash the DNA pellet with 70% ethanol, dry, and resuspend the DNA in sterile distilled H_2O to give a concentration of 100 ng/μl.
4. Set up the transcription reaction as follows:

Template DNA	2 μg (20 μl)
10 × transcription buffer	5 μl
ATP, CTP, GTP	2 μl of each
1.0 M DTT	2 μl
T7 RNA polymerase	10 units (1 μl)
RNase inhibitor	2 units (2 μl)
[^{35}S]UTP	2 μl
H_2O to 50 μl	

5. Incubate at 37 °C for 1 hour. Add an additional 10 units of T7 RNA polymerase and reincubate for 30 minutes.
6. Remove the DNA template by adding an RNase-free DNase. Incubate at 37 °C for 15 minutes.
7. Pass the transcription mixture through a G-50 spin column to remove the unincorporated labelled nucleotides. The probe is now ready for use in the Northern blot procedure.

Protocol 7

The Northern blot procedure

Equipment and reagents

- Agarose-formaldehyde gel (see *Protocol 5*)
- 50 mM NaOH
- 100 mM Tris-HCl pH 7.4
- Radiolabelled probe (see *Protocol 6*)
- Vacuum oven at 80 °C
- 0.2 μm Nytran (Schleicher and Schuell) or Genescreen (DuPont; NEN) membranes
- 20 × SSC: 3 M NaCl, 0.3 M trisodium citrate pH 7.0
- 2 × SSC, 1% SDS
- 100 × Denhardt's solution: 2% (w/v) bovine serum albumin, 2% (w/v) Ficoll (M_r 400 000), 2% polyvinylpyrrolidone (M_r 400 000)
- Hybridization solution: 50% (v/v) formamide, 5 × SSC, 8 × Denhardt's, 50 mM sodium phosphate pH 6.5, 0.1% SDS, 5 μg/ ml herring sperm DNA (denatured)
- Autoradiographic equipment and reagents

Protocol 7 continued

Method

1. Wash the agarose-formaldehyde gel for 20 minutes at room temperature in each of the following solutions, in this order:
   Double-distilled water
   50 mM NaOH
   100 mM Tris–HCl pH 7.4
   10 × SSC
2. Wet the Nytran or Genescreen membrane to which the RNA will be transferred, first in water, then in 10 × SSC.
3. Transfer the RNA to the membrane (see Chapter 1, *Protocol 6*).
4. Rinse the membrane with 2 × SSC for 5 minutes. Air-dry.
5. Bake the membrane in an 80 °C vacuum oven for between 30 minutes and 2 hours. The membrane can now be stored at room temperature for several months.
6. Pre-hybridize the membrane for 1 hour at 65 °C in hybridization solution.
7. Incubate the membrane with the radiolabelled probe in a minimum volume of hybridization solution. Hybridize overnight at 65 °C.
8. Wash the membrane for 30 minutes at room temperature in each of the following solutions, in this order:
   2 × SSC
   2 × SSC, 1% SDS
   0.1 × SSC
9. Air-dry the membrane and detect probe hybridization by autoradiography.

described here, two probes are required: one being of positive sense and able to hybridize to genomic RNA, the other being of negative sense that will hybridize to the anti-genome strand. In this instance, there are many possible regions of the genomic sequence that may be used to generate the probe, the N protein-coding region is as good as any. Incorporation of sequences from the terminal regions of the genome is not recommended since these regions exhibit strong complementarity to each other, and as such terminal probes will tend to bind to both positive- and negative-sense strands. A good size for a probe is between 500 and 1000 nucleotides, which will allow good radiolabel incorporation and easy purification if needed.

3 Introduction of site-specific mutations in the genome of negative-strand viruses

The component which is wholly responsible for the characteristics of a particular virus is the genome, and so the ability to alter the viral genome in a predetermined manner is crucial if the signals that reside within the genomic RNA are to be investigated. For many positive-strand RNA viruses, RNA which is simply transcribed from a full-length cDNA clone is capable of initiating a productive infection when introduced into permissive cells, and the seminal example of this is poliovirus (see Chapter 2). For these viruses, specific mutations can be incorporated into the viral genome simply by alteration of the corresponding cDNA template.

In contrast to this situation, genomic sense RNA transcribed from a full-length cDNA clone of a negative-strand RNA virus is not infectious. To be infectious, the genome must be in the form of the ribonucleocapsid complex. This requirement has necessitated the development of procedures, generally termed 'reverse genetics', that allow the assembly of a functional nucleocapsid structure from cDNA-encoded constituents (2) (see Chapter 9). The individual ribonucleocapsid components are synthesized within a cell-culture system, during which they self-assemble to form functional nucleocapsid structures. These structures are competent for RNA synthesis and thus are able to initiate the virus's infectious cycle. The individual components of the ribonucleocapsid are derived from manipulable plasmid cDNA templates, which may be introduced into cells by a variety of possible transfection methods. Expression of the nucleocapsid components is achieved by placing the virus-specific sequences under T7 RNA polymerase promoter control, and arranging for expression of this polymerase within the transfected cell. What is so attractive about this procedure is that by manipulating the corresponding cDNA templates, both the nucleotide sequence of the genomic RNA and also the amino acid sequences of the nucleocapsid components can be altered in a site-directed manner.

The way in which this system has been used to investigate the molecular biology of negative-strand viruses can be divided into two categories. The first category comprises experiments in which live, autonomously replicating virus is synthesized. These experiments require the use of cDNA plasmids that express the genomic RNA and all the mRNAs which encode the virus proteins. However, once assembled, the virus no longer depends upon cDNA plasmids to express necessary 'support'. The second category of experiments are those in which a truncated genome analogue is synthesized instead of the full-length virus genome. By including all the necessary *cis*-acting sequence signals, the analogues can be designed to perform all the stages of the virus life cycle, and their reduced size allows for both ease of manipulation and more efficient propagation. However, these genome analogues are not autonomous entities, in that their life cycle must always be supported by the necessary *trans*-acting protein components that their shortened genomes no longer encode.

Within the group of non-segmented negative-strand RNA viruses, the genome packaging mechanism is currently best characterized for the prototypic member of the group, which is VSV. This virus has been used as a model system for related viruses in many aspects of the virus life cycle such as RNA synthesis, glycoprotein maturation, and in many aspects of virus assembly and virion budding. Keeping with this trend, it is likely that the genome packaging process of VSV will also share characteristics with the packaging mechanisms of related viruses. The next section briefly describes the experimental approach that has been used to investigate the VSV packaging process. It is included as it is hoped that the approach will have relevance for related viruses, both in terms of the technicalities of the procedure and also in terms of the considerations that must be taken into account when designing an experiment of this type.

Clues to the nature of the VSV packaging mechanism were obtained by two important observations. First, that VSV DI particles having perfectly complementary 5′ genomic termini are packaged into particles, whereas DIs with complementary 3′ genomic ends are not. Second, a genome analogue, containing only the terminal regions of VSV surrounding sequences from bacteriophage ϕX174, was packaged into assembled particles when included in the reverse genetics procedure (2). These initial observations indicated that the packaging signals for VSV were located somewhere within the genome terminal sequences, and most likely within the 5′ end of the genome. Using this information, nucleotides from within the terminal sequences of a genomic analogue were gradually removed until an RNA template was created that could no longer direct the packaging of the resulting nucleocapsid structure (3). Importantly, this genome analogue was still fully able to perform the other functions of the original genome analogue, namely encapsidation, mRNA synthesis, and also replication. This indicated that the alterations had only affected the packaging ability of the genome. This last point is vital, as, for example, a mutation that prevented encapsidation would also indirectly prevent packaging. This consideration highlights a common problem that comes with introducing alterations to the terminal sequences of negative-strand RNA viruses, which is that these regions commonly contain overlapping *cis*-acting signals.

The protocol presented opposite is written as a general method, which can be adapted to investigate other aspects of VSV molecular biology other than packaging. The aspect of VSV molecular biology to be investigated depends largely on the design of the genome analogue incorporated in the system. As this procedure is complex and involves reagents that are not available from commercial sources, it is recommended that some technical assistance be sought from a laboratory in which this protocol is already in operation.

4 Influenza virus genome packaging

The Orthomyxoviridae family consists of four genera: influenza virus A, influenza virus B, influenza virus C, and the Thogotovirus genus. They are all enveloped viruses with virions that are pleomorphic and 80–120 nm in diameter. Depending on the genus, the genome consists of different numbers of negative-strand RNA molecules; influenza A and B viruses have eight segments, influenza C viruses and the tick-borne Dhori virus have seven and Thogotovirus, a virus of ticks, six. The segment lengths vary from approximately 890 to 2350 nucleotides and the genome size from 10.0 to 14.6 kb. Each RNA segment possesses conserved and partially complementary 5′ and 3′ terminal sequences. Most work on genome assembly has been carried out on influenza A viruses.

Protocol 8

The reverse genetics procedure applied to investigate the sequence signals required for genome packaging of a negative-strand RNA virus

Equipment and reagents

- Recombinant vaccinia virus VTF7-3, expressing T7 RNA polymerase
- cDNA plasmids encoding the five structural proteins of VSV under T7 polymerase promoter control
- cDNA plasmid(s) encoding the genome analogue RNA
- Lipofectin, liposome reagent (Gibco-BRL)
- 70% confluent BHK cells
- Serum-free GMEM cell-culture medium
- PBS
- 60 mm cell-culture dishes (Nunc)
- Microcentrifuge tubes (Eppendorf)
- 14 ml polycarbonate tubes (Falcon)

A. Primary transfection

1. Remove the medium from a 60 mm dish of 70% confluent BHK cells.
2. Rinse the cells with 2 ml of serum-free GMEM.
3. Infect the monolayer with a 200 μl volume of recombinant vaccinia virus VTF7-3 in PBS, to give an m.o.i. of 10. Incubate the dish for 60 minutes at 37 °C with occasional agitation.
4. While the cells are being infected, prepare a mixture of plasmid DNAs in a sterile 1.5 ml microcentrifuge tube containing the following:

N protein support plasmid	6.0 μg
P protein support plasmid	3.3 μg
L protein support plasmid	2.0 μg
M protein support plasmid	5.0 μg
G protein support plasmid	5.0 μg
Genome analogue-expressing plasmid	8.0 μg
GMEM	to 0.5 ml

5. Add 1.0 ml of GMEM to a sterile 14 ml polycarbonate tube, and to this add 22 μl of liposome preparation (Lipofectin reagent, Gibco-BRL).
6. Take the contents of the microcentrifuge tube and add it to the polycarbonate tube. Mix by gentle agitation, and leave for 15 minutes at room temperature.
7. After the correct incubation period (approximately 55 minutes), remove the inoculum of VTF7-3 from the infected BHK cells, and rinse the cells twice in serum-free GMEM.
8. Overlay the washed BHK cells with the transfection mixture contained in the polycarbonate tube. Incubate the cells for 4 hours at 37 °C.
9. Remove the transfection mixture, add 1.5 ml of GMEM with 2% foetal calf serum, incubate at 37 °C for between 30 and 42 hours.

Protocol 8 continued

B. Secondary transfection

1. Using the same procedure as described above, set up a 60 mm dish of BHK cells infected with VTF7-3, but transfected with only the support plasmids which express the N protein, the P protein, and the L protein. Transfected these new cells approximately 10 hours after the transfection of the first set of dishes. Incubate these cells at 37 °C for 4 hours to allow expressed VSV proteins to accumulate.
2. After the 30 to 42 hour incubation of the primary transfection cell monolayers is complete (part A, step 9) remove the supernatant (2 ml) from the cells and transfer it to sterile microcentrifuge tubes.
3. Centrifuge the tubes for 1 minute at 14 000 g in a benchtop centrifuge to pellet the unwanted cell debris.
4. Remove the culture medium from the secondary transfected cells, and apply the clarified primary transfection supernatant to the monolayer. Incubate for 6 hours at 37 °C.
5. Harvest and analyse the virus RNA to determine if infection has occurred.[a]

[a]An appropriate procedure for this analysis would be either by Northern blot analysis (see Section 2) or by directly labelling the RNA using [^{3}H]uridine, and then analysing the RNA by denaturing agarose-formaldehyde electrophoresis (see *Protocol* 5, this chapter).

Unlike gene analyses with viruses of the mononegavirales, including VSV, analysis of genetically modified influenza A viruses usually involves a helper virus to amplify and package modified virus genes or marker genes. Both virological and genetically modified virus have been used to study genome packaging, but a firm conclusion of the strategy for genome packaging has not been reached. The results from several studies of packaging reporter genome-like RNAs are consistent with the hypothesis that genome RNAs are not packaged in a segment-specific way (4, 5). But, in contrast, results from work with viruses having defective RNAs (which contain internal deletions of standard virus genes) support the alternate hypothesis that virion RNA packaging has segment-specificity (6, 7).

If genome packaging is not based on a segment-specific process, the hypothesis predicts that segments are packaged from the pool of virion RNPs present in the cell at random, and, therefore, to ensure a virus population with sufficient infectious particles the virus is expected to package significantly more than eight RNAs. A series of experiments in support of this hypothesis has been carried out based on the property of RNA segment 8 to encode two polypeptides. Through the construction of a plasmid containing a vRNA-like segment that encodes only the NS1 polypeptide but not the NS2 polypeptide it was possible to complement a mutant with a temperature-sensitive NS1 polypeptide. Plaque purification of some of the resulting virus population appeared to contain two copies of RNA segment 8. Thus, a virus with nine segments was possible (8).

It has long been established that it is possible to obtain viruses containing defective RNAs. These RNAs accumulate to high levels in infected cells. A prediction of random packaging of genome segments is that such a high level of defective RNAs would alter the genome profile of virus from cells infected with a defective virus, but it is likely that all segments would be affected equally. On the other hand, if genome packaging were segment-specific then a selective effect on packaging would be anticipated. In three studies, this segment-specific effect on the assembly of standard RNAs has been evident (6, 7, 9). In two cases, the effect could modify the packaging of non-defective virus in a mixed virus infection, or the segment-specific packaging of a genetically engineered reporter RNA in cells transfected with vRNA-like RNPs with terminal sequences derived from different RNA segments.

Many of the methods for analysing influenza virus RNA packaging are very similar to normal molecular virological techniques, and many of these have recently and thoroughly been described (10). This section will describe some methods used to study the packaging of influenza viruses. Many of the techniques, though, are general and have already been described in the section on the packaging of VSV. *Protocols 9–12* describe procedures we have used to investigate the nature of influenza virus genome packaging. In these experiments we have examined how defective RNAs derived from a single standard RNA segment affect the packaging of the segment from which they were derived, in comparison to how they affect the packaging of other RNA segments. Important consideration in these experiments has to be given to the ratio of defective virus to helper or standard virus. The multiplicity of infections of both defective and helper virus need to be high to ensure that a large proportion of cells are co-infected with both. It may not be possible to measure easily the amount of defective virus by standard virological techniques (11). We have used Northern blot analysis of the inoculum to estimate the relative concentration of the defective RNA compared to a standard segment, and both to that of a non-defective virus with a known infective titre. From these results and using the Poisson distribution, it is possible to estimate crudely the amount of defective virus being added to the culture as a proportion of helper virus. The Poisson distribution is described by the formula:

$$P_n = \frac{e^{-m}\, m^n}{n!}$$

It is used to estimate the probability (P_n) of a cell being infected with n infectious particles. In the formula, m is the mean multiplicity of infection and n is the number of particles in a cell. It is usual to estimate the proportion of cells in a culture that remain uninfected when infected at a particular multiplicity of infection. Examples are given in Table 1.

The protocols should be used as a guide to carrying out experiments on genome packaging, but there are many options. For example, *Protocol 10* describes the extraction of RNA from cells using hot phenol and low pH (pH 5.0). There are many other ways to make infected cell RNA and some may be con-

Table 1 The influence of the multiplicity of infection on the proportion of cells infected

Multiplicity of infection (p.f.u./cell)	Proportion of cells not infected (%)	Proportion of cells infected (%)
0.1	90.5	9.5
0.5	60.7	39.3
1.0	36.8	63.2
2.0	13.5	86.5
3.0	4.98	95.2
4.0	1.83	98.17
5.0	0.67	99.33

sidered attractive since they avoid phenol. Kits may be obtained from a large number of suppliers, including: Ambion Inc., Austin, TX; QIAGEN GmbH—Germany; Promega Inc., Madison WI; and Bioline, London, UK.

Protocol 9

The preparation of influenza virus from infected cells to examine its genome composition

Equipment and reagents

- 6 dishes of confluent monolayers of cultured cells (primary chick-embryo fibroblasts (CEF), BHK cells, MDCK cells) in 15-cm diameter tissue-culture dishes (e.g. Corning, cat. no. 430599) containing approximately 3×10^7 cells
- Virus inoculum (usually fresh allantoic fluid from infected hen's eggs (5 ml per dish, at least 3×10^8 p.f.u./ml))
- Maintenance medium: 2% bovine serum in M199 (for CEF cells), Dulbecco's Modified Eagle's Medium (for other cells) with 100 units/ml penicillin and 100 μg/ml streptomycin
- PBS (pH 7.2)
- 50 ml sterile plastic tubes

Method

1. Remove the medium from the dishes and rinse twice with PBS.
2. Add virus inoculum, 5 ml per dish, to cover the surface. Leave for 30 min, rocking occasionally.
3. Aspirate the inoculum and rinse the cells with PBS.
4. Remove the PBS and rinse the monolayer three more times with warmed (37 °C) maintenance medium. Finally replace the medium with 10 ml of maintenance medium and place in a 5% CO_2 incubator. Retain two plates as a control zero time point.
5. At 2 hours after transfer to 37 °C, take the dishes from the incubator, remove the medium, add fresh warmed maintenance medium, aspirate the medium, and replace again with 10 ml of maintenance medium.

Protocol 9 continued

6. At 4 hours' post-infection remove two dishes from the incubator to act as 4-hour time point for the extraction of intracellular RNA.
7. At 10 hours of incubation remove the remaining dishes from the incubator and transfer the medium into a 50 ml sterile plastic tube. Retain the dishes for the extraction of RNA from the cells.

Protocol 10

Extraction of RNA from cells with hot phenol

Equipment and reagents

- Sterile saline containing 10 mM Tris–HCl pH 7.4 (cooled on ice)
- Lysis solution: 50 mM NaOAc/HOAc pH 5.0, 1 mM EDTA, 0.5% SDS (cooled on ice).
- 3 M NaOAc/HOAc pH 5.2
- Water-saturated phenol
- Ethanol
- Water bath at 65 °C
- Benchtop centrifuge
- Sterile graduated 50 ml plastic centrifuge tubes
- Sterile Pasteur pipettes, some with the tip cut off and some drawn out
- Sterile rubber policeman (e.g. Corning; Fisher)

Method

1. Wash cell monolayers twice with Tris–saline (5–10 ml).
2. Add 5 ml lysis solution to each dish, ensuring an even wetting of the monolayer.
3. Scrape off the lysed monolayer with a sterile rubber policeman and transfer to a graduated 50 ml plastic centrifuge tube using a cut-off glass Pasteur pipette to ensure minimal shearing of DNA. Combine the cell extracts from the two dishes at each time point.
4. Add an equal volume of water-saturated phenol (10 ml) and mix gently in the 65 °C water bath until the phenol and water form a single milky phase.
5. Centrifuge at 2500 r.p.m. on a bench centrifuge at room temperature for 10 minutes to separate the phases.
6. Remove the top (aqueous) phase to a fresh tube and re-extract with an equal volume of phenol as above.
7. Remove the aqueous phase to a fresh tube add NaOAc (pH 5.2) to 0.3 M and add two volumes of ethanol. Mix and store at −20 °C, most conveniently overnight.
8. Centrifuge for 20 minutes at 1500 g at 4 °C.
9. Remove the ethanol and dissolve the RNA pellet in 200 μl sterile water.
10. Add NaOAc pH 5.2 to 0.3 M, two volumes of ethanol, and reprecipitate the RNA.

Protocol 10 continued

11. Centrifuge and remove the last drops of ethanol with a drawn-out Pasteur pipette.

12. Finally dissolve the RNA in 200 μl sterile water.

13. Determine the concentration of RNA by measuring the absorbance of a 1 to 100 dilution of the RNA in water at $A_{260\ nm}$ and $A_{280\ nm}$.[a]

14. Store RNA at −20 or −70°C.

[a]A 1 mg/ml solution of RNA has an $A_{260\ nm}$ of approximately 40 Absorbance units.

Protocol 11 describes the isolation of virus from the medium of infected cultured cells. It is our experience that a visible virus band can be detected using a torch in a darkened room. The virus is visible as a white band that scatters light equilibrated at a density of around 30–40% sucrose (1.13 to 1.18 specific gravity). The RNA from purified virus is then analysed by Northern blot analysis (*Protocols 5* and *7*) Typically, this protocol would be adapted as follows. Samples should be approximately 2 μg of intracellular RNA. Samples taken from time points through the replication cycle (usually 4 hours and 10 hours) and vRNA from virus purified from the medium are run with a sample from infected cells that has not been incubated at 37 °C (cells that have been harvested immediately after the room temperature period of virus adsorption to the cell surface). Following running the gel and transfer to a nylon membrane, probes need to be prepared (*Protocol 6*). Probes should be 200–400 nucleotides in length. For examining RNA segments 1, 2, and 3 separately, individual membranes will need to be used and these can be compared by including another probe, segment 7 RNA usually, but any segment that can be separated from the others would be suitable. All probes should be of similar specific activity determined by trichloroacetic acid precipitation. Following washing, bound radioactivity is determined most conveniently on a phosphoimager. Alternatively, it is possible to scan autoradiographs, which must not be overexposed and not underexposed. From the scan, the amount of radiolabel bound can be estimated by comparing it with a standard curve constructed from autoradiographs with known amounts of radiolabel.

An alternative way of estimating the relative amounts of RNA in cell cultures or in virus released from the cells is to use reverse transcription. Quantitative reverse transcription forms the basis for estimating RNA abundance by RT–PCR or RT–PCR with TaqMan technology (Perkin-Elmer Applied Biosystems). *Protocol 12* describes methods we have used for quantifying the products of reverse transcription. We have quantified directly by autoradiographic or phosphoimaging methods, but the general technique is suitable for adaptation to techniques that analyse amplified cDNA.

As discussed earlier, even the strategy used by influenza viruses for packaging their genome has not been unequivocally established. A complicating feature is the overlap of the genomic signals for transcription, replication, and packaging. Progress on the elucidation of the packaging problem has been slow but is a fundamental feature of the replication strategy of all viruses.

Protocol 11

Purification of virus from medium

Equipment and reagents

- 50 ml centrifuge tubes
- Ultracentrifuge and SW28, SW41 rotors
- Gradient maker
- PBS
- Beckman centrifuge tubes (SW28 polyallomer: 326823, ultra-clear SW28.1 tubes cat no. 344061; ultraclear SW41 tubes 344059) or equivalent
- Torch with a focusable beam
- 60% (w/v) sucrose in PBS
- 30% (w/v) sucrose in PBS
- 15% (w/v) sucrose in PBS
- Sonicator or sterile syringes and needles from 19 to 25 gauge
- Peristaltic pump
- Sterile saline containing 10 mM Tris–HCl pH 7.4

Method

1. Spin the medium from the cells in a 50 ml tube at 1000 g for 10 minutes at 4 °C.
2. Remove the supernatant to a Beckman SW28 tube and fill with cold PBS.
3. Centrifuge at 27 000 r.p.m. (100 000 g) for 90 minutes.
4. Decant the inoculum and resuspend the virus in 0.5 ml cold PBS by sonication or by passing the suspension through a series of sterile syringe needles from 19 to 25 gauge.
5. Prepare sucrose gradients in Ultraclear SW28.1 tubes with a linear gradient of 5 ml 60% sucrose in PBS and 5 ml 30% sucrose in PBS poured from a gradient maker and a peristaltic pump. The gradient is then carefully overlaid with 4 ml 15% sucrose in PBS.
6. Add the virus suspension to the top of the gradient and centrifuge at 27 000 r.p.m. (100 000 g) for 90 minutes at 4 °C.
7. The virus band can be visualized at about the position of 40% sucrose and can be visualized using the focused beam of a torch against a black background, preferably in a dark room.
8. Harvest the virus band with a syringe.
9. Dilute the sucrose at least two-fold with cold PBS, and pellet the virus in an SW28.1 tube for 90 minutes at 27 000 r.p.m. (100 000 g) or in an SW41 rotor for 60 minutes at 39 000 r.p.m. (190 000 g).
10. Take up the virus pellet in 0.2 ml cold saline containing 10 mM Tris–HCl pH 7.4 and store at 4 °C.

Protocol 12

Reverse transcription analysis of intracellular and released RNA

Equipment and reagents

- Human placental ribonuclease inhibitor (e.g. RNase Block, Stratagene)
- Moloney murine leukaemia virus reverse transcriptase (Life Technologies)
- Moloney murine leukaemia virus reverse transcriptase buffer (Life Technologies) (5 × buffer is 250 mM Tris–HCl pH 8.3, 375 mM KCl, 50 mM DTT, 15 mM $MgCl_2$)
- Deoxyribonucleoside triphosphates (100 mM dNTPs in Tris–HCl) pH 8.3
- Oligonucleotide primers as appropriate in water (for all influenza A virus vRNA segments, AGCAAAAGCAGG which is the complementary DNA sequence to the conserved terminal 3′ 12 nucleotides of each vRNA segment; for cRNA: AGTAGAAACAAGG, complementary to the conserved 13 nucleotides at the 3′ end of cRNA; or appropriate primers specific for each segment, for example AGCAAA-AGCAGGTAGATA for vRNA segment 7, complementary to the 3′ terminal 18 nucleotides of vRNA segment 7).
- RNA from virus or infected cells prepared as described in *Protocols 9–11*
- RNase-free water
- Boiling water bath
- [^{32}P]dATP (Specific activity 3000 Ci/mmol)
- 4% acrylamide gel (19:1 acrylamide: bisacrylamide) (Suitable gel dimensions are 40 cm × 20 cm × 0.35 mm)
- 10 × TBE buffer: Tris:Borate:EDTA (108 g, 55 g, 9.3 g) dissolve and make up to 1 l with water 89 mM Tris, 89 mM Boric acid, 2.5 mM EDTA)
- Gel buffer: 1 × TBE buffer pH 8.3 containing 8 M urea
- Gel apparatus
- Gel-Loader Tip (Eppendorf)
- Gel drier
- Gel fixative: 10% methanol, 10% acetic acid
- Formamide containing 0.025% (w/v) xylene cyanol and 0.025% (w/v) Bromophenol Blue

Method

1. Mix RNA (2 μg for intracellular vRNA estimation, 8 μg for intracellular cRNA estimation or approximately 10% of the yield of virus RNA from 4 × 10^7 cells) with 8 pmole of primer and make up the volume to 10 μl with RNase-free water.
2. Place the tubes in a boiling water bath for 2 minutes and then immediately put on ice.
3. In a separate tube, prepare a reaction cocktail consisting of:

dNTPs	0.5 mM
[^{32}P]dATP	10 μCi (1 μl)
Placental ribonuclease inhibitor	20 units

   2 × MMLV reverse transcriptase buffer (from the manufacturer)
   Moloney murine leukaemia virus reverse transcriptase 200 units.
4. Mix the denatured RNA and primer samples in 10 μl volumes with an equal volume of the reaction cocktail, and incubate at 37 °C for 60–90 minutes.[a]

Protocol 12 continued

5. Prepare a 4% acrylamide gel with the gel buffer.[b]
6. Dilute the sample of reverse transcribed cDNA with an equal volume of formamide and dyes.
7. Heat in a boiling water bath for 2 minutes, cool on ice, and load on to the prepared gel. Use a Gel-Loader Tip trimmed with a sharp scalpel to allow easy loading.
8. Carry out electrophoresis for a suitable length of time as required (for example until the xylene cyanol dye has reached to within 5 cm of the bottom of a 4% acrylamide gel).
9. Fix the gel in an aqueous solution of 10% methanol, 10% acetic acid for at least 15 minutes, but usually 30 minutes.
10. Dry on a gel drier.
11. Visualize by autoradiography or scan on a phosphoimager.

[a] These samples can be directly analysed by polyacrylamide gel electrophoresis.

[b] Methods for the preparation and running of polyacrylamide gels are described in many books and methods for analysis of nucleic acid sequencing (12).

References

1. Finke, S. and Conzelmann, K. K. (1997). *Journal of Virology*, **71**, 7281–8.
2. Pattnaik, A. K., Ball, L. A., LeGrone, A. W., and Wertz, G. W. (1992). *Cell*, **69**, 1011–20.
3. Whelan, S. P. J. and Wertz, G. W. W. (1997). *Seminars in Virology*, **8**, 131–9.
4. Luytjes, W., Krystal, M., Enami, M., Parvin, J. D., and Palese, P. (1989). *Cell*, **59**, 1107–13.
5. Neumann, G. and Hobom, G. (1995). *Journal of General Virology*, **76**, 1709–17.
6. Duhaut, S. D. and McCauley, J. W. (1996). *Virology*, **216**, 326–37.
7. Odagiri, T. and Tashiro, M. (1997). *Journal of Virology*, **71**, 2138–45.
8. Enami, M., Sharma, G., Benham, C., and Palese, P. (1991). *Virology*, **185**, 291–8.
9. Bergmann, M. and Muster, T. (1996). *Virus Research*, **44**, 23–31.
10. Paterson, R. G. and Lamb, R. A. (1993). In *Molecular virology, a practical approach* (ed. A. J. Davison and R. M. Elliott), p. 35. IRL at Oxford University Press, Oxford.
11. McLain, L., Armstrong, S., and Dimmock, N. J. (1988). *Journal of General Virology*, **69**, 1415–19.
12. Howe, C. J. and Ward, E. S. (ed.) (1989). *Nucleic acid sequencing, a practical approach*. IRL at Oxford University Press, Oxford.

Chapter 9

Reverse genetics of RNA viruses

A. Bridgen and R.M. Elliott
Division of Virology, Institute of Biological and Life Sciences, University of Glasgow, Church Street, Glasgow G11 5JR, U.K.

1 Introduction

RNA viruses comprise many of the most serious human pathogens. There are now possibly 500 million carriers of hepatitis C virus world-wide (1). Rotavirus infections are responsible for around 18 million cases of severe diarrhoea and nearly 1 million deaths in young children in developing countries annually (2). Measles is still also one of the leading causes of infant death in developing countries, and can induce the rare, but fatal, neurodegenerative disease subacute sclerosing panencephalitis (3). Many human respiratory infections are caused by RNA viruses such as influenza, human respiratory syncytial virus, coronaviruses, enteroviruses and, one of the most frequent virus pathogens of humans, the rhinoviruses (4). Hantaviruses, members of the *Bunyaviridae* family, are responsible for haemorrhagic diseases and form one of a number of virus groups whose incidence has increased greatly over the last few years, the so-called 'emerging viruses' (5). Perhaps the most frightening of all virus diseases, because of their high mortality and lack of effective prophylaxis and treatment, are those caused by the filoviruses Marburg and Ebola (6).

The severity of the diseases they cause means there is a great interest in understanding the biological processes governing the entry, replication, assembly, release, and transmission of RNA viruses. These processes have been studied in DNA viruses by mutating the genome and studying the effects of such mutations using infectious cloned virus DNAs in permissive cells. However, the difficulty of *in vitro* RNA manipulation makes such direct studies problematic for RNA viruses. For instance, site-directed mutagenesis of RNA to produce precisely defined nucleotide substitutions directly is not possible. The key to the study of these viruses lies in 'reverse genetics', the manipulation of cDNA copies of the virus genome, and an investigation of the effect of these alterations on the phenotype. The ideal is to produce or 'rescue' a virus from cDNA copies of the virus genome containing alterations to either the genes or non-coding sequences. But a great deal can also be learned from 'minigenomes' of the virus, comprising the terminal genomic sequences and a reduced number of the virus genes, with or without an additional reporter gene. Replication of these minigenomes is generally rather easier to achieve than rescue of the whole virus genome.

Such reverse genetic studies were first reported with positive-stranded RNA viruses (see below). The first virus genome to be rescued was the bacteriophage Qβ in 1978 (7). This was followed by the demonstration that cloned poliovirus cDNA was infectious (8), which allows the possibility of introducing mutations into the virus cDNA using standard recombinant DNA techniques and hence into infectious virus. Subsequently it was shown that infection can be initiated more efficiently from RNA transcribed *in vitro* from the cDNA template of the virus genome (reviewed in ref. 9).

To a large extent, RNA viruses have been classified on the nature of their genomes. Positive-stranded RNA viruses are those in which the virus RNA genome is in the same-sense as the message. Vertebrate families of positive-sense RNA viruses comprise the *Coronaviridae*, *Arteriviridae*, *Flaviviridae*, *Picornaviridae*, *Caliciviridae*, *Togaviridae*, and *Astroviridae* (space precludes a discussion of non-vertebrate viruses in this chapter). Reverse genetic studies have now been used to analyse the replication processes of six of these virus families (reviewed in refs 9, 10–12). The coronavirus family remains fully resistant to analysis using this approach owing to the large size of the genome (27–32 kilobases, some of the largest RNAs known). Changes have been introduced into the coronavirus genome, but by targeted RNA recombination rather than by transcription of an entire coronavirus genome (13).

Double-stranded RNA viruses, as their name implies, contain both positive- (message-sense) and negative-sense RNA. There are two families of double-stranded RNA viruses which infect vertebrates, invertebrates, and plants. Viruses belonging to the *Birnaviridae* family have genomes with two segments of RNA, while members of the *Reoviridae* (for example reoviruses, orbiviruses, and rotaviruses) have genomes with 10 or more segments of RNA. A rescue system has been developed for the birnavirus infectious bursal disease virus (14) and a helper virus-dependent rescue system has also been reported for reovirus, the prototype of the *Reoviridae* family (15).

Negative-strand RNA viruses, the genomes of which must be transcribed into mRNA for protein expression to occur, include viruses with either non-segmented RNA (*Filoviridae*, *Rhabdoviridae*, and *Paramyxoviridae*) or segmented RNA (*Orthomyxoviridae*, *Arenaviridae*, and *Bunyaviridae*) genomes. Reverse genetic systems with negative-strand RNA viruses took much longer to develop than those for positive-strand viruses, because the RNA of the former viruses is not infectious in either genome or anti-genome sense. The minimal infectious unit comprises the RNA genome encapsidated with the nucleoprotein, together with the virus-encoded RNA-dependent RNA polymerase protein(s) (16).

Two different approaches have been employed to derive a reverse genetic system for negative-stranded RNA viruses. They both involve the production of an intact ribonucleoprotein (RNP) complex, which can serve as a template for replication by the RNA-dependent RNA polymerase protein(s), but they differ in whether this complex is formed *in vitro* or *in vivo*. Both methods have been used to study the influenza A virus, which has eight segments of negative-stranded RNA in its genome. The first technique to be reported involved constructing an RNP complex *in vitro* using purified virus nucleocapsid (N) and polymerase (P) proteins and an *in vitro* transcribed RNA corresponding to one of the genome segments. This was then transfected into cells infected with wild-type influenza A virus to act as a helper, where it was replicated. The introduced segment could be incorporated into the influenza virus genome making use of the natural virus reassortment process (17). Selection for the modified segment or against the original virus segment allows production of a modified influenza A virus genome. More recently, an *in vivo* system allowing reconstitution of influenza virus ribonucleoprotein complexes within cells transfected with plasmids encoding virus segments has been described (18). However, this has not yet been used to produce all the virus segments simultaneously, so, again, rescue of a cDNA-derived genome segment requires co-infection with helper influenza virus.

Reverse genetic studies of the non-segmented paramyxoviruses and rhabdoviruses, and also of the segmented-genome Bunyamwera virus, have all been performed using the *in vivo* approach. All these viruses have been made without the aid of helper virus, thus alleviating problems of differentiating and separating the helper virus from the modified rescued virus. In this approach, cells are transfected with plasmids containing the full-length cDNA genome and the replication and nucleocapsid proteins, all under the control of the T7 promoter. Transcription and replication of these input cDNAs provides both the virus nucleic acid and proteins necessary for rescue of the virus. Researchers using this approach have found that it is more effective to use cDNA to the positive-sense anti-genome rather than the genome. Since 1994, two rhabdoviruses (rabies virus and vesicular stomatitis virus, VSV) (19–21) and a number of paramyxoviruses (Sendai virus, measles virus, respiratory syncytial virus, human parainfluenza virus type 3, rinderpest virus, and simian virus 5) (22–29) have all been rescued from cDNA clones. Thus far, Bunyamwera virus, the prototype bunyavirus with three RNA segments in the genome, is still the only segmented genome negative-strand virus to be rescued from cDNA without the use of a helper virus (30).[a]

2 Model systems for the manipulation of RNA viruses

Historically, model systems for the manipulation of RNA viruses have provided a first step in the study of the entire virus genome using reverse genetics. There are several advantages of starting with the analysis of a simpler model system:

(a) It is easier to ensure that a 'minigenome' of several hundred nucleotides in length contains the correct virus sequence than a full-length genome of perhaps 12 000 or 15 000 nucleotides. The correct sequence is important: a single nucleotide alteration in the cDNA clone of the arterivirus equine

---

[a] Note added in proof.

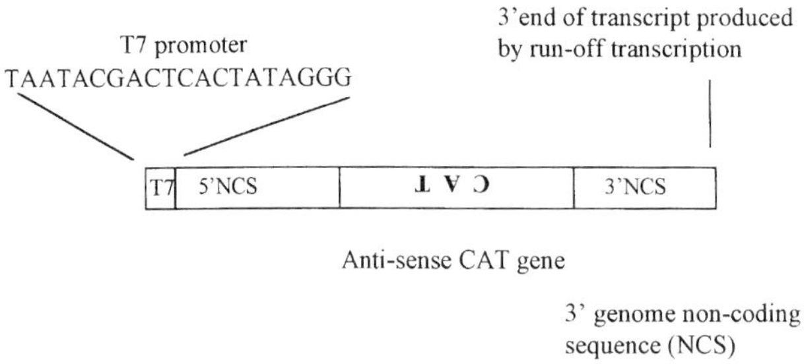

Figure 1 Composition of a typical minigenome used to analyse negative-strand virus genome replication. An anti-sense chloramphenicol acetyltransferase (CAT) gene is flanked by part or all of the terminal genomic sequences of the virus. Transcription is initiated at a T7 promoter, and terminated by run-off transcription.

arteritis virus caused an inactive replicase, and delayed construction of an infectious clone by 3 years (11).

(b) Small RNA molecules that mimic naturally occurring, defective interfering (DI) particles can replicate considerably more efficiently than larger ones, thus ensuring a level of replicated RNA in sufficient quantity to be able to analyse the products (31). Detection of replicated template can also be facilitated by inclusion of a reporter gene such as the beta-galactosidase, chloramphenicol acetyltransferase (CAT), or green fluorescent protein (GFP) genes flanked by the virus replication sequences (see *Figure 9.1*). Such experiments were first conducted with influenza A virus (32).

(c) Model systems using a reporter gene can yield quantitative data about the transcription or replication efficiency of wild-type or mutated promoter signals, which cannot otherwise be obtained (33). These results can also be obtained faster than those obtained by analysing virus particles produced from the rescue of whole virus genomes.

(d) Only cDNA clones corresponding to the virus replication proteins need to be introduced into the cell, and not cDNA encoding the envelope proteins, which are required for packaging the virus particles or other non-structural proteins (34).

For these reasons, model systems of studying virus transcription and replication are still being employed, even though methods are available to rescue complete virus genomes.

2.1 Transient expression system

The majority of model systems for manipulating RNA viruses have employed the transient vaccinia virus expression system (35), in which a recombinant vaccinia virus vTF7-3 expresses bacteriophage T7 RNA polymerase. Cells are initially infected with this virus, and then transfected with plasmids encoding the virus replication proteins under the control of a T7 promoter. Plasmids encoding the glycoprotein genes are also included if packaging of the minigenome is to be studied as well.

2.2 Transfection and electroporation techniques

Numerous techniques exist for the introduction of the virus cDNA or RNA into cells, including electroporation and transfection using calcium phosphate, DEAE dextran, or liposomes. Each method has its own advocates, but currently lipofection is probably the most widely used technique. The choice of lipofection reagents is increasing all the time, so recent molecular biological catalogues should be consulted, for example those of Boehringer Mannheim (now Roche Diagnostics), Gibco-BRL, and Stratagene. We include a protocol here for the synthesis of liposomes (*Protocol 1*), since it is considerably more economical to manufacture liposomes in the laboratory than to buy them ready-made.

Protocol 1

Preparation of cationic liposomes[a]

Equipment and reagents

- Dimethyldioctadecyl ammonium bromide (DDAB) (Sigma) (powder)
- 10 mg/ml dioleoyl-L-α-phosphatidyl ethanolamine (DOPE; Sigma) in chloroform
- Chloroform
- 20 ml glass Universal bottle
- 10 ml sterile distilled water
- Source of nitrogen gas
- Sonicator, ideally both a soniprobe and a sonicating water bath
- Vacuum desiccator

Method

1. Add 40 mg DDAB to 10 ml chloroform and shake to dissolve.
2. Remove 1 ml of the DDAB solution and add to a clean Universal bottle. If desired, freeze the remainder at −20 °C in 1 ml aliquots, with the lids of the Universals sealed with Parafilm®. It is important that exactly 4 mg DDAB is added to the DOPE.
3. Add 1 ml of the DOPE solution to the 1 ml DDAB solution.
4. Evaporate the chloroform using a stream of nitrogen gas; this takes about 5 min.
5. Dry the lipids for 2 h in a vacuum desiccator.
6. Add 10 ml sterile distilled water. Resuspend the lipids as far as possible in the water by sonicating in a water bath for 10 min.
7. Sonicate the lipids using a soniprobe. Sonicate at maximum power in 1 min bursts, keeping the Universal on ice, until the suspension clears (5–15 min according to the sonicator).
8. Store the lipids at 4 °C for up to 4 weeks; vortex briefly before use.

[a] This method is modified from ref. 36.

The linked *Protocols 2* and *3* describe a method for the expression of green fluorescent protein (GFP) using the vaccinia virus transient T7 system (see *Protocol 2*) and detection of the GFP (see *Protocol 3*). This allows the efficiency of transfection using liposomes made in the laboratory or bought from a commercial source to be tested. An assay for beta-galactosidase expression in fixed cells and tissues was included in an earlier volume of this series (37).

Protocol 2

Expression of green fluorescent protein (GFP) using the transient vaccinia virus T7 system

Equipment and reagents

- Titred vaccinia virus recombinant, expressing phage T7 polymerase, e.g. vTF7-3, MVA-T7[a]
- Plasmid encoding the GFP under the control of the T7 promoter
- Sterile plastic 60 mm cell-culture dishes (Nunc)
- Semi-confluent CV-1 or HeLa $T4^+$ cells in 60 mm dishes seeded 24 h previously
- Dulbecco's modified minimal essential medium (DMEM; Gibco-BRL) supplemented with 10% heat-inactivated fetal calf serum, L-glutamine (Gibco-BRL), and penicillin/streptomycin (Gibco-BRL)
- OptiMEM (Gibco-BRL)
- 14 ml plastic tubes (Falcon)
- Lipofection reagent (see *Protocol 1*)

Method

1. Remove the medium from the CV-1 or HeLa $T4^+$ cells in the 60 mm dishes; the cells should ideally be at around 70% confluence.
2. Wash the cells once with 1 ml OptiMEM.
3. Add 5×10^6 p.f.u. of vaccinia virus expressing T7 RNA polymerase in 0.5 ml OptiMEM. Incubate for 30 min at 37 °C, rocking the dishes occasionally.
4. Prepare the transfection mix. Add 50 ng of the GFP plasmid to 0.5 ml OptiMEM and vortex for a few seconds. Meanwhile add 50 μl lipofection reagent (see *Protocol 1*) to a further 0.5 ml in a second Falcon tube and vortex. Mix the contents of the two tubes, vortex again, then leave to stand for 5 min.
5. Remove the virus inoculum. Wash the cells once with 1 ml OptiMEM and add the transfection mix. Return the cells to 37 °C.
6. After a minimum of 2 h, add 4 ml DMEM to each dish and incubate for 24 h to 48 h.

[a] See Section 4 for a discussion of the different T7 polymerase-expressing vaccinia viruses.

Protocol 3

Detection of green fluorescent protein following expression with the transient vaccinia virus T7 system

Equipment and reagents

- Complete phosphate-buffered saline (PBS)
- 4% formaldehyde freshly diluted from a 40% stock into PBS (2 ml per dish)
- 1 μg/ml propidium iodide (Sigma), made up in PBS (2 ml per dish)
- PBS/glycerol solution (Citifluor)
- Fluorescence microscope
- Square, 25 mm coverslips

Method

1. At 24 or 48 hours' post-transfection, remove the medium, wash the cells once with 4 ml PBS, and fix for 30 min in 5 ml of the 4% formaldehyde solution.
2. Wash the cells with 5 ml of PBS and stain cells with the propidium iodide solution for 5–10 min.[a]
3. Wash the cells 3 times with PBS, add 2 drops of PBS/glycerol solution and cover with two coverslips. Cut down the sides of the dishes to aid viewing under the microscope, and view under the UV settings of a fluorescence microscope.[b]

[a] Propidium iodide stains nucleic acids, both DNA and RNA.

[b] Transfected cells appear green, while cells not expressing the GFP protein have an orange background staining from the propidium iodide.

2.3 Choice of expression system

For positive-strand RNA viruses, a source of virus RNA or cDNA is sufficient to initiate infection in susceptible cells (see Chapter 2). The pros and cons of RNA versus cDNA as the source of virus nucleic acid for these viruses are discussed in ref. 9. Up until now, *in vitro* transcribed RNA has been used more commonly than cDNA clones, but this may change as the technique of using a self-cleaving ribozyme to delineate the 3′ terminus of the RNA becomes more popular (38).

As discussed in the introduction, production of an infectious clone for negative-sense RNA viruses requires a source of both replication proteins and virus genome to form a functional RNP. These have been constructed *in vitro* for influenza A virus, but all other negative-strand RNA viruses that have been rescued into infectious virus have been done so using an *in vivo* method. Both the plasmids encoding the genome and the replication proteins are expressed under the control of a suitable promoter. The vaccinia virus T7 expression system (see Section 2.1) has been used extensively for viruses that replicate in the cytoplasm. Orthomyxoviruses such as influenza A virus, which replicate in the nucleus of infected cells, may require a different promoter. The cytomegalovirus immediate–early promoter, which is expressed constitutively in the nuclei of most eukaryotic cells, is a suitable candidate. Use has also been made of the murine polymerase I promoter (39). Other investigators used a mouse hydroxymethylglutaryl-coenzyme A reductase promoter and a truncated human RNA polymerase I promoter (positions −250 to −1) to express the influenza A virus replication proteins and a CAT-containing minigenome, respectively (18).

2.4 Generation of RNAs containing authentic 5′ and 3′ termini

The effects of 5′ and 3′ extensions and the m^7GpppG cap on the infectivity of positive-strand RNA virus transcripts has been reviewed by Boyer and Haenni (9). In general, 5′ extensions are not tolerated but 3′ extensions of up to seven nucleotides are. A cap structure is required for some virus families. For negative-strand RNA viruses, the picture is somewhat different—5′ extensions of one or two nucleotides are tolerated; viruses differ, however, in whether these extra nucleotides are subsequently cleaved off the transcript (40, 41), 3′ extensions are not generally tolerated at all (40). The following techniques have been utilized to generate exact termini:

(a) Run-off transcripts can be made to terminate on the final 3′ nucleotide. Some restriction enzymes, for example *Bbs*I, *Alw*I, *Bbv*I, and *Ear*I cut at a distance from their recognition sites and are therefore particularly useful for such purposes. Molecular biology catalogues such as that of New England Biolabs list enzymes with non-palindromic recognition sequences (http://www.neb.com/neb/products/REs/RE×frame.html).

(b) For polyadenylated viruses, restriction endonucleases such as *Mlu*I have a recognition site (A ↓ CGCGT) in which cleavage occurs after the initial A residue, thus it can be used in run-off transcription experiments to produce products terminating in runs of A residues (42).

(c) The correct 5′ terminus can generally be achieved by cloning immediately downstream of the promoter. Use of the polymerase chain reaction (PCR) in cloning facilitates this, since hybrid primers containing both the promoter sequences (such as T7, SP6, and T3) and the 5′ virus sequence can be employed. When the complete T7 RNA polymerase promoter (TAATACG-ACTCACTATAGGG) is used, three G residues are generated at the 5′ end of the synthesized RNA. The promoter can be truncated to omit these residues, but this reduces transcription levels.

(d) Self-cleaving ribozymes have been used to generate exact 5′ and 3′ end transcripts. Both hammerhead and hepatitis delta ribozymes (38) have been used in reverse genetic experiments.

3 Synthesis of RNA templates

RNA templates are generally either synthesized *in vitro* by run-off transcription or *in vivo* using ribozymes to delineate the 3′ terminus. Generation of the constructs used to synthesize RNA has already been discussed in Section 2. *Protocol 4* describes the production of *in vitro* RNA transcripts.

Protocol 4

In vitro transcription of virus RNA

Equipment and reagents

- DNA sample
- Appropriate restriction enzyme
- 1% agarose gel
- Electrophoresis equipment and reagents
- Ethidium bromide
- Geneclean kit (BIO-101)
- 15 U/μl T7 RNA polymerase (Promega)
- 40 U/μl RNasin (Promega)
- 0.1 M dithiothreitol (DTT; Promega)
- 10 mM mix nucleotide triphosphates (from 100 mM stocks, Pharmacia)
- 5 × transcription buffer (Promega, supplied with the enzyme)
- 1 U/μl RQ DNase I (Promega)
- Sterile distilled water

Method

1. Linearize DNA with the appropriate restriction enzyme and purify by agarose gel electrophoresis and the Geneclean kit.
2. Resuspend the linearized DNA at a concentration of about 200 ng/μl in sterile distilled water.
3. Set up the transcription reaction at room temperature and wear gloves to avoid possible RNase contamination. Add the reagents in the following order:

Sterile distilled water	52.5 μl
5 × transcription buffer	20.0 μl
DTT	10.5 μl
Linearized DNA (1 μg)	5.0 μl
RNasin	2.5 μl
T7 RNA polymerase	5.0 μl
4 × 10 mM NTP mix (ATP, CTP, GTP, UTP)	5.0 μl

4. Vortex mix after adding each reagent and briefly spin in a microcentrifuge.
5. Incubate at 37 °C for 2 h.
6. Check that the transcription reaction has worked by running 4 μl on a 1% agarose gel containing 1 μg/ml ethidium bromide and view on a UV light-box.
7. To the remainder of the reaction mix add 5 μl RQ DNase I, vortex mix, and spin down in a microcentrifuge briefly. Incubate for 20 min at 37 °C, then store the tube on ice until used. Use 10 μl of the reaction to transfect a 35 mm dish of cells.

4 Purification/synthesis of virus proteins required for replication

Purification/synthesis of virus proteins required for replication is really only an issue for negative-sense RNA viruses, since the proteins can be translated directly from the genome of positive-sense RNA viruses (see Chapter 2). Initially, virus rescue of negative-sense RNA viruses employed the vTF7-3 system described for the minigenomes (35). However, the toxicity of this virus gives rise to problems for complete virus rescue and the vaccinia virus has to be separated from the rescued virus by some means. More recently, the less pathogenic fowl pox and the attenuated Ankara strain of vaccinia virus have been used to express the bacteriophage T7 polymerase (43–45). Some authors have found that vaccinia virus inhibitors aid the recovery of rescued virus (25). A recombinant baculovirus has also been made which expresses T7 RNA polymerase, but, as yet, this has not been used in any rescue experiments (46).

Protocol 5

Generation and selection of cell lines expressing T7 RNA polymerase

Equipment and reagents

- Plasmid expressing the T7 RNA polymerase gene under the control of a promoter suitable for the cell line in question and an appropriate selectable marker, e.g. neomycin gene
- Selection reagent (e.g. geneticin; Sigma). Make up to 80 mg/ml in distilled water, filter-sterilize, and store at −20 °C.
- Baby hamster kidney (BHK) cells (or other cell line as appropriate)
- Growth medium (Glasgow modified Eagle's Medium, GMEM)
- Lipofection reagent and OptiMEM (see *Protocols* 1 and 2)
- 14 and 50 ml plastic Falcon tubes (Falcon)
- 60 mm sterile plastic cell culture dishes (Nalge Nunc)
- 1:4 trypsin/versene solution for disruption of cell monolayer (trypsin is 0.25% (w/v) in Tris-buffered saline, versene is 0.02% (w/v) in PBS)
- Sterile narrow-tip forceps
- Whatman 3MM paper cut into 2 mm squares and autoclaved
- Plastic 24-well plate (Nalge Nunc)

Method

1. Seed 10^6 BHK cells in a 60 mm dish. Incubate overnight in a CO_2 incubator at 37 °C.
2. Transfect the cells using 2 μg of the selectable plasmid, as described in steps 4–6 of *Protocol 2*.[a] Incubate the cells until the monolayer is confluent (1 or 2 days' post-transfection).
3. Trypsinize cells and dilute with 300 ml GMEM.
4. Plate 5 ml of cells per 60 mm dish. Incubate overnight at 37 °C.

Protocol 5 continued

5. Remove the medium. Add 3 ml fresh medium containing the selection reagent (800 μg/ml for geneticin).
6. Incubate at 37 °C until individual colonies of cells start to appear (about 2 weeks), changing the medium every 3–4 days.
7. Trypsinize the individual colonies. For best results, soak the small squares of sterile 3MM paper in the trypsin/versene solution, place them on the colonies, and then shake off the cells into 1 ml medium containing geneticin in separate wells of a 24-well plate.
8. Continue expanding the selected colonies. The concentration of geneticin in the medium can be reduced to 200 μg/ml once the cells are grown in a 60 mm dish.
9. Test for the presence of the T7 RNA polymerase in the cell lines according to *Protocol 5*.

[a] Linearization of the plasmid was performed in ref. 22, but this is not essential. Note that no helper vaccinia is required for this protocol.

Protocol 6

Screening cell lines for the expression of T7 RNA polymerase by chloramphenicol acetyltransferase (CAT) assay

Equipment and reagents

- 10^6 BHK cells/35 mm dish
- Acetyl coenzyme A, sodium salt (Sigma). Prepare 50 mM solution in sterile distilled water and store in aliquots at −20 °C.
- [^{14}C]chloramphenicol (Dupont NEN)
- TNE buffer: 40 mM Tris–HCl, 150 mM NaCl, 1 mM EDTA
- 0.25 M Tris–HCl pH 7.5
- Thin-layer chromatography (TLC) plate
- Whatman 3MM paper
- 100 ml of 95:5 chloroform/methanol
- Ethyl acetate
- Sterile plastic scrapers (Falcon)
- X-ray film

Method

1. For each cell line to be tested, seed approximately 1 × 10^6 BHK cells stably expressing the T7 RNA polymerase gene in a 35 mm dish. Incubate overnight in a CO_2 incubator at 37 °C or until the cells reach confluence.
2. Remove 1 ml of the medium. Resuspend the cells in the remaining medium using a sterile plastic scraper and transfer to microcentrifuge tube.
3. Spin for 1 min and remove the supernatant.

Protocol 6 continued

4. Resuspend the pellet gently in 250 μl TNE, spin for 30 sec, and remove the supernatant.
5. Resuspend the pellet in 70 μl of 0.25 M Tris–HCl pH 7.5.
6. Freeze–thaw in dry ice and a 37 °C water bath three times, for 5 min in each.
7. Pellet the cell debris for 5 min in a microcentrifuge and transfer the supernatant to a new tube.
8. Set up the CAT assay (per reaction):

50 mM acetyl CoA	1 μl
[^{14}C]chloramphenicol	1 μl
0.25 M Tris–HCl pH 7.5	13 μl

9. Mix 25 μl of the cell extract with the 15 μl reaction mix and incubate at 37 °C for 2 h.
10. Spin the tubes for 5 sec, add 250 μl ethyl acetate, and vortex well for 15 sec. Allow to stand for 2 min, then centrifuge for 5 min. Remove the upper phase to a new tube and lyophilize under vacuum until all traces of ethyl acetate are removed.
11. Add 25 μl of ethyl acetate to each tube, vortex well for 20 sec, and spin briefly in a microcentrifuge.
12. Spot all samples on to the TLC plate on spots drawn 20 mm above the bottom of the plate, add 5 μl of each sample at a time.
13. Place the TLC plate in an equilibrated tank, lined with Whatman 3MM paper, containing the chloroform/methanol. Run until the solvent front is nearly at the top of the plate (about 15 min).
14. Air-dry the TLC plate and expose to X-ray film overnight.

An alternative approach, used to rescue measles virus, was to construct a cell line that expresses the virus N and P proteins in addition to T7 RNA polymerase (22). This latter method is gaining acceptance as a more convenient approach than the vaccinia virus transient expression system because of the toxicity problems associated with vaccinia virus, so a protocol is given here for deriving a cell line producing T7 RNA polymerase (see *Protocol 5*). This protocol can be modified to create cell lines expressing more than one protein by co-transfection with multiple plasmids at the first stage; many cells become stably transfected with all the plasmids used, and this can be tested by Western blotting. A further protocol for screening cell lines for the presence of T7 RNA polymerase using the CAT assay is also given (see *Protocol 6*). Screening can alternatively be performed using a luciferase gene expressed under the T7 RNA polymerase promoter.

The first reverse genetics experiments using influenza A virus were performed using purified nucleocapsids (32). Virions were purified, then detergent-treated, and nucleocapsids isolated by caesium-chloride density centrifugation. *In vitro*-transcribed RNAs were added to the purified nucleocapsids, which were then transfected into influenza A-infected cells. In an alternative method, micrococcal nuclease was added to the purified nucleocapsids to digest the wild-type virus RNAs; *in vitro*-transcribed RNA was then added (47).

5 Rescue of infectious virus

Virus rescue experiments are generally performed in a similar manner to those for minigenomes, except that a source of the complete genome is needed instead of a minigenome. The first researchers in the field of negative-strand virus reverse genetics found that the larger the genome size, the harder it was to rescue the construct, so the complete virus may well be rescued with a tenth or a hundredth the efficiency of the minigenome. This therefore means that all processes need to be carried out at optimal efficiency, so transfection reagents should always be tested for efficiency before starting with rescue experiments.

For the rescue of positive-sense RNA viruses, the basic techniques have already been described, since transfection of the RNA genome transcribed *in vitro* or of the cDNA equivalent both lead to the generation of infectious virus. Transfectant virus can be detected by direct visualization of virus plaques aided by immunofluorescence staining. Difficulties encountered in rescue vary with the virus family. For example, many flavivirus cDNAs are reported to be unstable in bacteria, so an infectious virus may have to be made by *in vitro* ligation of cDNA fragments followed by *in vitro* transcription of the complete cDNA. This approach has been used successfully to generate infectious yellow fever virus and dengue type 2 (48, 49). The large size of coronavirus genomes, coupled with the instability of the frame-shift region cDNA in bacterial hosts pose particular problems. Use of specific targeted RNA recombination or of modified defective-interfering particle cDNAs is overcoming these difficulties (13).

Rescue of negative-sense RNA viruses is inevitably more complex, because of the need for a source of both virus replication proteins and genome. The precise protocols tend to be very specific to the virus family being studied, so no single generalized protocol can be provided. Instead, we provide an example of the rescue procedure we use for Bunyamwera virus (see *Protocols* 7 and 8) (30). This is followed by *Protocol* 9, the growth of Bunyamwera virus from plaques produced in the rescue experiments for further analysis. Other specific rescue protocols, such as that for VSV, are included in Chapter 8 of this volume.

The use of constructs containing silent mutations (see the next section for the generation of mutant viruses) is a useful means of ensuring that the rescued virus is, in fact, a rescued virus and not a laboratory contaminant with the wild-type virus. For example, a genetic tag of three nucleotides was incorporated into measles virus (22). One of the first experiments we performed with the BUN rescue was to add a unique, silent restriction endonuclease site to both the small (S) and long (L) genome segment cDNAs of Bunyamwera virus (30).

Protocol 7

Rescue of Bunyamwera virus from cDNA clones—Part 1: transfection procedure

Equipment and reagents

- 100 mm plastic cell-culture dishes (Corning)
- Liposomes (see *Protocol 1*)
- Plasmid DNAs
- Titred stock of vaccinia virus vTF7-3
- HeLa $T4^+$ cells
- Dulbecco's modified minimal essential medium (DMEM; Gibco-BRL)
- 14 ml plastic tubes (Falcon)
- OptiMEM (see *Protocol 2*)
- Lipofection reagent (see *Protocol 1*)

Method

1. Seed 1×10^6 HeLa T^{4+} cells in DMEM in each of six 100 mm dishes and incubate in a CO_2 incubator at 37 °C for 48 h (cells should reach 70% confluence). Check the efficiency of transfection (although not essential, it is advised) by performing a GFP control transfection at the same time according to the instructions given in *Protocol 2*.
2. Wash the cells once with 2 ml OptiMEM per dish.
3. Add 4×10^6 p.f.u. of vaccinia virus expressing T7 RNA polymerase in 1.5 ml OptiMEM. Incubate for 30–60 min at 37 °C, rocking the dishes occasionally.
4. Prepare the transfection mix. For each dish, add 20 μg of pTF7-5BUNS, 4 μg of pTF7-5BUNM, and 10 μg of pTF7-5BUNL in 1 ml OptiMEM and vortex for a few seconds. Meanwhile add 100 μl of the lipofection reagent (see *Protocol 1*) to a further 1 ml OptiMEM in a second Falcon tube and vortex mix. Mix the contents of the two tubes, vortex again, and leave to stand for 5–10 min. These plasmids provide the source of virus proteins.
5. Remove the virus inoculum. Wash the cells three times with 4 ml OptiMEM and add the transfection mix. Return the cells to 37 °C.
6. After 2–3 h, prepare the second transfection mix. This time use 1 μg of pT7riboS+, 4 μg of pT7riboBUNM+, and 10 μg of pT7riboBUNL+ and mix with the liposomes as described in step 4. Remove the first transfection mix and add the second. These plasmids provide the source of virus genome RNA.
7. After 2–3 h, add 9 ml DMEM to each dish and incubate for 48 h.

Protocol 8

Rescue of Bunyamwera virus from cDNA clones—Part 2: amplification of rescued virus and assay for virus rescue

Equipment and reagents

- C3/36 *Aedes albopictus* cells grown in L-15 medium (Gibco-BRL)
- Sterile, cell scrapers (Falcon)
- Ice bucket containing dry ice
- 50 ml plastic tubes (Falcon)
- 60 mm and 100 mm cell culture dishes (Nalge Nunc, Corning)
- L-15 medium
- BHK cells
- Glasgow modified Eagle's Medium (GMEM)
- Overlay medium: 31 ml Eagles A solution, 6.5 ml Eagles B solution, 2 ml new-born calf serum pre-warmed to 42 °C
- 2% (w/v) HSA agarose in Eagles A solution for the overlay (Park Scientific Ltd)
- 4% formaldehyde in PBS
- Giemsa stain (BDH, Gurr® solution)
- 50 mg/ml phosphonoacetic acid (PAA) solution in distilled water, filter-sterilized (Sigma)

A. Amplification of rescued virus

1. On the same day as the initial transfection (see *Protocol 7*), seed six 100 mm dishes of C6/36 cells using 3×10^6 cells per dish and culture at 30 °–33°C (without CO_2).
2. At 48 hours' post-transfection, harvest the transfected HeLa $T4^+$ cells into a 50 ml Falcon tube using the sterile cell scraper.
3. Freeze–thaw three times, alternately using the dry ice and a 37 °C water bath.
4. After the third thaw, clarify the cell extract by centrifugation at 2500 r.p.m. for 5 min.
5. Add 4 ml of this clarified cell extract to the C6/36 cells (having first removed the medium from the cells). Allow to adsorb for 2 h at 30 °–33°C.
6. Wash the C6/36 cells three times in 9 ml of L-15 medium. Overlay with another 9 ml of L-15 medium and incubate at 30 °–33°C for 6 days.
7. Harvest the C6/36 cell supernatant medium after 6 days and assay for the presence of Bunyamwera virus.

B. Titration of Bunyamwera virus

1. Seed 60 mm dishes with 1×10^6 BHK cells and incubate at 37 °C for 48 h or until cells reach 100% confluence.
2. Make serial tenfold dilutions of the clarified C6/36 supernatant in GMEM up to a 10^{-7} dilution.
3. Remove most of the medium from the BHK cells, leaving approximately 300 μl per dish. Add 100 μl of the virus dilution per dish, adsorb for 1 h at 37 °C.
4. Microwave the overlay agar and allow to cool to approximately 60 °C.

Protocol 8 continued

5. Remove the virus dilution. Add 12 ml of the agar to the overlay solution together with 0.25 ml of the PAA solution and immediately add 5 ml of the overlay solution per dish, allow to set.
6. Incubate the dishes at 37 °C for 4 days. Fix in formaldehyde solution for 2 h and stain with Giemsa. Count the virus plaques.

Protocol 9

Growth of Bunyamwera virus (BUN) from plaques

Equipment and reagents

- Semi-confluent BHK cells
- Sterile 1.5 ml cryotubes (Sarstedt)
- 35 mm cell-culture dishes (Nalge Nunc)
- 75 cm^2 cell-culture flasks (Nalge Nunc)
- Glasgow modified Eagle's Medium (GMEM) (Gibco-BRL)
- 50 ml sterile plastic tube (Falcon)

Method

1. Pick a single, discrete plaque of BUN virus into 0.5 ml of GMEM in a sterile plastic cryotube, vortex.
2. Take a semi-confluent 35 mm dish of BHK cells seeded the previous day using approximately 3×10^5 cells and remove the medium. Add 0.25 ml of the plaque-picked virus to the dish and adsorb for 1 h at 31–33 °C.
3. Remove the inoculum. Add 2 ml of fresh GMEM and incubate at 31–33 °C until the cells show a cytopathic effect (c.p.e., 2–3 days).
4. Harvest the supernatant from the dish. Use 0.5 ml to infect a 75 cm^2 cell-culture flask of BHK cells set up 24–48 h previously (again, cells should ideally be 70% confluent). Adsorb for 1 h in 1 ml GMEM and then add a further 20 ml GMEM.
5. Incubate until the cells show 100% c.p.e. (usually 48–72 h).
6. Harvest the cell supernatant and remove the cells by centrifugation in 50 ml Falcon tubes at 2500 r.p.m. for 5 min.
7. Repeat steps 4–6 to make sequential virus passages, using 0.1 ml (approximately 10^6 p.f.u.) for infection each time.

6 Creation of mutant viruses

Since the purpose of reverse genetics is to study the biology of the virus, it is important to be able to construct virus genomes that are modified in the coding or non-coding regions of the genome. Modification of virus genes can generally be achieved using standard molecular biological techniques such as PCR mutagenesis (see *Protocols 10* and *11*) and replacement or insertion of restriction endonuclease fragments.

Protocol 10

Addition of several mutations by overlapping PCR

Equipment and reagents

- Plasmid encoding the virus genome, digested with restriction enzyme and dephosphorylated
- 4 primers (see Method)
- Sequence analysis software, e.g. University of Wisconsin GCG package
- Thermocycler
- Geneclean® III reagent or equivalent (BIO-101)
- 0.5 ml plastic PCR tubes
- 10 mM dNTPs (diluted from 100 mM stock, Pharmacia)
- Long wavelength UV transilluminator
- Agarose gel electrophoresis equipment and reagents
- Scalpels for excising agarose gel slices
- Pfu DNA polymerase (Stratagene)

Method

1. Choose the mutations required.[a]
2. Design 4 PCR primers, 25–40 nucleotides in length—two should cover the region to be mutated and should overlap by 18 or more nucleotides.[b]
3. Perform the first two PCR reactions, from each outer primer to the appropriate inner primer. Use a high-fidelity polymerase such as Pfu polymerase (Stratagene); this enzyme does not add additional nucleotides at the 3′ end of the amplified sequence. Use no more than 20 cycles of PCR (95 °C 40 sec, 50 °C 60 sec, 74 °C 90 sec) and around 100 ng of plasmid template.
4. Separate the PCR products from any remaining primers by agarose gel electrophoresis. Excise the amplified fragments and purify the DNA by Geneclean or an equivalent protocol.
5. Mix molar ratios of the two PCR products and amplify for a further 20 or so cycles using the two outer primers.
6. Either clone directly into the plasmid to be mutated or subclone and reclone. Confirm the new plasmid by restriction digestion and sequence analysis.

[a] Silent mutations can be devised using the UWGCG command 'map' with delimiter/silent. The output using the command 'type' lists existing restriction sites in capitals, and those which could be introduced by mutation in small type. When deleting a gene, it is advisable not only to remove the ATG codon but also to add stop codons within the reading frame to avoid the possibility of back mutation.

[b] Subsequent cloning is easier if the outer primers lie outside unique restriction endonuclease sites in the plasmid to be amplified.

Epitope tags or additional promoter regions can both be added by inserting appropriately tailored linkers into restriction sites (*Protocol 11*). For example, an epitope tag can be inserted at the C terminus of a virus protein which has a unique restriction site just before the end of the gene by designing linkers that contain:

(a) nucleotides to complete the 3′ end of the upstream restriction site;

(b) nucleotides to complete the 3′ end of the virus gene less the stop codon;

(c) nucleotides that encode the epitope tag;

(d) stop codon(s) after the epitope tag;

(e) nucleotides to complete the 5′ end of the downstream restriction site.

Protocol 11

Insertion of linker sequence into a restriction site

Equipment and reagents

- Plasmid encoding the virus gene, digested with restriction enzyme, and dephosphorylated
- Two complementary primers containing the additional sequences to be cloned
- Heating block/PCR machine/water baths
- 10 U/μl T4 polynucleotide kinase (e.g. New England Biolabs)
- 5 mM ATP solution for the kinase reaction (from 100 mM stock, Pharmacia)
- 400 U/μl T4 DNA ligase (e.g. New England Biolabs)
- 1 × PCR buffer (provided with Pfu DNA polymerase, see *Protocol 10*)
- Competent *E. coli* cells

Method

1. Design primers as required (see text), such that the two primers are completely complementary if they are to be cloned into a blunt restriction site. If the primer is to be added at a restriction site having cohesive ends, design appropriate 5′ or 3′ overhangs.
2. Phosphorylate the 5′ ends of the primers using 0.5 μl T4 polynucleotide kinase in a total of 10 μl for 30 min at 37 °C.[a]
3. Anneal 200 ng of the complementary primers in 1 × PCR buffer. Heat for 5 min at 95 °C and anneal for 60 min at 55 °C.
4. Ligate to 200 ng of the dephosphorylated plasmid.
5. Transform competent *E. coli* with the ligated plasmid.

[a] The New England Biolabs enzyme comes complete with 10 × kinase buffer, but ATP must be added to 500 μM.

Single nucleotide changes, replacements of amino acids, or amino acid insertions can conveniently be made in double-stranded DNA plasmids using the Stratagene QuikChange™ Site Directed mutagenesis kit. This works by amplifying both strands of the entire plasmid using the high-fidelity Pfu DNA polymerase enzyme. Complementary primers containing the desired mutations are designed for the two DNA strands of the plasmid. Following multiple rounds of

Table 1 Selection systems for transfectant influenza viruses

Gene	Basis of selection	Reference
PB2[a]	Host-cell restriction	Subbarao *et al.* (50)
HA	Monoclonal antibody	Barclay and Palese (51); Horimoto and Kawaoka (53)
NA	Host-cell restriction	Enami *et al.* (17)
NP	Temperature-sensitive helper virus	Li *et al.* (54)
M	Temperature-sensitive helper virus	Yasuda *et al.* (55)
M2	Drug (amantadine) resistance	Castrucci and Kawasaki (56)
NS1	Temperature-sensitive helper virus	Enami *et al.* (57)

[a]PB2, virus polymerase; HA, haemagglutinin; NA, neuraminidase; NP, nucleoprotein; M, M2, matrix; NSI, non-structural.

amplification, the DNA is digested with *Dpn*I endonuclease, which selectively digests methylated or hemi-methylated DNA. This results in selective degradation of the original DNA, since this will have been methylated by the *dam* methylase present in most laboratory strains of *E. coli*. Following the final annealing stage, complementary PCR-generated strands will become annealed, leaving a non-overlapping gap in each strand. Transformation of supercompetent *E. coli* XL-1 cells with the nicked PCR-generated DNA allows religation and amplification of the mutated DNA.

Modification of influenza A virus genes has generally been performed on one segment at a time. Rescue of this modified segment is then followed either by positive selection for this segment, or negative selection against the unmodified segment. *Table 1* summarizes the approaches that have been employed.

The insertion of additional genes requires knowledge of the biological processes of transcription and replication of the virus being studied. Techniques used include:

(a) Incorporation of an additional internal virus promoter. The gene encoding GFP has been introduced into influenza A virus using a second, internal virus promoter (58).
(b) Expression of an additional gene in an ambisense orientation, such as has been achieved with rabies virus (59).
(c) Use of an internal ribosomal entry site (IRES). A mammalian IRES sequence, derived from the 5′ non-coding region of the human immunoglobulin heavy-chain binding protein mRNA, has been used to create a bicistronic neuraminidase gene in influenza A virus (60). The mammalian IRES is shorter than the equivalent picornavirus sequence and can be utilized in influenza A virus-infected cells.
(d) Use of a protease cleavage signal to separate products at the protein level. For example, reverse genetics was used to produce a fusion protein in influenza A virus between the full-length CAT protein, the 2A protease of foot-and-mouth disease virus, and the influenza neuraminidase, which was subsequently cleaved by the 2A protease (61).

(e) Creation of an additional virus segment. An influenza A virus with nine, as opposed to the normal eight, segments has been created in an extension of the technique described above for the rescue of modified influenza virus segments. One of the virus segments encodes two genes, *NS1* and *NS2*, in overlapping reading frames. A recombinant segment containing the *NS1* gene alone was rescued into a virus containing a temperature-sensitive lesion in this gene (57).

In addition to adding complete extra genes, there are many examples of the creation of fusion products. For example, the neuraminidase gene of influenza A virus is relatively flexible in the stalk region, and a number of different epitopes of up to 80 amino acids have been added here. Epitopes have also been added at several points in the haemagglutinin molecule of influenza A virus (reviewed in ref. 16).

Increasing use is being made of insertion cassettes, in which unique restriction endonuclease sites are introduced by a suitable mutagenesis technique either side of the virus genome, or of a gene or gene region. The gene or region of interest can then be removed and then replaced with modified or alternate genes. Such techniques have been in use for some time with positive-sense RNA viruses (see Chapter 2). Evans has described their use for poliovirus reverse genetics (42). This author highlights particular features of a pCAS7 vector, so designed that it cannot be used to create infectious virus unless additional sequences are added between the unique *Sal*I and *Dra*I sites, because the virus reading frame would be incorrect. Insertion cassettes are currently being used for negative-sense RNA viruses such as measles (62).

7 Analysis of mutant phenotypes

Once a mutant virus has been obtained through reverse genetic experiments, it is imperative to first determine whether the genome does indeed correspond to the input genetic material, and, second. what is the phenotype of the mutant virus.

7.1 Genomic studies of the virus mutants

The genome is generally studied initially by the techniques of reverse transcription followed by polymerase chain reaction (RT-PCR). Mutant and wild-type viruses can often be distinguished from each other purely by the size of the RT-PCR products. For example, gene deletions lead to smaller RT-PCR products, and mutations incorporating restriction sites give rise to RT-PCR products that can be distinguished from those of the authentic virus following restriction enzyme digestion. *Protocols 12* and *13* describe the preparation of virus RNA from cell-culture supernatant fluids, and also the synthesis of first-strand cDNA from this RNA. PCR-amplification from this cDNA is not described; the reader is referred to standard molecular biology texts for this technique and to Chapter 6, *Protocol 4*.

Protocol 12

Preparation of RNA from transfectant virus using Trizol™

Equipment and reagents

- 175 cm^2 cell-culture flasks
- Subconfluent BHK cells
- 50 ml Falcon tubes
- Trizol™ (Gibco-BRL)
- Chloroform, AnalaR
- Isopropyl alcohol
- 75% ethanol
- Sterile 1.5 ml microcentrifuge tubes
- 35 ml polyallomer ultracentrifuge tubes (Sorvall), ultracentrifuge, and SW28 rotor
- Disposable gloves
- Sterile distilled water
- 1 ml Gilson pipette and sterile tips

Method

1. Culture the virus as described in *Protocol 9*. To prepare a bulk stock of RNA for multiple analysis infect two 175 cm^2 flasks of subconfluent BHK cells in 40 ml of medium, and incubate until the cells show a 100% cytopathic effect (approximately 48 h).
2. Pour the cells and supernatant into one 50 ml Falcon tube per flask (gently tap the flask to generally ensure that all cells are removed from the flask). Centrifuge for 5 min at 2500 r.p.m. to remove cell debris.
3. Pellet the virus from the clarified supernatant, for example 20 000 r.p.m. for 2 h at 4 °C in a SW28 rotor of a Sorvall ultracentrifuge.
4. Remove all medium from the ultracentrifuge tubes. Remove the final traces of medium using a dry tissue, taking care not to disturb the virus pellet.
5. Add 0.5 ml Trizol™ reagent per virus pellet using a Gilson pipette. Pipette up and down until the pellet is largely dissolved. Incubate for 5 min at room temperature.
6. Transfer the samples to a sterile 1.5 ml microcentrifuge tube. Add 0.1 ml chloroform, shake the tubes vigorously for 15 sec, and incubate at room temperature for 2–3 min.
7. Centrifuge the samples at 12 000 g for 15 min at 4 °C.[a]
8. Transfer the colourless, upper, aqueous phase to a new tube. Precipitate the RNA using 0.25 ml of isopropyl alcohol. Incubate the samples at room temperature for 10 min, then centrifuge at 7500 g for 10 min at 4°C.
9. Remove the supernatant. Wash the pellet once with 75% ethanol. Store the pellets in 75% ethanol at −20 °C if the samples are not to be used immediately.
10. For immediate use, microcentrifuge the samples for 5 min at 7500 g, air-dry the RNA pellet, and resuspend the pellet in 50 μl of sterile distilled water on ice.

[a] Following centrifugation, the mixture separates into: a lower, red, phenol phase; an interphase; and a colourless, upper, aqueous phase of approximately 300 μl containing the RNA.

Protocol 13

Reverse transcription of virus RNA

Equipment and reagents

- Sterile 1.5 ml microcentrifuge tubes
- Disposable gloves
- Sterile distilled water
- Gilson pipette and sterile tips
- Superscript™ II (Gibco-BRL)[a]
- RNasin (Promega)
- Primer for reverse transcription
- 10 mM dNTPs (diluted from 100 mM stock, Pharmacia)

Method

1. Prepare virus RNA as described in *Protocol 12*.
2. Add 2 μg of RNA to a fresh microcentrifuge tube with 0.5 μl RNasin.
3. Add 1 μg specific primer to the RNA and make up to a total of 12 μl with distilled water. Heat at 70 °C for 10 min and place on ice.
4. Collect the contents of the tube by brief centrifugation and add (in order):

5 × first strand buffer (supplied with enzyme)	4 μl
0.1 M dithiothreitol (DTT, supplied with enzyme)	2 μl
10 mM dNTP mix (10 mM each of dATP, dCTP, dGTP, dTTP)	1 μl

   Mix contents of the tube gently and incubate at 42 °C for 2 min.
5. Add 1 μl Superscript™ II, mix by pipetting.
6. Incubate at 42 °C for 70 min, then inactivate the enzyme at 75 °C for 10 min.

[a] The first-strand buffer and DTT are supplied with the enzyme.

Where restriction endonuclease sites have been introduced into the mutant genome, this can be examined by restriction digestion of the RT–PCR products. *Figure 2* shows *Xho*I digests of the BUN RT–PCR products from the S segments of authentic virus and transfectant virus produced by reverse genetics.

Final confirmation that the virus genotype is identical to that of the input cDNA can be provided by DNA sequencing the RT–PCR products, or occasionally by direct RNA sequencing. The latter is trickier to perform, but it does guarantee absolute results without the risks of possible contamination of the PCR assay. The former approach is the one generally used by researchers in this field, and this approach is normally acceptable for publication so long as it is backed up by other data to confirm the nature of the altered genotype.

Armed with the knowledge that the mutant virus is indeed that which was being constructed, the biological and biochemical properties of the mutant virus should also be examined. In addition, it is standard practice to passage the mutant virus at low multiplicity of infection for 10 or more passages to ensure

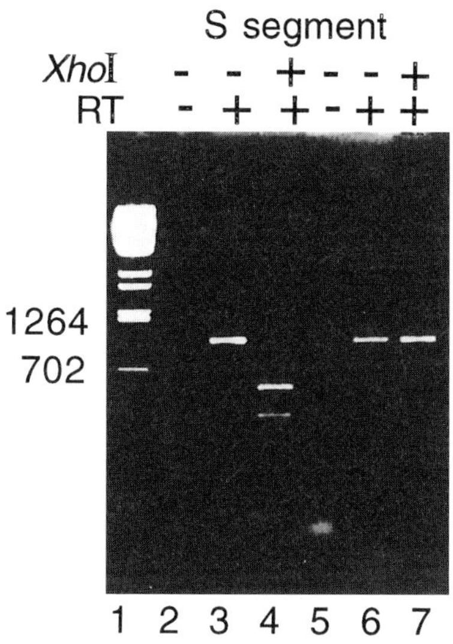

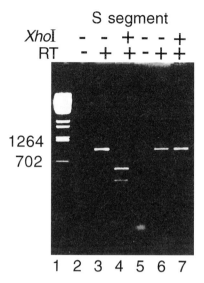

Figure 2 Confirmation of the identity of the transfectant Bunyamwera virus (BUN) by RT–PCR and restriction enzyme digestion. Virion RNA was extracted from the supernatant of cells infected with rescued or authentic BUN virus. Full-length, S segment cDNA was reverse-transcribed and amplified by PCR using terminal primers. Aliquots of the PCR products were digested with *Xho*I as indicated. Control reactions were set up omitting the reverse transcriptase in the first step (lanes 2 and 5). Authentic virus yielded an RT–PCR product of 961 bp that was resistant to *Xho*I digestion (lanes 6 and 7). In contrast, the transfectant virus, which contained a genetic tag including an added *Xho*I site, yielded an RT product of 961 bp (lane 3) that was cleaved into two products of 566 and 395 bp after *Xho*I digestion (lane 4). (Reproduced from ref. 30.)

that the virus mutation is stable. Many introduced mutations can be deleterious to the virus and may therefore be lost on passage, or the virus may develop compensatory mutations.

7.2 Phenotypic studies

The phenotypic studies performed on mutant viruses comprise:

(a) *Virological techniques*, including:
- plaque size and morphology (for instance, is there still fusion in a fusogenic virus?);
- one-step growth curves;
- virus yield;
- virus pathogenesis;
- host range (generally in cell culture);
- electron microscopy studies.

(b) *Protein studies*. If the mutation in the genome leads to an altered protein profile, either by an alteration to a protein or its removal by deletion of a non-essential gene from the virus, this can easily be seen by PAGE of [^{35}S]methionine-labelled virus proteins. Cellular localization of the virus or of particular virus products can be visualized by indirect immunofluorescence. The increased popularity of confocal microscopy allows the co-localization of two virus-encoded proteins, or of the virus and host proteins to be examined. In some cases, the resolution of the confocal microscope may be insufficient, and immunoelectron microscopy may be required.

(c) *RNA studies*. The ratios and amounts of different messenger RNAs can be studied by Northern blot analysis.

(d) *Biochemical studies*. The analysis used would depend on the particular mutation introduced into the virus. Specific studies could include:
- protein trafficking and processing;
- protein–protein interactions and protein–RNA interactions;
- antibody binding;
- activity of additional marker genes, e.g. GFP, beta-galactosidase, CAT.

No protocols are provided for this section, because these techniques are standard and are described elsewhere. For example, ref. 63 includes many of them. Other methods are to be found in standard virology or molecular biology texts such as *Current protocols in molecular biology* (64).

8 Technical and ethical issues

Most of this chapter has been addressing the question of 'how' to utilize reverse genetics. Numerous technical questions to which we have only limited answers remain, including:

- What are the size constraints of the mutant virus?
- Are particular sequences refractory to virus rescue?
- Do extra G residues at the 5′ end of negative-sense RNA viruses hinder rescue? Or are they removed subsequent to rescue?
- To what extent is it a problem that less fit viruses may not be rescued?

The question of 'what' reverse genetics is where science meets ethics. Since reverse genetics creates novel viruses that would not otherwise exist, due thought should be given to the nature of these new viruses before the experiments are undertaken. At a recent discussion on 'Genetic Rescue of enveloped RNA viruses: potentials and consequences' at the Novartis foundation (London) it was also suggested that, given our inability to predict the outcome of particular gene manipulations, changes to genes that might affect tissue tropism or host range should be conducted under higher containment facilities than those required for handling the parental virus (65).

9 Perspectives

Application of the reverse genetic techniques are still at an early stage, but a number of interesting experiments have already been performed. An excellent recent review of reverse genetics experiments performed with negative-sense RNA viruses is given in ref. 66. Non-essential genes have been deleted from the genomes of a number of viruses, and the effect on the virus phenotype observed. It was discovered that adding a further transcription signal to both paramyxoviruses and rhabdoviruses led to the production of another virus transcript. This information was quickly utilized to determine the minimum transcription signal, and thence to add numerous reporter genes (GFP, CAT, beta-galactosidase, and luciferase) to these viruses. In VSV, the level of gene expression is generally related to the relative position of the gene on the genome, and so the *CAT* gene has been inserted at various locations to confirm this relationship between genomic position and gene expression. Additional transcription signals have been added to the rabies virus anti-genome as well as to the genome, thus giving rise to an ambisense virus.

More recently, genes have been added that alter the biological properties of the virus. The *CD4* gene, which encodes the cellular human immunodeficiency virus (HIV) receptor, together with additional HIV co-receptor genes were added to both rabies virus and to VSV. Cells infected with the modified VSV or rabies virus express these HIV receptors at the cell surface, and thus cause the HIV to target the VSV or rabies-infected cells, where the HIV infection becomes swamped by the more rapidly growing rhabdovirus. The influenza virus haemagglutinin gene has been added to VSV and used to provide protection from influenza virus challenge (all reviewed in ref. 66). The possibilities are endless, and will provide an exciting area of work for the foreseeable future.

References

1. Clarke, B. (1997). *J. Gen. Virol.*, **78**, 2397–410.
2. Shaw, R. D. and Greenberg, H. B. (1994). In *Encyclopedia of virology* (ed. R. G. Webster and A. Granoff), pp. 1274–81. Academic Press, London.
3. Griffin, D. E. and Bellini, W. J. (1996). In *Virology* (ed. B. N. Fields, D. M. Knipe, and P. M. Howley), pp. 1267–312. Lippincott–Raven Press, New York.
4. White, D. O. (1994). In *Encyclopedia of virology* (ed. R. G. Webster and A. Granoff), pp. 1219–26. Academic Press, London.
5. Elliott, R. M. (1997). *Mol. Med.*, **3**, 572–7.
6. Klenk, H-D., Slenczka, W., and Feldmann, H. (1994). In *Encyclopedia of virology* (ed. R. G. Webster and A. Granoff), pp. 827–832. Academic Press, London.
7. Taniguchi, T., Palmieri, M., and Weissmann, C. (1978). *Nature* **274**, 223–8.
8. Racaniello, V. R. and Baltimore, D. (1981). *Science*, **214**, 916–19.
9. Boyer, J-C. and Haenni, A-L. (1994). *Virology*, **198**, 415–26.
10. Sosnovtsev, S. and Green, K. Y. (1995). *Virology*, **210**, 383–90.
11. vanDinten, L. C., den Boon, J. A., Wassenaa, A. L. M., Spaan, W. J. M., and Snijder, E. J. (1997). *Proc. Natl Acad. Sci. USA*, **94**, 991–6.
12. Geigenmuller, U., Ginzton, N. H., and Matsui, S. M. (1997). *J. Virol.*, **71**, 1713–17.
13. Koetzner, C. A., Parker, M. M., Ricard, C. S., Sturman, L. S., and Masters, P. S. (1992). *J. Virol.*, **66**, 1841–8.
14. Mundt, E. and Vakharia, V. N. (1996). *Proc. Natl Acad. Sci. USA*, **93**, 11131–6.
15. Roner, M. R., Sutphin, L. A., and Joklik, W. K. (1990). *Virology*, **179**, 845–52.
16. Palese, P., Zheng, H., Engelhardt, O. G., Pleschka, S., and Garcia-Sastre, A. (1996). *Proc. Natl Acad. Sci. USA*, **93**, 11354–8.

17. Enami, M., Luytjes, W., Krystal, M., and Palese, P. (1990). *Proc. Natl Acad. Sci. USA*, **87**, 3802–5.
18. Pleschka, S., Jaskunas, S. R., Engelhardt, O. G., Zurcher, T., Palese, P., and Garcia-Sastre, A. (1996). *J. Virol.*, **70**, 4188–92.
19. Schnell, M. J., Mebatsion, T., and Conzelmann, K-K. (1994). *EMBO J.*, **13**, 4195–203.
20. Lawson, N. D., Stillman, E. A., Whitt, M. A., and Rose, J. K. (1995). *Proc. Natl Acad. Sci. USA*, **92**, 4477–81.
21. Whelan, S. P. J., Ball, L. A., Barr, J. N., and Wertz, G. T. W. (1995). *Proc. Natl Acad. Sci. USA*, **92**, 8388–8392.
22. Radecke, F., Spielhofer, P., Schneider, H., Kaelin, K., Huber, M., Dotsch, C., Christiansen, G., and Billeter, M. A. (1995). *EMBO J.*, **14**, 5773–84.
23. Collins, P. L., Hill, M. G., Camargo, E., Grosfeld, H., Chanock, R. M., and Murphy, B. R. (1995). *Proc. Natl Acad. Sci. USA*, **92**, 11563–7.
24. Garcin, D., Pelet, T., Calain, P., Roux, L., Curran, J., and Kolakofsky, D. (1995). *EMBO J.*, **14**, 6087–94.
25. Kato, A., Sakai, Y., Shioda, T., Kondo, T., Nakanishi, M., and Nagai, Y. (1996). *Genes Cells*, **1**, 569–79.
26. Baron, M. D. and Barrett, T. (1997). *J. Virol.*, **71**, 1265–71.
27. Hoffmann, M. A. and Banerjee, A. K. (1997). *J. Virol.*, **71**, 4272–7.
28. Durbin, A. P., Hall, S. L., Siew, J. W., Whitehead, S. S., Collins, P. L., and Murphy, B. R. (1997). *Virology*, **235**, 323–32.
29. He, B., Paterson, R. G., Ward, C. D., and Lamb, R. A. (1997). *Virology*, **237**, 249–60.
30. Bridgen, A. and Elliott, R. M. (1996). *Proc. Natl Acad. Sci. USA*, **93**, 15400–4.
31. Conzelmann, K. K. and Schnell, M. (1994). *J. Virol.*, **68**, 713–19.
32. Luytjes, W., Krystal, M., Enami, M., Parvin, J. D., and Palese, P. (1989). *Cell*, **59**, 1107–13.
33. Neumann, G. and Hobom, G. (1995). *J. Gen. Virol.*, **76**, 1709–1717.
34. Dunn, E. F., Pritlove, D. C., Jin, H., and Elliott, R. M. (1995). *Virology*, **211**, 133–43.
35. Fuerst, T. R., Niles, E. G., Studier, F. W., and Moss, B. (1986). *Proc. Natl Acad. Sci. USA*, **83**, 8122–6.
36. Rose, J. K., Buonocore, L., and Whitt, M. A. (1991). *Biotechniques*, **4**, 520–5.
37. Stoker, A. W. (1993). *Molecular virology. a practical approach* (ed. A. J. Davison and R. M. Elliott), pp. 171–97. IRL Press, Oxford.
38. Perrotta, A. T. and Been, M. D. (1991). *Nature*, **350**, 434–6.
39. Neumann, G., Zobel, A., and Hobom, G. (1994). *Virology*, **202**, 477–9.
40. Pattnaik, A. K., Ball, L. A., LeGrone, A. W., and Wertz, G. W. (1992). *Cell*, **69**, 1011–20.
41. Samal, S. K. and Collins, P. L. (1996). *J. Virol.*, **70**, 5075–82.
42. Evans, D. (1993). *Molecular virology: a practical approach* (ed. A. J. Davison and R. M. Elliott), pp. 199–226. IRL Press, Oxford.
43. Britton, P., Green, P., Kottier, S., Mawditt, K. L., Penzes, Z., Cavanagh, D., and Skinner, M. A. (1996). *J. Gen. Virol.*, **77**, 963–7.
44. Wyatt, L. S., Moss, B., and Rozenblatt, S. (1995). *Virology*, **210**, 202–5.
45. Sutter, G., Ohlmann, M., and Erfle, V. (1995). *FEBS Lett.*, **371**, 9–12.
46. Polkinhorne, I. and Roy, P. (1995). *Nucl. Acids Res.*, **23**, 188–91.
47. Seong, B. L. and Brownlee, G. G. (1992). *J. Gen. Virol.*, **186**, 247–60.
48. Rice, C. M., Grakoui, A., Galler, R., and Chambers, T. J. (1989). *New Biol.* **1**, 285–96.
49. Kapoor, M., Zhang, L., Mohan, P. M., and Padmanabhan, R. (1995). *Gene*, **162**, 175–80.
50. Subbarao, E. K., Kawaoka, Y., and Murphy, B. R. (1993). *J. Virol.*, **67**, 7223–8.
51. Barclay, W. S. and Palese, P. (1995). *J. Virol.*, **69**, 1275–9.
52. Enami, M. and Palese, P. (1991). *J. Virol.*, **65**, 2711–13.
53. Horimoto, T. and Kawaoka, Y. (1994). *J. Virol.*, **68**, 3120–8.

54. Li, S., Xu, M., and Coelingh, K. (1995). *Virus Res.*, **37**, 153–161.
55. Yasuda, J., Bucher, D. J., and Ishihama, A. (1994). *J. Virol.*, **68**, 8141–6.
56. Castrucci, M. R. and Kawaoka, Y. (1995). *J. Virol.*, **69**, 2725–8.
57. Enami, M., Sharma, G., Benham, C., and Palese, P. (1991). *Virology*, **185**, 291–8.
58. Flick, R., Hoffmann, E., Azzah, M., Neumeier, E., and Hobom, G. (1997). *10th International Conference on Negative Strand Viruses* (abstract).
59. Finke, S. and Conzelmann, K. K. (1997). *J. Virol.*, **71**, 7281–8.
60. Garcia-Sastre, A., Muster, T., Barclay, W. S., Percy, N., and Palese P. (1994). *J. Virol.*, **68**, 6254–61.
61. Percy, N., Barclay, W. S., Garcia-Sastre, A., and Palese, P. (1994). *J. Virol.*, **68**, 4486–92.
62. Singh, M., Hangartner, L., Cornu, T., Christiansen, G., and Billeter, M. A. (1997). *10th International Conference on Negative Strand Viruses* (abstract).
63. Paterson, R. G. and Lamb, R. A. (1993). *Molecular virology: a practical approach* (ed. A. J. Davison and R. M. Elliott), pp. 35–73. IRL Press, Oxford.
64. Ausubel, F. M., Brent, R., Kingston, R. E., Moore, R. E., Seidman, J. G., Smith, J. A., and Struhl, K. (ed.) (1995). *Current protocols in molecular biology*. Wiley, New York.
65. Randall, R. E. (1998). *SGM Quart.*, **25**, 95.
66. Roberts, A. and Rose, J. K. (1998). *Virology*, **247**, 1–6.

Since writing this chapter, the efficient rescue of influenza A virus entirely from cDNA clones has been reported by two groups, which represents a significant improvement over previous helper-independent systems.

Generation of influenza A viruses entirely from cloned cDNAs. Neumann, G., Watanabe, T., Ito, H., Watanabe, S., Goto, H., Gao, P., Hughes, M., Perez, D. R., Donis, R., Hoffmann, E., Hobom, G. and Kawaoka, Y. (1999). *Proc., Natl. Acad. Sci (USA)* **96**, 9345–9350.

Rescue of influenza A virus from recombinant DNA. Fodor, E., Devenish, L., Engelhardt, O. G., Palese, P., Brownlee, G. G., GarciaSastre, A. (1999). *J. Virol.* **73**, 9679–9682.

Chapter 10

Development of RNA virus vectors for gene delivery

B.A. Usmani, A. Fassati, and G. Dickson
School of Biological Sciences, Division of Biochemistry, Royal Holloway College, University of London, Egham, Surrey TW20 0EX, U.K.

1 Introduction

Experimental gene transfer and clinical gene therapy approaches rely on the efficient delivery of genes to desired target cells. The first approved human gene transfer study began almost a decade ago using an RNA virus vector derived from a murine oncoretrovirus to deliver a bacterial marker gene to patients' cells (1). This was rapidly followed by protocols to treat severe combined immunodeficiency disease (SCID) and advanced melanoma. There are now hundreds of human gene therapy trials ongoing to treat many different classes of illness including inherited genetic disorders, cancer, virus infections, and cardiovascular disease, the majority of which use modified retrovirus vectors as gene delivery vehicles.

The functional organization of a number of RNA virus genomes and the relative simplicity of their structure have made retrovirus-based systems attractive. The RNA virus genome is converted into a double-stranded DNA provirus element, which is then stably integrated into host chromosomal DNA in a predictable manner that generally leaves host gene expression patterns unaffected. Virtually all human cells can be successfully infected by at least one type of retrovirus vector, and once integrated retrovirus promoters and enhancers can be used to drive the high-level expression of foreign transgenes. Finally, in terms of vector construction, the relative simplicity of retrovirus genomes allows proteins necessary for all aspects of infection to be provided entirely *in trans*. Thus it is possible to generate particles that are infectious, but replication-defective and which do not produce virus antigens, making them potentially very safe and controlled agents for gene therapy.

Retrovirus vectors derived from C-type mammalian oncoretroviruses, notably murine leukaemia virus (MLV), have been the prototype gene delivery system utilizing RNA virus biology (2–4), and this system is described in detail in Section 2. However, a major problem with these type-C retrovirus vectors is the requirement for proliferation in target cells for provirus integration. More recently, interest has developed in the use of vectors derived from the lentivirus family of RNA viruses, such as human immunodeficiency virus type-1 (HIV-1), that can infect non-dividing cells *in vitro* and *in vivo* (5–7). These lentivirus vectors are pseudotyped with the vesicular stomatitis virus G glycoprotein (VSV-G), and hence can transduce a broad range of tissues and can be concentrated to high titres. The production and properties of lentivirus vectors is described in Section 3.

2 Vectors based on murine retroviruses

Vectors based on the murine oncoretroviruses, and in particular on the Moloney murine leukaemia virus (MoMLV), have been the most extensively used tools for gene transfer so far. The relative simplicity of the functional organization of the MLV genome has allowed the generation of many different types of vectors with different cell tropisms, suitable for a large variety of gene transfer purposes. The advantages of such vectors are several. On infection, the retrovirus genome is stably integrated into the host-cell genome in a predictable orientation, whilst the normal cell functions are unaffected by the infection process (8). In principle, this feature is very valuable when genetic transduction of the stem cells of a particular tissue is required. The gene of interest, once inserted into the stem cells, should then be expressed by large populations of more differentiated cell types.

The simplicity of the MoMLV genome allows proteins necessary for infectious particle production to be provided *in trans* (9). As such, retrovirus vectors themselves do not invoke an immune response against the infected cells for they do not express any virus proteins. This is important for *in vivo* gene therapy applications where the immune system plays an important role in limiting the efficiency of stable gene transfer. There are, however, limitations in the use of retrovirus vectors. For instance, these vectors can only infect cells that undergo at least one round of division, for the integration of the virus genome is strictly dependent on the breakdown of the nuclear envelope (10, 11). In addition, retroviruses can accommodate only relatively small genomes (up to 10.5 kb), therefore transduction of large genes and their regulatory sequences (untranslated regions, large introns, etc.) is not possible. Retroviruses are also prone to recombination and rearrangements, so great care is needed to avoid the generation of replication-competent virus, especially when large stocks of vector are required. Finally, a problem of these vectors is the potential for insertional mutagenesis and accidental oncogene activation due to the randomness of integration in the host genome. However, the generation of vectors with tissue-specific tropism should reduce these risks substantially.

Retrovirus vectors can be used in a variety of experimental settings, including somatic gene therapy, cell-lineage analyses in developing animals (12), studies on the functional organization of tissues and organs (13), and the study of retrovirus biology. An outline of the features of existing vectors and methodologies will be given in the next sections of this chapter, to guide investigators in the choice of the retrovirus vector and packaging system best suited for the individual research goals.

2.1 Design and choice of MoMLV retrovirus vectors

The study of the acutely oncogenic retroviruses has provided the basis for the development of retrovirus vectors. The genome of oncogenic retroviruses is partially replaced by an oncogene. Since these replication-defective viruses cannot encode for most virus proteins, they are incapable of independent replication. However, if the virus proteins are provided *in trans* by concomitant infection of the wild-type virus, the defective virus genome can be rescued and infect new cells. A closer analysis of the acutely oncogenic retroviruses has revealed that the only sequences required *in cis* for retrovirus survival are the long terminal repeats (LTRs), the primer binding site, the polypurine tract, and the packaging signal. Therefore, recombinant retroviruses could be engineered to carry genes other than oncogenes, and serve as vectors with the proteins necessary for replication being provided *in trans* by the so-called packaging cells (9). A variety of different retrovirus vectors have been developed in recent years.

2.1.1 Single-gene MoMLV vectors

In these simple vectors, the coding regions for *gag-pol* and *env* are deleted and substituted with the gene of interest. Generally, a polylinker is present after the packaging signal to allow easy cloning of the cDNA in the vector backbone. Two such vectors are N2 and MFG (14, 15). A few simple rules should be borne in mind when using these vectors. The size of the insert should not be over 8 kb to avoid instability of the vector and low titres. Polyadenylation sites and introns (but there are exceptions) should be avoided. Promoters other than the LTR can be used, independent of orientation, but specific transcription cannot be guaranteed depending on the circumstances of individual cases (8). The expressed genes should be easy to detect in transfected/infected cells and these vectors are best suited when the target cells can be easily selected, i.e. by a fluorescent marker and fluorescence-activated cell sorting (FACS) analysis.

2.1.2 MoMLV vectors expressing multiple genes

The expression of both the gene of interest and a selectable marker within the same vector allows simple cloning of transduced cells or permits the fate of the transduced cells to be followed *in vivo* and their later recovery for analyses *in vitro*. There are two kinds of vectors widely employed that are capable of expressing two or more genes: double promoter vectors and internal ribosome entry site (IRES)-based vectors. In the double promoter vectors, one gene is expressed from the LTR and the other gene from an internal promoter placed downstream. The internal promoter can be either of virus origin, like the simian virus SV40 or cytomegalovirus (CMV) (12, 16, 17), or of cellular origin (18).

Sometimes the presence of two distinct promoters within the same vector can result in interference and partial suppression of one of them. In particular, the expression from the LTR can be suppressed when there is selection for the downstream promoter and vice-versa, although this effect is much stronger in spleen necrosis virus- (SNV) based vectors than in MoMLV-based vectors (19–22).

In any case, for practical purposes, it is important to check for the expression of both genes when using the double promoter vectors. An alternative strategy is to use IRES-based vectors to express polycistronic mRNAs. IRES elements allows translational initiation within an RNA transcript in a 5′ cap independent manner. IRES sequences have been inserted into retrovirus vectors between the genes of interest, so that the level of translation of the two genes is very close (23). IRES-based vectors containing up to three genes have also been successfully designed (24). In the IRES-based vectors the virus LTRs drive transcription of the bicistronic provirus. As such, the selection for the expression of one gene should ensure expression of the second gene at very similar levels.

2.1.3 Tropism of MoMLV retrovirus vectors

Choice of the right vector tropism is essential to obtain high levels of transduction and to restrict infection to the desired subset of cells. The tropism of the retrovirus vector is decided at two different levels: the type of *env* glycoproteins present on the virus surface and the enhancer and promoter present in the virus LTRs.

The *env* glycoproteins present on the virus envelope are responsible for the initial recognition and attachment to the cell receptor and subsequent virus internalization (25). Naturally occurring *env* glycoproteins of MLV can be ecotropic, amphotropic, xenotropic, and polytropic. The ecotropic and amphotropic *env* glycoproteins are the ones most commonly used for gene transfer procedures. The ecotropic *env* allows infection of mouse cells only (but murine hepatocytes seems to be refractory) (26), while the amphotropic *env* allows infection of mouse, rat, rabbit, dog, primates, and human cells (25). The xenotropic *env* has been used to successfully infect human hepatic cells *in vivo* (26), while the polytropic *env* allows infection of some human cell lines but not others.

Virus vectors can be pseudotyped, i.e. *env* glycoproteins of different viruses can be artificially incorporated into MoMLV particles. The incorporation of the VSV-G proteins into MoMLV vectors has broadened the virus tropism to cell types previously refractory to infection (27), and has allowed the efficient concentration of virus stocks by ultracentrifugation due to their high stability (28). MoMLV vectors can also be pseudotyped by chimeric *env* glycoproteins containing regions that recognize specific cell ligands to re-target virus infection (29, 30). However, the manipulation of the *env*-glycoprotein to redirect virus tropism is not a trivial procedure and incorporation of at least low levels of wild-type *env* seems to be necessary to restore infectivity. More recently, SNV-based vectors have been pseudotyped with chimeric *env*-glycoproteins to obtain specific cell targeting. Incorporation of the wild-type SNV *env* was still necessary to promote virus–cell membrane fusion and achieve productive infection. However, since the receptor for SNV virus in human cells has a very low affinity for its ligand, the resulting vectors pseudotyped with the chimeric SNV *env* glycoproteins maintained a specific tropism (31).

The enhancer and promoter in the virus LTRs are active in almost all cell types, although there are differences in the levels of expression. An exception is represented by embryonic carcinoma cells in which the LTR is not transcriptionally active (32, 33). Transcriptional activity of the virus LTR is not always predictable (particularly *in vivo*) and can often be short-term, e.g. in haematopoietic stem cells. The virus LTRs can also be modified to change the virus transcriptional tropism. Generally, the 3′ virus enhancer and promoter are deleted to obtain a self-inactivating vector (SIV) (34). Then a tissue-specific enhancer is inserted to substitute the virus enhancer. On infection, the tissue-specific enhancer in the virus 3′ LTR is copied to the 5′ LTR and the resulting vector has a tissue-specific transcriptional tropism. Many variants of this type of vectors are available, some of them contain tetracycline or steroid-inducible cassettes (35). The problem with these vectors is the relatively low titres. More recently it has been shown that the HTLV-I *tax*-responsive element can be inserted in the LTR of an SIV together with a tissue-specific enhancer. *Trans*-complementing packaging cells expressing *tax* allowed the production of high-titre virus stocks. In the absence of *tax*, however, the vector showed tissue-specific expression (36).

2.2 MoMLV packaging cell lines

The choice of appropriate packaging cell lines should be based on the following considerations:

- their ability to produce high-titre recombinant virus;
- their tropism; and
- their safety.

There are a number of well-characterized packaging cell lines which can produce high-titre vectors (see *Table 1*). In our hands, the most reliable ecotropic cell lines have proven to be AmpliGPE (37) and GP+E86 (38), and the most reliable amphotropic packaging cells proved to be PA317 (39) and the GP+envAm12 (40). AmpliGPE, GP+E86, GP+envAm12 are so-called 'split' packaging cells in which the *gag-pol* and *env* genes are encoded by separate plasmids stably transfected in a sequential manner. As such, three recombination events would be required to produce replication-competent virus. Many packaging cell lines are derived from murine fibroblasts and a potential problem using murine packaging cell lines is the co-packaging of endogenous murine retrovirus sequences with the vector, which can occasionally lead to the formation of replication-competent virus (41).

More recently, packaging cell lines have been developed using highly transfectable 293 cells (43, 44). The high transfection efficiency of 293 cells allows the production of high-titre vectors by transient transfection (45, 46). Moreover, 293 cells are of human origin and do not contain murine endogenous C-type retrovirus elements which could be co-packaged with the vector, further reducing the risk of contamination with replication-competent virus. For gene therapy applications requiring the *in vivo* administration of retrovirus vectors in primates or humans, a series of packaging cells producing virus-resistant to complement-mediated lysis have also been developed (44, 47).

Table 1 Retrovirus packaging cell lines

Name	Type	Host range	Maximum titer (c.f.u./ml)	Reference
ψ-2	y-Deletion	Ecotropic	10^7	9
PA317	Multiple deletions	Amphotropic	10^7	39
ψ-CRE	Split coding regions with deletions	Ecotropic	10^6	42
ψ-CRIP	Split coding regions with deletions	Amphotropic	10^6	42
GP+E86	Separation of coding regions	Ecotropic	4×10^6	38
GP+envAM12	Separation of coding regions	Amphotropic	10^6	40
AmpliGPE	Separation of coding regions, LTRs deleted	Ecotropic	5×10^6	37
Bosc-23	Separation of coding regions	Ecotropic	4×10^6	45
FLY	Separation of coding regions	Amphotropic/FeLV	10^7	47
ProPak-A	Separation of coding regions	Amphotropic	5×10^6	44

2.3 MoMLV vector production

To produce helper-free retrovirus vector stocks, plasmid DNA containing the provirus form of the vector is introduced into a packaging cell line. The packaging cell line provides all the proteins essential for virus replication *in trans*, while the plasmid DNA containing the specific virus packaging signal provides the RNA genome to be packaged into the recombinant particle. Virus stocks can be generated transiently or stably. The transient production of virus vectors has the advantage of being fast, so a large variety of constructs can be tested. Vectors expressing toxic genes can also be produced at high titres by transient transfection methods. The stable production of virus has the advantage of higher titres, a better characterization of the progeny virus vector, and the availability of an almost indefinite source of virus vector stocks. The essential requirement for the transient production of virus is the good transfectability of the packaging cell line. The BOSC/BING and the AmpliGPE lines offer two such examples. *Protocol 1* describes the transient production of high-titre virus stocks from AmpliGPE packaging cells.

To make stable producer cell lines, the packaging cells must be transfected or infected with the plasmid vector and individual clones selected. *Protocol 2* describes the steps to make stable producer cell lines.

Once the retrovirus producer cell clone with the highest titre has been identified, cells should be expanded and frozen in multiple aliquots. Continuous passaging of producer cells tends to reduce the titres, and as such, early passage stocks and continued monitoring are very essential. If the titres of the producer cell clone obtained by transfection are unsatisfactory, other packaging cells with a different *env* tropism can be infected with the recombinant virus stock. This procedure is described in *Protocol 3* and generally results in titres $>10^6$ c.f.u./ml.

Protocol 1

Transient production of MoMLV vectors from AmpliGPE cells

Equipment and reagents

- AmpliGPE packaging cell (37)
- Lipofectamine (Gibco-BRL)
- Dulbecco's modified Eagle's medium (DMEM) + 10% fetal calf serum (FCS) + 2 mM L-glutamine
- 0.5–1 mg/ml purified retrovirus vector plasmid in water
- Serum-free DMEM + 2 mM L-glutamine
- 60 mm dishes
- 0.45 μm filter

Method

1. Plate 1.3×10^4 cells/cm^2 in a 60 mm dish. Make sure that the cells are evenly dispersed in the dish.
2. The next day, transfect the cells by exposure to lipofectamine–DNA complexes by incubating the cells at 37 °C for 6 hours in the presence of 3 ml of DMEM (without FCS) containing a mixture of 10: μg/ml lipofectamine and 2: μg/ml supercoiled plasmid DNA prepared according to the manufacturer's instructions.
3. Add 3 ml of fresh DMEM + 20% FCS and incubate for 18–24 h.
4. Change the medium to 3 ml DMEM + 10% FCS.
5. Collect the supernatant after 36–48 h, when the cells are 100% confluent, and filter through a 0.45 μm filter. Store at −70 °C or use immediately to infect target cells.

Protocol 2

Isolation of stable MoMLV virus producer cell clones by DNA transfection

Equipment and reagents

- Packaging cell line and medium
- Lipofectamine (Gibco-BRL) or $CaPO_4$
- Selection antibiotic (e.g. G418 or hygromycin B)
- Trypsin
- 96-well microtitre cloning olates
- 100 mm dishes
- PCR or Southern blot equipment and reagents
- Equipment and reagents for Western blot analysis

Protocol 2 continued

Method

1. Thaw the packaging cells and apply the specific selection media to boost the expression of both *gag-pol* and *env* genes for 7–10 days.
2. Transfect the plasmid vector into the packaging cells as a $CaPO_4$-precipitate or lipofectamine complex.
3. At 72 h after transfection, trypsinize the cells and seed them into 96-well microtitre cloning plates at limiting dilutions. Generally, prepare at least three plates containing 1, 3 and 10 cells/well, respectively. Use selective medium (i.e. 400 μg/ml G418 for neomycin-resistant cells, 70 μg/ml hygromycin B for hygromycin-resistant cells).
4. After growth and selection, isolate resistant clones (at least 10–15) by trypsinization, expand them and divide in 3 × 100 mm dishes. One dish must be frozen, one dish used to isolate genomic DNA, and one dish used to perform Western blot analysis for the relevant transgene.
5. Check the integrity of the transfected provirus by PCR or Southern blot analysis.
6. Select those clones with an intact integrated provirus and evaluate them for expression of the gene of interest by Western blot analysis.
7. Collect the supernatant culture medium when the producer cells are 100% confluent, and use it to evaluate individual clones for virus production titres and the absence of helper virus (see *Protocol 5*).

Protocol 3

Isolation of stable MoMLV virus producer cell clones by cross-transduction between packaging cells of different tropism

Equipment and reagents

- Packaging cell line
- 60 mm dishes
- Virus stock
- DMEM culture medium (Gibco-BRL)
- 0.45 μm filter
- 24-well plates
- Polybrene

Method

1. Seed target packaging cells of different tropisms in 60 mm dishes (i.e. amphotropic if the virus stock is ecotropic or vice versa) so that 16–18 h later they will be about 50% confluent
2. After 16–18 h infect the cells with 5 ml of the virus stock in the presence of 8 μg/ml polybrene
3. Change the media 24 h later. After a further 24 h, passage the cells and re-plate them at the same density as in step 1.
4. A further 16–18 h later re-infect the cells as in step 2.

Protocol 3 continued

5. Repeat the procedure three to five times
6. Collect the supernatant when the cells are 100% confluent, filter (0.45 μm) and check virus titres.
7. If the virus titres are in the range of 10^5 c.f.u./ml, isolate individual clones by dilution-cloning (see *Protocol 2*). Isolate 20–30 clones, plate them in a 24-well plate and quickly screen them for virus titres
8. Isolate, expand, and freeze selected clones in multiple aliquots at −70°C.

In *Protocol 3* there is no need to screen the producer cell clones for the integrity of the provirus or expression of the proteins since this has already been done in *Protocol 2*. Infection of a second packaging cell line can also be performed using virus vector stocks obtained by transient transfection procedures. To do this, combine *Protocols 1* and *3*. It should be noted that in this case all the screening procedures described in *Protocol 2* must be performed on the producer cell clones isolated after infection with transiently produced virus.

2.4 Titration of MoMLV virus vector stocks

Direct determination of virus vector titres can be obtained by transduction of target cells and subsequent counting of the number of productive infection events. NIH 3T3 fibroblasts are the most commonly used target cells providing a standard for the measurement of virus titres. They are susceptible to amphotropic and ecotropic virus infection, but not to xenotropic virus or to virus pseudotyped with Feline leukaemia virus (FeLV) *env* glycoproteins. After infection of the target cells with serial dilutions of the virus stock, selection for the appropriate marker is applied and the number of resistant colonies counted. Virus titres are then expressed as colony-forming units per ml of virus-containing supernatant (c.f.u./ml). If the virus vector contains a readily detectable marker gene such as *lac-Z* or the green fluorescent protein (GFP) gene, an *in-situ* histochemical staining or a FACS analysis will allow the titration directly. Two procedures for virus titration are described in *Protocol 4*.

Protocol 4

Titration of MoMLV vector preparations

Equipment and reagents

- NIH 3T3 or other target cells
- DMEM culture medium (Gibco-BRL)
- 40-mm, 6-well cluster plates (NUNC)
- Polybrene
- Selective drug
- PBS pH 7.3
- Fixative solution: PBS, 1 mM $MgCl_2$, 0.5% glutaraldeyde
- X-Gal stain: 1 mM $MgCl_2$, 150 mM NaCl, 30 mM $K_3Fe(CN)$, 30 mM $K_4Fe(CN)_6$, 0.1% X-Gal in PBS

Protocol 4 continued

A. For vectors expressing a selectable marker

1. Plate 1×10^4 cells/cm^2 in culture media in 40 mm, 6-well cluster dishes.
2. After 16–18 h infect the cells with tenfold serial dilutions of vectors-containing supernatant in the presence of 8 μg/ml of polybrene.
3. Allow infection to occur for 24 h, then passage the cells and split 1:4–1:8. Apply drug selection 24 h later.
4. Wait for 7–10 days and then count the number of drug-resistant colonies and multiply by the dilution factor to yield the titre.

B. For vectors expressing expressing Lac-Z

1. Infect the cells as described in steps 1 and 2 above.
2. After a 48-h infection, wash the cells in PBS, add the fixing solution to cover the cells and incubate at room temperature for 10 min.
3. Wash the cells three times for 5 min each time with PBS.
4. Add X-Gal staining solution and incubate in an humidified atmosphere at 37 °C for 2–24 hours.
5. Count the number of blue-staining cells and multiply by the dilution factor.

Protocol 5

Concentration of ecotropic and amphotropic MoMLV retrovirus vectors

Equipment and reagents

- 150 mm dishes
- Hepes-buffered DMEM (Gibco-BRL)
- TNE buffer: 10 mM Tris–HCl pH 7.4, 100 mM NaCl, 1 mM EDTA
- Culture medium
- Selective agent
- 0.45 μm filter
- 70% sucrose in TNE buffer
- Producer cells
- Ultracentrifuge and Beckman SW27 or SW28 rotors (or equivalent)
- Beckman ultraclear tubes (25 × 89 mm)
- Centricon 100 or 500 filters (Amicon)

Method

1. Plate producer cells in 150 mm dishes in the presence of the selective agent.
2. When cells are 80-90% confluent (24–36 hours) replace the culture medium with 15 ml of fresh Hepes-buffered DMEM without the selective drug.
3. Incubate the cells for 16–18 h at 32 °C instead of 37 °C and then harvest the supernatant. Filter the supernatant through a 0.45 μm filter.

Protocol 5 continued

4. Place the virus-containing supernatant (30 ml) at 4 °C in a Beckman 25 × 89 mm ultra-clear tube underlaid with 3 ml 70% sucrose in TNE buffer.
5. Spin in a SW27 or SW28 rotor in a Beckman ultracentrifuge at 18 000 r.p.m. for 4 h at 2 °C.
6. Pierce a hole in the bottom of the tube and let the contents drain dropwise by gravity flow. Virus, which is located in the third and fourth ml, is then collected. If sucrose is a problem for subsequent applications, dilute the fractions 3–5 times in TNE buffer and concentrate by spinning at 4 °C in Centricon 100 or 500 filters (Amicon) following the manufacturer's instructions.

2.5 Concentration of MoMLV virus vector stocks

For *in vivo* gene transfer procedures it is often useful to concentrate the virus to small injection volumes. Unfortunately, amphotropic and ecotropic *env* glycoproteins are very sensitive to mechanical stress and infectious virus is often degraded during ultracentrifugation steps. However, when gentle conditions are used it is still possible to concentrate the virus 10–20-fold. If very concentrated stocks are needed, then the vector should be pseudotyped with VSV-G glycoprotein and ultracentrifuged (28). Using this procedure a 1000-fold concentration can be achieved with a high yield of infectious particles. VSV-G glycoproteins are toxic and must be expressed transiently in the packaging cells. Attempts to make packaging cells expressing VSV-G in a tightly inducible way have met with limited success so far. *Protocol 5* describes a method to concentrate ecotropic and amphotropic retrovirus vectors.

2.6 Detection of replication-competent MoMLV (RCR) vector contamination

Replication competent retroviruses (RCR) can arise by recombination between the homologous region of the provirus vector and the *gag-pol* and *env* plasmid transgenes in the packaging cells. They may also arise by recombination during reverse transcription if homologous or partially homologous sequences are co-packaged. All newly made retrovirus producer cells must be screened for the presence of RCR. Moreover, virus vector stocks must be regularly checked because RCR can arise while the producer cells are passaged in culture. The importance of stringent screenings for the presence of RCR must not be underestimated, even when split packaging cell lines are used for the production of the vector (indeed, RCR has recently been detected in a split producer cell line) (48). RCR can cause T-cell neoplasm in primates (49) and, in principle, could recombine with endogenous C-type retrovirus sequences giving rise to new virus variants.

One of the most sensitive methods for the detection of RCR is the mobilization assay. In this assay virus stocks are used to infect a population of indicator cells containing an integrated recombinant provirus expressing an easily detectable marker (i.e. nuclear localizing *Lac-Z*). The infected indicator cells are passaged for 1–2 weeks and the supernatant collected and used to infect fresh target cells such as NIH 3T3 fibroblasts. If RCR is present in the original virus stock it will infect the indicator cells and mobilize the provirus RNA containing the marker. Subsequent infection of fresh target cells will reveal the mobilized provirus by histochemical staining. *Protocol 6* describes the mobilization assay.

Protocol 6

Assay for replication-competent MoMLV virus

Equipment and reagents

- Indicator cells with an integrated provirus expressing nuclear localizing LacZ
- NIH 3T3 fibroblasts or other target cells
- DMEM cell culture medium (Gibco-BRL)
- 100 mm dishes
- 0.45 μm filter
- Fixing solution and X-Gal staining solution (see *Protocol 4*)
- PBS

Method

1. Plate 5×10^5 indicator cells/100 mm dish, and incubate at 37 °C.
2. After 16–18 h infect the indicator cells with 1–5 ml of the virus-containing stock.
3. Incubate for 48 h, then split the cells 1:4–1:8, let them reach 100% confluency and split again 1:4–1:8. Passage the cells in this way for 1–2 weeks.
4. Collect the supernatant, filter through a 0.45 μm filter, and infect fresh NIH 3T3 cells or other target cells which were plated 24 h previously in 100 mm dishes.
5. Replace the culture medium 16–18h later
6. The following day fix and then stain with X-Gal (see *Protocol 4*).

2.7 Infection of target cells *in vitro* with MoMLV vectors

Once high-titre producer cell clones with the appropriate tropism are available, different strategies can be used to infect target cells with high efficiency. As a general rule, primary cells isolated from animal tissues are less infectable than established cell lines, and repeated rounds of infection with the virus vector or co-cultivation with producer cells is necessary to achieve high-level transduction. This effect is likely to depend on lower levels of expression of the virus receptor and/or on lower rates of cell cycling in primary cell cultures. Since retrovirus integration is dependent on mitosis, the right conditions to maximize cell proliferation *in vitro* must be established. Often, a particular combination of substrate to coat the culture dish and optimization of serum and/or growth factors in the culture medium is required to give the best results. The density at which the cells should be plated must be decided empirically, and this varies from cell type to cell type. It is important to infect cells during the exponential growth phase to ensure maximum transduction levels. The infection procedure can be repeated several times, although the infected cells should be monitored for early signs of toxicity. For the infection procedure, refer to *Protocol 3*.

Co-cultivation of the target cells with the producer cells is the most effective procedure for obtaining high levels of transduction. However, the target cells must be passaged for at least 1 week and some primary cell types (e.g. hepatocytes) cannot be used. *Protocol 7* describes the co-cultivation procedure. Moreover, the conditions in which the target cells grow well must also be tested for the maintenance of producer cells to ensure continued high virus vector output.

Protocol 7

Target cell transduction by co-cultivation with MoMLV producer cells

Equipment and reagents

- Suitable culture medium (e.g. DMEM, 10% FCS)
- Vector producer cells
- Target cells
- 100 mm dishes
- Trypsin (0.025% in PBS)
- Polybrene
- Mitomycin C (Sigma)

Method

1. Plate 5×10^5 producer cells/100 mm dish in DMEM, 10% FCS containing 2 g/ml mitomycin C. Take care not to expose mitomycin C to light. Incubate the cells for exactly 16 h.
2. Trypsinize the producer cells and wash once by centrifugation in DMEM. Trypsinize the target cells and mix with the producer cells at 3:1, 1:1, and 1:3 ratios. Plate on to 100 mm dishes in the presence of 8 μg/ml polybrene.
3. Passage the cells for 1–2 weeks in suitable medium.[a]
4. Check the levels of transduction of the target cells by appropriate histochemical or immunocytochemical staining, or by enzymatic assay on an aliquot or replicate culture.

[a] The producer cells treated with mitomycin C will slowly die over this period.

Drug selection can be applied to the infected cells to enrich for the transduced population. It should be noted, however, that drug selection may change some of the properties of the primary target cells. For example, selection of transduced primary myoblasts with G418 greatly reduces fusion and myotube formation. The dose of mitomycin C used to treat producer cells, as well as the time of exposure to the drug, can be optimized depending on the individual needs. Higher doses of mitomycin C will result in faster cell death, which sometimes may be useful if the target cells cannot be passaged for 2 weeks. A titration curve for the dose of mitomycin C on NIH 3T3-based producer cells can be found in Fassati *et al.* (50).

2.8 Direct transduction of target cells *in vivo* with MoMLV vectors

Transduction of the target cells *in vivo* with retrovirus vectors is a much more problematic goal than *in vitro* approaches since high virus titres are required ($>10^7$ c.f.u./ml). Retrovirus particles have a relatively large diameter (100 nm) and cannot diffuse freely outside the vascular bed. The liver can be perfused more easily with retrovirus vectors due to the presence of capillaries with large pores (the so-called fenestrated capillaries), and solid tumours can sometimes be infected by systemic intravenous injection of vector particles, probably due to the aberrant structure of neoplastic capillaries. In general, however, intravenous injection of virus yields only poor transduction efficiencies. It is possible to inject the concentrated virus vector stock directly into the tissue of interest, by-passing the vascular bed. In this case, the number of dividing cells at the time of infection is the critical factor since the half-life of retroviruses *in vivo* is probably very short. Mouse serum contains neutralizing antibodies against ecotropic *env* glycoproteins (51) and the complement in primate and human sera can lyse MLV particles very quickly (52). Much higher transduction efficiencies can be obtained *in vivo* by injecting mitotically inactivated, retrovirus producer cells (53–55). The producer cells injected *in vivo* survive a few days while releasing virus. As such, a higher number of dividing cells can be targeted. However, immunosuppressive strategies must be designed to avoid the immunorejection of the implanted cells and other adverse inflammatory reactions. *Protocol 8* describes this procedure.

Protocol 8

Implantation of MoMLV producer cells for *in vivo* transduction of target cells

Equipment and reagents

- Producer cells
- DMEM culture medium, 10% FCS, 2 mM L-glutamine
- Hepes-buffered culture medium (serum-free)
- 100 mm dishes
- Mitomycin C (Sigma)
- Versene (5 mM EDTA in PBS) pH 7.3
- Trypsin (Gibco-BRL)

Method

1. Plate 1×10^6 producer cells/100 mm dish in culture medium containing 1–1.5 μg/ml mitomycin C. Take care to minimize exposure to light. Incubate the cells in the dark for 16 h.

Protocol 8 continued

2. Rinse the cells in culture medium then twice briefly in versene (5 mM EDTA in PBS) and add 1 ml/100 mm dish of 0.025% trypsin in versene. Resuspend the trypsinized cells in 20 ml/100 mm dish with serum-free culture medium and pellet the cells.
3. Resuspend the cells (1×10^6 cells/100 μl) in Hepes-buffered culture medium (serum-free) and keep on ice. The cells are now ready to be injected into the tissue of interest.

Cells can survive a few hours on ice and should survive a few days after implantation. We recommend a temporary immunosuppressive regimen to avoid inflammatory responses, which should be started 1 day before treatment. In our hands, mice responded very well to 6 days of treatment with 2.5 mg/kg tacrolimus (FK506, Fushisawa Pharmaceuticals) injected intraperitoneally starting 1 day before cell implantation (53).

An alternative strategy for *in vivo* retrovirus-mediated transduction, which has been proposed recently, is to engineer macrophages or other migratory blood-cell types that have the natural ability to extravasate, to become transient retrovirus producer cells. The engineered cells can be injected intravenously and may then be able to home into sites of inflammation/degeneration and deliver vector particles *in situ* (56). Essential to the success of this strategy is the design of virus vectors with a tropism restricted to specific cell populations.

3 Retrovirus gene transfer vectors based on lentiviruses

Until recently, MoMLV-based retrovirus vectors have been most frequently used as vehicles for the transport of foreign genes of interest to targeted cellular sites. Section 2 of this chapter described in detail the use of MoMLV-based retrovirus technology in relation to their efficient transfer, stable integration, and relatively long-term expression of foreign genes. However, a major drawback of these vectors is their inability to infect post-mitotic cells, which are proliferatively inactive. Such cell types include neurons, hepatocytes, muscle fibres, and non-dividing haematopoietic cells (10, 57, 58). Therefore, one might legitimately ask if the potentially dangerous lentiviruses hold any distinctly favourable advantages over the clinically used oncoretroviruses, as gene delivery vehicles. In general, retroviruses have a strong requirement for dividing cells since the synthesis of virus DNA and integration of the provirus genome is dependent on the breakdown of the nuclear envelope (10, 59). Lentiviruses, on the other hand, replicate efficiently in non-dividing cells both *in vitro* and *in vivo* (5, 6).

The lentivirus genome (see *Figure 1*), like the MoMLV genome, contains the three structural genes, *gag, pol*, and *env*, typical of all retroviruses. In addition,

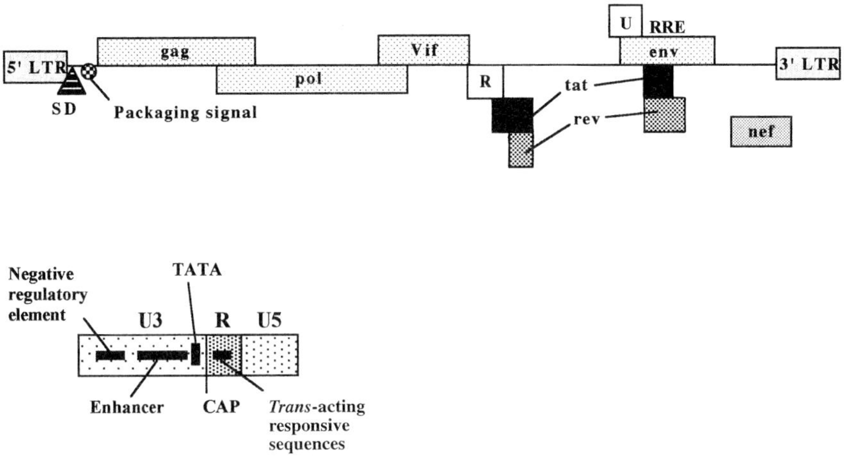

Figure 1 The genomic organization of the lentivirus. Rectangles (apart from the LTRs) are open reading frames (ORFs) with respective protein products indicated in each ORF. Bottom left shows a 5′ LTR comprising a U3 region (containing enhancer, promoter, and regulatory elements), an R region (containing the cap site which is the 5′ terminus of all virus RNAs), and a U5 region.

lentiviruses have a number of unique small open reading frames (ORFs) located between *pol* and *env* and at the 3′ terminus. These ORFs code for regulatory proteins. HIV-1 has six such genes, *vif, vpr, vpu, tat, rev,* and *nef* (60, 61). The gene *vif* (virion infectivity factor) facilitates virion maturation; *vpr* (virus protein regulatory) arrests cell proliferation and contains a nuclear localization signal (NLS); *vpu* (virus protein unknown) downregulates CD4 and promotes virus budding; *tat* (trans-activator of virus transcription) binds tat-responsive RNA element (TAR) and upregulates virus transcription; *rev* (regulator of virus expression) binds rev-responsive RNA element (RRE) and regulates virus RNA transport and splicing; *nef* (negative factor) downregulates CD4, binds cellular kinases, and is essential for virus disease induction.

Thus, any HIV-based minimal vectors for gene therapy purposes can only be safe and efficient if they lack the pathology-associated accessory proteins without affecting production of the vector or efficiency of infection. Investigators have therefore concentrated on eliminating non-essential proteins, whilst at the same time broadening the tropism of the virus.

Lentivirus vectors have been made with a transgene enclosed between the LTRs and injected into rodent brain, muscle, liver, eye, or pancreatic-islet cells with sustained expression over 6 months without the transcriptional 'shut-off' of the transgene commonly observed in MoMLV retrovirus vectors. Injection of up to 10^7 infectious units does not evoke a cellular immune response at the site of injection and, furthermore, there is no potent antibody response (6, 62). So, lentivirus vectors seem to offer an alternative mechanism for *in vivo* gene delivery with sustained expression.

Table 2 Lentiviruses

Virus	Host species
Equine infectious anaemia virus (EIAV)	Horse
Maedi-visna virus	Sheep
Progressive pneumonia virus	Sheep
Caprine arthritis-encephalitis virus (CAEV)	Goat
Bovine immunodeficiency virus (BIV)	Cattle
Simian immunodeficiency virus (SIV)	Macaque
Simian immunodeficiency virus (SIV)	African monkeys and baboons
Human immunodeficiency virus-type 1 (HIV-1)	Human
Human immunodeficiency virus-type 2 (HIV-2)	Human

(Taken from ref. 7)

3.1 Lentivirus vector sources and design

The ultimate goal of vector choice and design is to be able to develop vectors that can enter specific cells defective in a particular gene(s), integrate precisely into the genome of the host cell, and provide a regulated gene product. HIV, and other lentiviruses, can infect non-dividing cells (see *Table 2*). However, because of the safety issues surrounding the use of AIDS-related virus, a number of efforts have been made to develop HIV type 1-based packaging systems which would generate a replication-defective vector (63). Safe HIV type 1 vector systems have been constructed to contain minimal essential genes and exclude any auxiliary genes in addition to the three common retrovirus genes *gag*, *pol* and *env*. Unlike retrovirus systems, efficient packaging cell lines for lentivirus vectors have been difficult to establish (64, 65) and subsequently, three-plasmid expression systems have been used to generate replication-deficient but infective provirus after transient transfection *in vitro*.

HIV-based lentivirus vector production utilizes a three-plasmid expression system. The HIV provirus sequence (see *Figure 2*) is de-constructed, and its various components expressed from individual plasmids to construct a replication-deficient virus harbouring the transgene of interest.

Plasmid pCMVΔR9, the packaging construct (see *Figure 3*), forms the backbone of the virus system. It contains the human cytomegalovirus (hCMV)

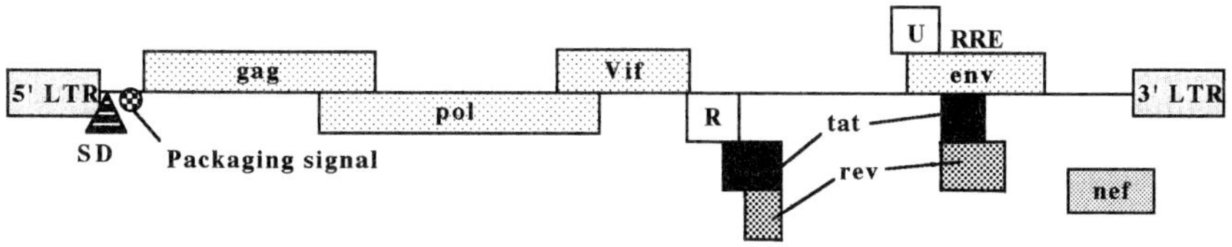

Figure 2 The HIV provirus. The coding region of the virus proteins including the accessory proteins is shown. The splice donor site (SD) and the packaging signal are indicated. The three-plasmid expression system takes the provirus and uses its various parts to construct a replication deficient virus with the appropriate insert sequence.

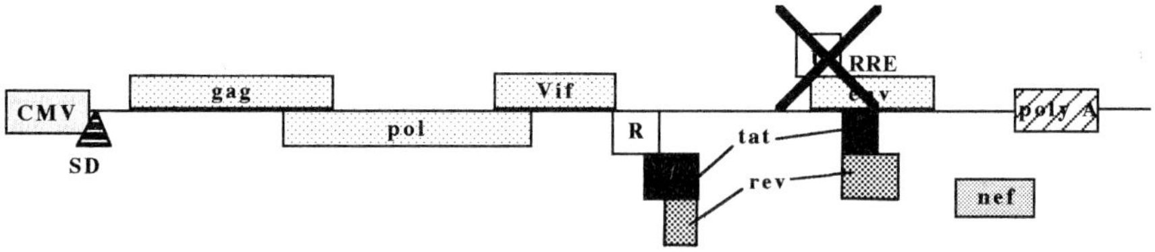

Figure 3 The packaging construct. The 5′ packaging signal has been removed and the 3′ LTR has been replaced with a polyadenylation signal. The reading frames of *env* and *vpu* are blocked (X). In this way, the packaging parts of the virus have been separated from the sequences that activate them, so that the construct can only produce the necessary envelope proteins to package the virus in the presence of the signal located on another vector.

immediate-early promoter, which drives the expression of virus proteins *in trans*. The 5′ untranslated region has been changed to preserve the 5′ splice donor site and delete the packaging signal (Ψ) and adjacent sequences. This virus is defective for the production of the virus envelope and *vpu*. The 3′ LTR sequence has been substituted by a polyadenylation signal (poly A) at the end of the *nef* reading frame (66). This particular design has also eliminated *cis*-acting sequences, thereby separating the packaging parts of the virus DNA from the sequences that activate them (67).

The second plasmid of the system, which broadens the tropism of the vector, encodes for a heterologous envelope protein. There are two variants of the envelope used (see *Figure 4*). The first of these transcribes the amphotropic envelope of MoMLV. The other codes for the VSV-G glycoprotein (68). These serve to pseudotype the particles generated by pCMVΔR9, and the latter envelope offers an additional advantage of high stability allowing efficient particle concentration by ultracentrifugation.

The third plasmid is the transfer or transducing vector (see *Figure 5*). This plasmid includes both the 5′ and 3′ LTRs of the original provirus, along with *cis*-acting sequences required for packaging, reverse transcription, and integration. There are also unique restriction sites enabling cloning insertion of a desired gene of interest. Nearly 350 base pairs of the *gag* sequences as well as the Rev

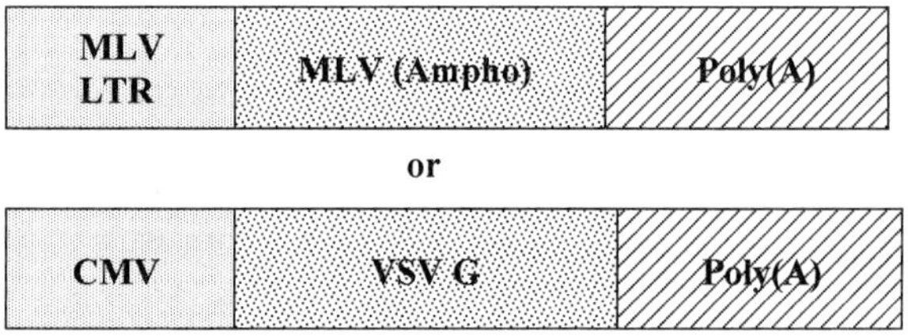

Figure 4 The envelope-coding plasmids. In the *env*-coding plasmid, the VSV-G coding region is flanked by the CMV promoter and a poly(A) site. The amphotropic envelope of MLV is flanked by an MLV LTR and a SV40 poly(A) site. These serve both to bring together the particles generated by the packaging construct and to provide stability to the vector produced.

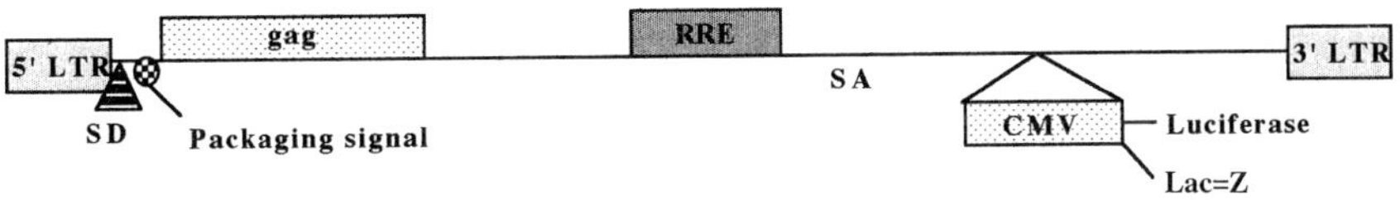

Figure 5 The transfer vector incorporates both the 5′ and 3′ LTR sequences removed from the original provirus when making the packaging construct. The promoter before the *gag* sequences has been removed along with most of the original envelope proteins. The *gag* gene is truncated and out of frame, and the internal CMV promoter is used to drive expression of a reporter gene such as lac-Z or luciferase. The RRE and splice acceptor site (SA) are indicated.

response element (RRE) of the *env* sequences, flanked by splice signals, were included to increase packaging efficiency (64, 69, 70).

Productive assembly and release of replication-deficient virus is obtained by co-transfection of the three plasmids on 293T cells (71). The use of a three-plasmid combination, heterologous envelope pseudotyping, and elimination of *cis*-acting sequences from the packaging vector, makes it highly unlikely that a replication-competent recombinant virus could emerge. The HIV-based vectors can be successfully produced for the transduction of dividing as well as non-dividing cells and long-term transgene expression. This HIV-based lentivirus vector system has been compared with conventional MoMLV-derived transducing vectors (63, 72).

3.2 Packaging and pseudotyping constructs for lentivirus vectors

In lentiviruses, there is still doubt over the exact regions necessary and sufficient for packaging (the Ψ region). There is a recognized Ψ component in the untranslated leader region between the splice donor (SD) and the *gag* initiation codon. However, the 5′ and 3′ extent of this are not known and a minimal essential packaging signal has not been defined. Instead, most investigators choose to package with subfragments of the region (64). Efforts to establish stable producer cell lines have been made (73, 74), and one group of workers have managed to generate stable HIV-1 packaging lines that constitutively express high levels of HIV-1 structural proteins (65).

It has been long been known that co-infection of a cell with retrovirus and vesicular stomatitis virus (VSV), a member of the Rhabdoviridae family, produces progeny pseudotypes in which the genome of one virus is encapsidated by the envelope protein of the other virus (75). The host range of the pseudotyped virus is that of the virus donating the envelope protein. VSV-G pseudotyped vectors can withstand the shearing forces encountered during ultracentrifugation, and by this process can be efficiently concentrated to titres of $>10^9$ c.f.u./ml (76), without significant loss of infectivity (77, 78). This ability to concentrate VSV-G pseudotyped vectors has facilitated gene therapy model studies and other gene transfer experiments that require direct delivery of vectors *in vivo*. Use of pseudotyped vectors to transduce non-dividing neuronal cells *in vivo* has been achieved, including sustained long-term expression in adult rat brain (63). Other advantages of pseudotyping with VSV-G are a broad host range tropism (79), and elimination of potential homologous recombination events that may generate replication-competent virus.

3.3 Production and concentration of recombinant vector

To produce lentivirus vector stocks, a three plasmid co-transfection system is used on 293T cells. In the absence of packaging cell lines, infective vector is produced by transient transfection of an envelope-coding plasmid, a *gag/pol*-coding plasmid, and a plasmid expressing the relevant genome. For 293T cells, the calcium phosphate method of transfection is suitable, although other suitable polycation-based reagents such as Transfast (Promega), lipofectamine (Gibco), etc. may also be used. Transfection with co-precipitates of calcium phosphate and DNA give peak transient gene expression after 48 hours, and the procedure is described in *Protocol 9*.

Protocol 9

Calcium phosphate-mediated transfection of 293T cultured cells with minimal lentivirus expression vectors

Equipment and reagents

- Lentivirus envelope, transfer, packaging plasmid DNAs
- Dulbecco's modified Eagle's medium (DMEM) containing 10% (v/v) FCS, 2 mM L-glutamine
- 100 mm dishes
- Promega Profection kit (calcium phosphate transfection kit) (this contains nuclease-free water, 2 M $CaCl_2$, 2 × Hepes-buffered saline (HBS))
- 1 M sodium butyrate
- 293T cells (human embryonic kidney cells containing SV40 T-antigen)
- 50 ml syringe
- 0.45 μm filter
- Beckman 15 ml and 50 ml centrifuge tubes
- PBS
- 20% (w/v) sucrose in PBS
- Beckman SW28 and SW40 rotors

Method

1. Seed 10 × 100 mm dishes with 3 × 10^6 cells and incubate overnight at 37 °C.
2. Three hours prior to transfection, remove the medium from the cells and replace with fresh growth medium.
3. Prepare a three plasmid mixture for transfection in the ratio of 200 μg (genome): 150 μg (*gag/pol*):60 μg (*env*).
4. Add nuclease-free water (from the Promega Profection Kit) to 4.38 ml total volume.
5. Add 620 μl of the 2 M $CaCl_2$ (from the Promega Profection Kit), to yield a total volume of 5 ml DNA/$CaCl_2$ solution.
6. In two lots, add a maximum 2.5 ml $CaCl_2$/DNA solution to 2.5 ml 2 × HBS (from the Promega Profection Kit). Add slowly in a dropwise manner whilst gently vortexing, to allow a precipitate to form.[a]

Protocol 9 continued

7. Leave the transfection mix for 30 min at room temperature to allow the precipitate to form.
8. Vortex the solution again prior to adding it to the cells. Add 1 ml of the transfection mix per 10 cm dish and swirl the dish to allow even distribution of the precipitate over the cells.
9. Leave the dishes for 6 h at 37 °C; remove the transfection mixture and replace with 5 ml of fresh medium
10. Leave the dishes overnight at 37 °C. **NB**: Always pipette the medium down the side of the dish so as not to flush cells off the plate surface.
11. The following day, add 50 μl of 1 M sodium butyrate[b] directly into each dish.
12. Leave the dishes for 4 h at 37 °C, then replace the spent medium with 7 ml fresh medium.
13. Leave the dishes overnight at 37 °C.
14. Collect the medium from multiple plates with a 50 ml syringe, place a 0.45 μm filter on the syringe nozzle and filter into Beckman 50 ml centrifuge tubes. Put aside 1 ml of un-concentrated virus for titration.
15. Place 7 ml of fresh media on to Petri dishes for the second harvest tomorrow, and return dishes to incubator.
16. Underlay 5 ml of a 20% (w/v) sucrose in PBS cushion per tube by placing the tip of a pipette to the bottom of the tube and dispensing slowly.
17. Top up the tubes with PBS and balance the tubes.
18. Spin for 90 min at 20 000 r.p.m. and 4 °C in a Beckman SW28 swing-out rotor.
19. Remove the tubes and aspirate the supernatant leaving a small volume over the pellet which may be difficult to visualize.
20. Resuspend the pellet in 1 ml of PBS and leave on ice for 1 h.
21. Transfer the resuspended virus to a Beckman 15 ml centrifuge tube; rinse the original 50 ml tube with 1 ml PBS and pool into the 15 ml tube. Top up the tubes with PBS and balance.
22. Spin for 75 min at 20 000 r.p.m. and 4 °C in a Beckman SW40 swing-out rotor.
23. Decant the supernatant and resuspend the final virus pellet in 200 μl PBS; the pellet should be visible at this stage.
24. Leave on ice for 1 h. Resuspend the pellet every 15 min using a micropipette to get a good suspension.
25. Store in 10 μl aliquots at −80 °C. Do not thaw and re-freeze.
26. Repeat steps 1–12 for the second harvest.

[a] The solution should appear slightly opaque due to the formation of a fine calcium phosphate-DNA co-precipitate.

[b] Sodium butyrate induces promoter activity and gives higher transgene expression.

The titre of the virus vector stocks can be determined by infecting the target cells and subsequently counting the cells or derived colonies carrying a selectable or detectable marker. Virus is routinely titred on 293T cells (human embryonic kidney), D17 cells (dog osteosarcoma), or HeLa cells. Serial dilutions of concentrated virus stocks are placed on appropriate cells to allow for infection. Added polybrene can aid in virus infection. Selection is then applied for the marker protein and the number of cells or colonies staining positive are counted. Virus titres are expressed as the number of positive infected cells/colonies per ml of stock. Selectable markers include Lac-Z and green fluorescent protein (GFP) genes. Direct determination of the vector can also be performed *in situ* if antibodies to the vector-encoded transgene product are available. Below is the protocol for using a Lac-Z marker gene.

Protocol 10

Titration of lentivirus vectors stocks

Equipment and reagents

- 293T cells or other target cells
- 12-well tissue-culture dishes
- DMEM culture medium (Gibco-BRL)
- PBS pH 7.3
- 4 mg/ml polybrene stock in PBS
- Formaldehyde (37%)
- X-Gal staining solution: 33 mM $KFe_3(CN)_6$, 33 mM $KFe_4(CN)_6$, 5 mM $MgCl_2$, 1 mg/ml X-Gal in PBS

Method

1. Seed 1×10^5 cells/well in a 12-well tissue culture dish overnight at 37 °C.
2. Next day, prepare serial dilutions of virus stock to be titred in culture medium containing 8 μg/ml polybrene.
3. Remove the multiwell dish containing the cells from the incubator and aspirate the medium.
4. Overlay the cells with 500 μl of the respective virus dilution.
5. Replace the multiwell dish in the incubator for a minimum of 4 h (overnight achieves better infection efficiency).
6. Remove the dish, add 1 ml fresh medium per well, and leave for 48 h at 37 °C.
7. Aspirate the medium and wash the cells once with 1 ml PBS.
8. Add 2 ml of 4% formaldehyde in PBS (v/v) per well and leave for 5 min.
9. Aspirate and wash the cells twice with 1 ml PBS.
10. Add 2 ml of the X-Gal staining solution per well and leave overnight at 37 °C.
11. Next day, score the most convenient dilution for the number of Lac-Z forming units (blue cells), multiply by the dilution factor, and express the result as the number of Lac-Z forming units/ml.

3.4 *Ex vivo* and *in vivo* transduction

For *ex vivo* transduction, primary cells can be removed from an animal under sterile conditions, established in culture, then transduced. Also, transduced primary cultures can be used as a method of gene delivery for gene therapy by returning them back to the host. In line with the interests of the authors, the following protocols describe the transduction of a muscle cell line C2C12, and the culture and transduction of primary myogenic cells. Primary cultures derived from skeletal muscle consist mainly of myoblasts and fibroblasts. Myoblasts eventually become the dominant population after 1–2 weeks of passaging and selective growth in culture, with nearly 100% staining positive for the myoblast-specific protein, desmin (80). Primary myoblasts can be isolated from mice of any age, but greater yields of myogenic cells are obtained from neonatal mice.

Protocol 11

In vitro transduction of myoblast and myotube cultures from an established C2C12 mouse muscle cell line

Equipment and reagents

- Mouse C2C12 myoblast and myotube cultures
- 12-well tissue-culture plates
- DMEM + 10% FCS + 2 mM L-glutamine
- Lentivirus vectors of choice
- 4 mg/ml polybrene
- Fetal calf serum (FCS)
- PBS
- 4% formaldehyde in PBS
- X-gal staining solution: 33 mM $KFe_3(CN)_6$, 33 mM $KFe_4(CN)_6$, 5 mM $MgCl_2$, 1 mg/ml X-Gal in PBS

Method

1. Seed a 12-well tissue culture plate with 5×10^4 cells/well (50–80% confluent) and leave overnight at 37 °C.
2. Next day, prepare 10-fold serial dilutions of a previously titred lentivirus virus stock in DMEM (undiluted virus titre should be $> 10^6$/ml).
3. In each dilution, supplement the lentivirus mixture with polybrene to a final concentration of 8 μg/ml.
4. Overlay the cells with 500 μl of the respective virus dilution per well, and in triplicate.
5. Follow steps 5–10 as described in *Protocol 10* above.
6. After infection with the virus carrying the Lac-Z reporter gene, score the number of blue cells.
7. (Optional) As a positive control, use an MLV-based retrovirus to compare retrovirus and lentivirus transduction efficiencies.

Protocol 12

Ex vivo transduction of primary myoblasts and myotubes

Equipment and reagents

- Primary myoblast and myotube cultures
- Lentivirus vectors of choice
- 12-well tissue-culture plates
- DMEM + 10% FCS + 2 mM L-glutamine
- 4 mg/ml polybrene
- Fetal calf serum (FCS)
- PBS
- 4% formaldehyde in PBS
- X-gal staining solution: 33 mM $KFe_3(CN)_6$, 33 mM $KFe_4(CN)_6$, 5 mM $MgCl_2$, 1 mg/ml X-Gal in PBS

Method

As described in *Protocol 11*, steps 2–6.

The transduced myoblasts can be injected back into the skeletal muscle to look at *in vivo* expression (81). Alternatively, muscle tissue can be transduced directly with infective provirus. The injection of infective lentivirus is an alternative to myoblast transplantation for the introduction of exogenous genes into skeletal muscle (81). The injected lentivirus will infect host skeletal muscle fibres via its ecotropic envelope. After the transfer vector is constructed with the desired transgene, or, using a reporter gene, 293T cells are transfected with three plasmids and the packaged virus is collected. The virus is concentrated to achieve titres of $> 10^7$ infectious units/ml, then injected into the host animal. Efficient delivery of a transgene using an HIV-based expression system has been described (63, 82). This vector showed little toxicity at the site of injection, and sustained expression for up to 6 months.

3.5 Safety considerations in the use of lentivirus vectors

Are HIV and other lentiviruses a good idea for gene therapy? HIV is a major human pathogen currently propagating the AIDS epidemic, and any intentional introduction of HIV or related lentivirus vectors into human patients provokes serious ethical issues. The major concern is the generation of replication-competent retroviruses (RCRs) during the process of vector preparation or indeed within the host subjects. With MoMLV-based vectors, homologous and/or non-homologous recombination can occasionally occur between retrovirus vectors and virus *trans*-complementing genes to generate RCRs (83). For this reason, the use of the HIV envelope in vector particles is likely to be unacceptable. On the other hand, although the use of a heterologous envelope would eliminate the risk of homologous recombination, there remains a potential to generate RCRs by non-homologous recombination (84). As yet, such recombination has not been observed in a laboratory setting or elsewhere, although it might be more likely in the large scale-ups required for clinical trials. Furthermore, the use of

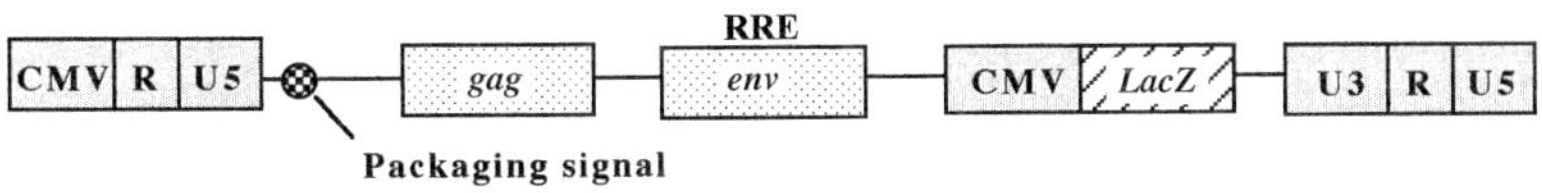

Figure 6 HIV-1 based vector constructed with the potent hCMV promoter replacing U3 of the 5′ LTR. The *tat* expression is eliminated due to a deletion encompassing the first 50 bp of the *tat* gene. A frameshift mutation prevents translation of *gag* sequences present and remaining *gag-pol* sequences have been deleted. A second deletion has removed part of *env*. The remaining HIV-1 sequences in the vector include RRE and *rev*. The β-galactosidase reporter gene is expressed from an internal hCMV promoter.

VSV-G is of some concern since it can encapsidate foreign RNA to generate pseudoviroids (85). Also, Naldini *et al.* (63), in order to produce their high-titre virus stocks, used a transient transfection system, but this might be unsuitable for clinical grade material since it increases the rate of recombination. Yu *et al.* (73) tried a different approach using an inducible system employing a tetracycline-repressible promoter, but the virus titres obtained were low.

HIV-1 *tat* is a strong transcriptional activator and is essential for virus replication. It functions through a *tat* activation response element located downstream of the transcription initiation site (see *Figure 1*). Kim *et al.* (1998) have shown that in single-cycle infection (transduction), *tat* is dispensable if the U3 of the 5′ LTR is replaced with the hCMV promoter and, if any transgene is expressed from a strong internal promoter (see *Figure 6*). Srinivasakumar *et al.* (65) have generated HIV-1 packaging cell lines that constitutively expressed the HIV-1 structural proteins and efficiently packaged vector RNA without *rev* expression. This was done using a small fragment from the Mason–Pfizer monkey virus (MPMV) genome called the constitutive transport element (CTE), which seemed to substitute completely for *rev*-RRE function and allowed HIV structural protein synthesis (86). Virus titres obtained were essentially similar to a parallel *rev*-containing system. In contrast, however, virus titres after several rounds of CTE-containing HIV replication were always reduced at least 10-fold compared to viruses utilizing *rev* and the RRE.

Recently, Miyoshi *et al.* (82) have constructed a new series of self-inactivating (SIN) HIV-1 vectors by replacing the U3 region of the 5′ LTR with a CMV promoter and deleting 133 bp in the U3 region of the 3′ LTR, including the TATA box and transcription factor-binding sites. The deletion is transferred to the 5′ LTR during replication eliminating any *cis*-acting effects of the 5′ LTR in the provirus, and also overcoming the possibility of insertional activation of cellular oncogenes. All accessory proteins including *tat* and *rev* have been eliminated (see *Figure 7*). These SIN vectors still maintain high levels of expression and should be safer for gene therapy. In this case, no theoretical scenario of homologous recombination can be constructed that will generate RCRs.

Perhaps the way forward is not to use HIV-based vectors, but, instead, to use non-human lentiviruses such as FIV (feline immunodeficiency virus) (5), EIAV (equine infectious anaemia virus) (7), bovine immunodeficiency virus (BIV) (7), and caprine arthritis-encephalitis virus (CAEV) (87) as gene therapy vehicles in

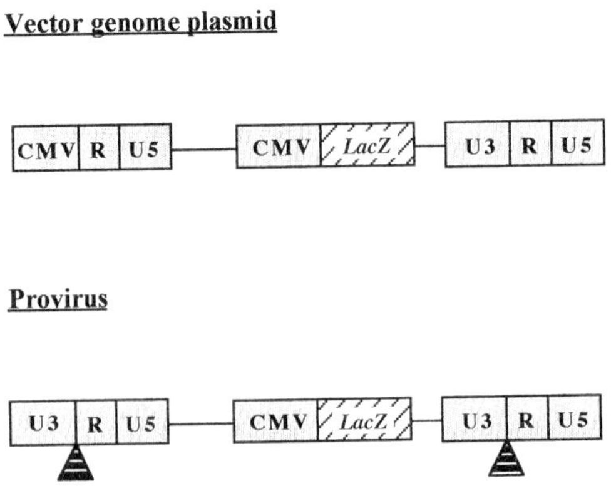

Figure 7 Structure of HIV SIN vector construct and corresponding provirus. The U3 region of the 5′ LTR was replaced with the CMV promoter, resulting in *tat*-independent transcription but still maintaining high levels of expression. The vector construct is co-transfected with the envelope and packaging constructs into 293T cells. Virus transcription is initiated at the 5′ LTR and terminates at the R/U5 border in the 3′ LTR. SIN vector has a 133 bp deletion in the U3 region (represented by triangles). The deletion is transferred to the 5′ LTR after reverse transcription and integration in infected cells, resulting in the transcriptional inactivation of the LTR in the provirus.

humans. Although these approaches will provide for greater safety, they simultaneously generate uncertainty because significant differences in genome organization between HIV and the animal lentiviruses exist, thus making it difficult to transpose what has been learned from the HIV vectors to other lentiviruses (5).

Gradually, however, as HIV- and other lentivirus-based vectors are explored for gene therapy purposes in the laboratory, it seems quite possible that HIV or a related virus will gain acceptance as an attractive vector system to target gene therapy to genetic disorders such as haemophilia, hypercholesterolaemia, and cystic fibrosis, as well as for cancer and neurological, neuromuscular, cardiovascular, and infectious diseases.

References

1. Rosenberg, S. A., Aebersold, P., Cornetta, K., Kasid, A., Morgan, R. A., Moen, R., Karson, E. M., Lotze, M. T., Yang, J. C., Topalian, S. L., Merino, M. J., Culver, K., Miller, A. D., Blaese, R. M., and Andersson, W. F. (1990). *New Engl. J. Med.*, **323**, 570.
2. Anderson, W. F. (1998). *Nature*, **392**, 25.
3. Strauss, M. and Barranger, J. A. (ed.) (1997). *Concepts in gene therapy*. Walter de Gruyter, Berlin.
4. Dickson, G. (ed.) (1995). *Molecular and cell biology of human gene therapeutics.* Chapman and Hall, London.
5. Poeschla, E. M., Wong-Staal, F., and Looney, D. J. (1998). *Nature Med.*, **4**, 354.
6. Bl`mer, U., Naldini, L., Kafri, T., Trono, D., and Verma, I. M. (1997). *J. Virol.*, **71**, 6641.
7. Narayan, O. and Clements, J. E. (1990). *Virology* (2nd edn) (ed. B. N. Fields, D. M. Knipe, P. M. Howley), p. 1571. Raven Press, New York.
8. Weiss, R. A., Teich, N., Varmus, H. E., and Coffin, J. M. (ed.) (1985). *Molecular biology of*

tumor viruses: RNA tumor viruses (2nd edn). Cold Spring Harbor Laboratory, Cold Spring Harbor, NY.

9. Mann, R., Mulligan, R. C., and Baltimore, D. (1983). *Cell*, **33**, 153.
10. Miller, D. G., Adam, M. A., and Miller, A. D. (1990). *Mol. Cell Biol.*, **10**, 4239.
11. Lewis, P., Hensel, M., and Emerman, M. (1992). *EMBO J.*, **11**, 3053.
12. Price, J., Turner, D., and Cepko, C. (1987). *Proc. Natl Acad. Sci. USA*, **84**, 156.
13. Overturf, K., Al-Dhalimy, M., Tanguay, R., Brantly, M., Ou, C-N., Finegold, M., and Grompe, M. (1996). *Nature Genet.*, **12**, 266.
14. Armentano, D., Yu, S-F., Kantoff, P. W., von Ruden, T., Anderson, W. F., and Gilboa, E. (1987). *J. Virol.*, **61**, 1647.
15. Jaffee, E. M., Dranoff, G., Cohen, L. K., Hauda, K. M., Mulligan, R. C., and Pardol, D. (1993). *Cancer Res.*, **53**, 2221.
16. Osborne, W. R. and Miller, A. D. (1988). *Proc. Natl Acad. Sci. USA*, **85**, 6851.
17. Morgernstern, J. P. and Land, H. (1990). *Nucl. Acids Res.* **18**, 3587.
18. Bowtell, D. D. L., Cory, S., Johnson, G. R., and Gonda, R. J. (1988). *J. Virol.*, **62**, 2464.
19. Emerman, M. and Temin, H. M. (1984). *J. Virol.*, **50**, 42.
20. Emerman, M. and Temin, H. M. (1984). *Cell*, **39**, 459.
21. Emerman, M. and Temin, H. M. (1986). *Mol. Cell Biol.*, **6**, 792.
22. Emerman, M. and Temin, H. M. (1986). *Nucl. Acids Res.*, **14**, 9381.
23. Adam, M. A., Ramesh, N., Miller, A. D., and Osborne, W. R. A. (1991). *J. Virol.*, **65**, 4985.
24. Morgan, R. A., Couture, L., Elroy-Stein, O., Ragheb, J., Moss, B., and Anderson, W. F. (1992). *Nucl. Acids Res.*, **20**, 1293.
25. Weiss, R. A. (1992). In *The Retroviridae*, Vol. 2 (ed. J. A. Levy), Plenum Press, New York.
26. Adams, R. M., Soriano, H. E., Wang, M., Darlington, G., Steffen, D., and Ledley, F. D. (1992). *Proc. Natl Acad. Sci. USA*, **89**, 8981.
27. Lin, S., Gaiano, N., Culp, P., Burns, J. C., Friedmann, T., Yee, J. K., and Hopkins, N. (1994). *Science*, **265**, 666.
28. Yee, J. K., Miyanohara, A., La Porte, P., Bouic, K., Burns, J. C., and Friedmann T. (1994). *Proc. Natl Acad. Sci. USA*, **91**, 9564.
29. Cosset, F-L. and Russell, S. J. (1996). *Gene Ther.*, **3**, 946–56.
30. Dornburg, R. (1997). *Biol. Chem.*, **378**, 457–68.
31. Chu, T. H. and Dornburg, R. (1997). *J. Virol.*, **71**, 720.
32. Linney, E., Davis, B., Overhauser, J., Chao, E., and Fan, H. (1984). *Nature*, **308**, 470.
33. Gorman, C. M., Rigby, P. W. J., and Lane, D. (1985). *Cell*, **42**, 519.
34. Yu, S-F., von Ruder, T., Kantoff, P. W., Garber, C., Seiberg, M., Ruther, U., Anderson, W. F., Wagner, E. F., and Gilboa, E. (1986). *Proc. Natl Acad. Sci. USA*, **83**, 3194.
35. Miller, N. and Whelan, J. (1997). *Hum. Gene Ther.*, **8**, 803.
36. Fassati, A., Bardoni, A., Sironi, M., Wells, D. J., Bresolin, N., Scarlato, G., Hatanaka, M., Yamaoka, S., and Dickson, G. (1998). *Hum. Gene Ther.*, **9**, 2459.
37. Takahara Y., Hamada, K., and Housman, D. (1992). *J. Virol.*, **66**, 3725.
38. Markovitz, D., Goff, S., and Bank, A. (1988). *J. Virol.*, **62**, 1120.
39. Miller, A. D. and Buttimore, C. (1986). *Mol. Cell. Biol.*, **6**, 2895.
40. Markovitz, D., Goff, S., and Bank, A. (1988). *Virology*, **167**, 400.
41. Patience, C., Takeuchi, Y., Cosset, F. L., Weiss, R. A. (1998). *J. Virol.*, **72**, 2671.
42. Danos and Mulligan (1982).
43. Grignani, F., Kinsella, T., Mencarelli, A., Valtieri, M., Riganelli, D., Grignani, F., Lanfrancone, L., Peschle, C., Nolan, G. P., and Pelicci, P. G. (1998). *Cancer Res.*, **58**, 14.
44. Rigg, R. J., Chen, J., Dando, J. S., Forestell, S. P., Plavee, I., and Bohnlein, E. (1996). *Virology*, **218**, 290.
45. Pear, W. S., Nolan, G. P., Scott, M. L., and Baltimore, D. (1993). *Proc. Natl Acad. Sci. USA*, **90**, 8392.

46. Kinsella, T. M. and Nolan, G. P. (1996). *Hum. Gene Ther.* **7**, 1405.
47. Cosset, F-L., Takeuchi, Y., Battini, J. L., Weiss, R. A., and Collins, M. K. (1995). *J. Virol.* **69**, 7430.
48. Chong, H., Starkey, W., and Vile, R. G. (1998). *J. Virol.*, **72**, 2663.
49. Donahue, R. E., Kessler, S. W., Bodine, D., *et al.* (1992). *J. Exp. Med.*, **176**, 1125.
50. Fassati, A., Wells, D. J., Walsh, F. S., and Dickson, G. (1996). *Hum. Gene Ther.*, **7**, 595.
51. Fassati, A., Wells, D. J., Walsh, F. S., and Dickson, G. (1995). *Hum. Gene Ther.*, **6**, 1177.
52. Takeuchi, Y., Porter, C. D., Strahan, K. M., Preece, A. F., Gustafsson, K., Cosset, F. L., Weiss, R. A., and Collins, M. K. (1996). *Nature,* **379**, 85.
53. Fassati, A., Wells, D. J., Sgro-Serpente, P. A., Walsh, F. S., Brown, S. C., Strong, P. N., and Dickson, G. (1997). *J. Clin. Invest.*, **100**, 620.
54. Hurford, R. K., Dranoff, G., Mulligan, R. C., and Tepper, R. I. (1995). *Nature Genet.,* **10**, 430.
55. Culver, K. W., Ram, Z., Wallbridge, S., Ishii, H., Oldfield, E. H., and Blease, R. M. (1992). *Science,* **256**, 1550.
56. Parrish, E. P., Cifuentes-Diaz, C., Li, Z. L., Vicart, P., Paulin, D., Dreyfus, P. A., Peschanski, M., Harris, A. J., and Garcia, L. (1996). *Gene Ther.,* **3**, 13.
57. Roe, T., Reynolds, T. C., Yu, G., and Brown, P. O. (1993). *EMBO J.*, **12**, 2099.
58. Lewis, P. F. and Emerman, M. (1994). *J. Virol.*, **68**, 510.
59. Lewis, P., Hensel, M., and Emerman, M. (1992). *EMBO J.*, **11**, 3053.
60. Subbramanian, R. A. and Cohen, E. A. (1994). *J. Virol.*, **68**, 6831.
61. Trono, D. (1995). *Cell,* **82**, 189.
62. Miyoshi, H., Takahashi, M., Gage, F. H., and Verma, I. M. (1997). *Proc. Natl Acad. Sci. USA,* **94**, 10319.
63. Naldini, L., Bl`mer, U., Gallay, P., Ory, D., Mulligan, R., Gage, F. H., Verma, I. M., and Trono, D. (1996). *Science,* **272,** 263.
64. Kaye, J. F., Richardson, J. H., and Lever, A. M. L. (1995). *J. Virol.*, **69**, 6588.
65. Srinivasakumar, N., Chazal, N., Helga-Maria, C., Prasad, S., Hammarskjøld, M. L., and Rekosh, D. (1997). *J. Virol.*, **71**, 5841.
66. Trono, D., Feinberg, M. B., and Baltimore, D. (1989). *Cell,* **59**, 113.
67. Kim, H-J., Lee, K., and O'Rear, J. J. (1994). *Virology,* **198**, 336.
68. Page, K. A., Landau, N. R., and Littman, D. R. (1990). *J. Virol.*, **64**, 5270.
69. Richardson, J. H., Kaye, J. F., Child, L. A., and Lever, A. M. L. (1995). *J. Gen. Virol.*, **76**, 691.
70. Berkowitz, D. R., Hammarskjøld, M-L., Helga-Maria, C., and Goff, S. P. (1995). *Virology,* **212**, 718.
71. Chen, C. and Okayama, H. (1987). *Mol. Cell. Biol.*, **7**, 2745.
72. Scharfmann, R., Axelrod, J. H., and Verma, I. M. (1991). *Proc. Natl Acad. Sci. USA,* **88**, 4625.
73. Yu, H., Rabson, A. B., Kaul, M., Ron, Y., and Dougherty, J. P. (1996). *J. Virol.*, **70**, 4530.
74. Carroll, R., Lin, J-T., Dacquel, E. J., Mosca, J. D., Burke, D. S., and St. Louis, D. C. (1994). *J. Virol.*, **68**, 6047.
75. Zavada, J. (1972). *J. Gen. Virol.*, **15**, 183.
76. Burns, J. C., Friedmann, T., Driever, W., Burrascano, M., and Yee, J-K. (1993). *Proc. Natl Acad. USA,* **90**, 8033.
77. Akkina, R. K., Walton, R. M., Chen, M. L., Li, Q-X., Planelles, V. and Chen, I. S. Y. (1996). *J. Virol.*, **70**, 2581.
78. Reiser, J., Harmison, G., Stahl, S. K., Brady, R. O., Karlsson, S., and Schubert, M. (1996). *Proc. Natl Acad. USA,* **93**, 15266.
79. Marsh, M. and Helenius, A. (1989). *Adv. Virus Res.*, **36**, 107.
80. Rando, T. and Blau, H. M. (1994). *J. Cell Biol.*, **125**, 1275.

81. Dracopoli, N. C., Haines, J. L., Korf, B. R., Moir, D. T., Morton, C. C., Seidman, C. E., Seidman, J. G., and Smith, D. R. (ed.). (1998). *Current protocols in human genetics.* Wiley, New York.
82. Miyoshi, H., Bl`mer, U., Takahashi, M., Gage, F. H., and Verma, I. M. (1998). *J. Virol.*, **72**, 8150.
83. Chong, H., Starkey, W., and Vile, R. G. (1998). *J. Virol.*, **72**, 2663.
84. Reiprich, S., Gundlach, B. R., Fleckenstein, B., and Uberla, K. (1997). *J. Virol.*, **71**, 3328.
85. Rolls, M. M., Webster, P., Balba, N. H., and Rose, J. K. (1994). *Cell*, **79**, 497.
86. Bray, M., Prasad, S., Dubay, J. W., Hunter, E., Jeang, K. T., Rekosh, D., and Hammarskjøld, M-L. (1994). *Proc. Natl Acad. Sci. USA*, **91**, 1256.
87. MselliLakhal, L., Favier, C., Teixeira, M. F. D., Chettab, K., Legras, C., Ronfort, C., Verdier, G., Mornex, J. F., and Chebloune, Y. (1998). *Arch. Virol.*, **143**, 681.

List of suppliers

Ambion, 2130 Woodward Street, Austin, Texas 78744-1832, USA
Tel: (512) 651-0200 Fax: (512) 651-0201 Web site: www.ambion.com
CliniSciences, 147, Avenue Henri, Ginoux, 92120 Montrouge, France
Tel: 33 1 42 53 14 53 Fax: 33 1 46 56 97 33
AMS Biotechnology (UK) Ltd., 185A & B, Milton Park, Abingdon, Oxfordshire OX14 4SR, UK
Amersham Pharmacia Biotech, Amersham Place, Little Chalfont, Buckinghamshire HP7 9NA, UK. Web site: www.apbiotech.com/
AMICON *Millipore (U.K.) Ltd*, The Boulevard, Blackmoor Lane, Watford, Hertfordshire WD1 8YW, UK; Millipore Corporation, 80 Ashby Road, Bedford, Massachusetts, 01730-2271, USA.
AMICON Inc., 72 Cherry Hill Drive, Beverly, MA 01915, USA
Amicon Ltd, Upper Mill, Stonehouse, Gloucestershire, GL10 2BJ, UK
Amresco, 30175 Solon Industrial Parkway, Solon, OH 44139, USA
Anderman and Co. Ltd, 145 London Road, Kingston-upon-Thames, Surrey, KT2 6NH
Tel: 0181 541 0035 Fax: 0181 541 0623
Applied Scientific, 154 W Harris Avenue, South San Francisco, CA 94080, USA
BDH Laboratory Supplies, Poole, Dorset BH15 1TD, UK. Web site: www.bdh.com/
Beckman Coulter Inc., 4300 N Harbor Boulevard, PO Box 3100, Fullerton, CA 92834-3100, USA
Tel: 001 714 871 4848 Fax: 001 714 773 8283 Web site: www.beckman.com
Beckman Coulter (U.K.) Limited, Oakley Court, Kingsmead Business Park, London Road, High Wycombe, Buckinghamshire, HP11 1JU
Tel: 01494 441181 Fax: 01494 447558 Web site: www.beckman.com
Beckman Instruments (UK) Ltd, Oakley Court, Kingsmead Business Park, London Road
Becton Dickinson and Co., 21 Between Towns Road, Cowley, Oxford, OX4 3LY
Tel: 01865 748844 Fax: 01865 781627 Web site: www.bd.com
Becton Dickinson, 1 Becton Drive, Franklin Lakes, New Jersey 07417-1883, USA.
Tel: 001 201 847 6800 Web site: www.bd.com
BIO 101 Inc., 1070 Joshua Way, Vista, CA 92083, USA.
Bio 101 Inc., c/o Anachem Ltd, Anachem House, 20 Charles Street, Luton, Bedfordshire, LU2 0EB
Tel: 01582 456666 Fax: 01582 391768 Web site: www.anachem.co.uk
Bio 101 Inc., PO Box 2284, La Jolla, CA 92038-2284, USA
Tel: 001 760 598 7299 Fax: 001 760 598 0116 Web site: www.bio101.com

BioDiscovery, 11150 W Olympic Blvd. Ste. 805E, Los Angeles, CA 90064, USA

Bio-Rad Laboratories Ltd., Bio-Rad House, Maylands Avenue, Hemel Hempstead, Hertfordshire, HP2 7TD

Tel: 0181 328 2000 Fax: 0181 328 2550 Web site: www.bio-rad.com

Bio-Rad Laboratories Ltd., Division Headquarters, 1000 Alfred Noble Drive, Hercules, CA 94547, USA

Tel: 001 510 724 7000 Fax: 001 510 741 5817 Web site: www.bio-rad.com

Boehringher Mannheim: See Roche

Cartesian Technologies Inc., 17781 Sky Park Circle, Irvine, CA 92614, USA

Citifluor Ltd, 18 Enfield Cloisters, Fanshaw St, London N1 6LD, UK.

CLONTECH Laboratories, Inc., 1020 East Meadow Circle, Palo Alto, CA 94303, USA

CLONTECH Laboratories, UK Ltd, Unit 2, Intec 2, Wade Road, Basingstoke, Hampshire RG24 8NE, UK

Corning Costar UK, One the Valley Centre, Gordon Road, High Wycombe, Bucks HP13 6EQ, UK

Corning Incorporated, Science Products Division, 45 Nagog Park, Acton MA 01720, USA

CP Instrument Company Ltd, PO Box 22, Bishop Stortford, Hertfordshire, CM23 3DX

Tel: 01279 757711 Fax: 01279 755785 Web site: www.cpinstrument.co.uk

Difco: See Becton Dickinson

Dupont (UK) Ltd, Industrial Products Division, Wedgwood Way, Stevenage, Herts, SG1 4QN

Tel: 01438 734000 Fax: 01438 734382 Web site: www.dupont.com

Dupont Co, (Biotechnology Systems Division), PO Box 80024, Wilmington, DE 19880-002, USA

Tel: 001 302 774 1000 Fax: 001 302 774 7321 Web site: www.dupont.com

Eastman Chemical Company, 100 North Eastman Road, PO Box 511, Kingsport, TN 37662-5075, USA

Tel: 001 423 229 2000 Web site: www.eastman.com

Fisher Scientific UK Ltd, Bishop Meadow Road, Loughborough, Leicestershire, LE11 5RG

Tel: 01509 231166 Fax: 01509 231893 Web site: www.fisher.co.uk

Fisher Scientific U.S. Headquarters, Pittsburgh, PA (412) 490-8300, USA

Tel: 1-800-766-7000 Fax: 1-800-926-1166 Web site: www.fishersci.com

European Headquarters Fisher Scientific, Janssen Pharmaceuticalaan 3A, B-2440 Geel, Belgium

Tel: 32 14 57 5284 Fax: 32 14 57 5283

Fisher Scientific, Fisher Research, 2761 Walnut Avenue, Tustin, CA 92780, USA

Tel: 001 714 669 4600 Fax: 001 714 669 1613 Web site: www.fishersci.com

Flowgen Instruments Ltd., Lynn Lane, Shenstone, Lichfield, Staffordshire WS14 0EE

Web site: www.philipharris.co.uk/flowgen

Fluka, PO Box 2060, Milwaukee, WI 53201, USA

Tel: 001 414 273 5013 Fax: 001 414 2734979 Web site: www.sigma-aldrich.com

Fluka Chemical Company Ltd, PO Box 260, CH-9471, Buchs, Switzerland

Tel: 00 41 81 745 2828 Fax: 00 41 81 756 5449 Web site: www.sigma-aldrich.com

Fushisawa Pharmaceutical Co., Fujisawa Healthcare, Inc, Parkway Center North, Three Parkway North, Deerfield, IL 60015-2548, USA.

GeneMachines, PO Box 2048, Menlo Park, CA 94026, USA

General Scanning Inc., 500 Arsenal Street, Watertown, MA 02472, USA
Genetic Microsystems Inc., 34 Commerce Way, Woburn, MA 01801, USA
Genetix Ltd., Unit 1, 9 Airfield Road, Christchurch, Dorset BH23 3TG
Genomic Solutions Inc., 4355 Varsity Drive Ste. E, Ann Arbor, MI 48108, USA
Genome Systems Inc., 4633 World Parkway Circle, St. Louis, MO 63134, USA
Genosys Biotechnologies Inc., Lake Front Circle, Suite 185, The Woodlands, TX 77380, USA
Gibco-BRL See Life Technologies
Hybaid Ltd, Action Court, Ashford Road, Ashford, Middlesex, TW15 1XB
Tel: 01784 425000 Fax: 01784 248085 Web site: www.hybaid.com
Hybaid US, 8 East Forge Parkway, Franklin, MA 02038, USA
Tel: 001 508 541 6918 Fax: 001 508 541 3041 Web site: www.hybaid.com
HyClone Laboratories, 1725 South HyClone Road, Logan, UT 84321, USA
Tel: 001 435 753 4584 Fax: 001 435 753 4589 Web site: www.hyclone.com
ICN Radiochemicals Inc., Biomedical Research Products, 3300 Hyland Avenue, Costa Mesa, CA 92626, USA
Tel: 800-854-0530 Fax: 800-334-6999
ICN Biomedicals Ltd., Unit 18, Thame Park Business Centre, Wenman Road, Thame, Oxfordshire OX9 3XA, UK
Tel: 0800-282474 Fax: 0800-614735
Imaging Research Inc., Brock Univesity, 500 Glenridge Avenue, St. Catherines, Ontario L2S 3A1, Canada
Intelligent Automation Systems, 149 Sidney Street, Cambridge, MA 02139, USA
Invitrogen BV, PO Box 2312, 9704 CH Groningen, The Netherlands
Tel: 00800 5345 5345 Fax: 00800 7890 7890 Web site: www.invitrogen.com
Invitrogen Corporation, 1600 Faraday Avenue, Carlsbad, CA 92008, USA
Tel: 001 760 603 7200 Fax: 001 760 603 7201 Web site: www.invitrogen.com
IVEE Development AB, Forsta Langgata 26, SE-413 28 Goteborg, Sweden
Life Technologies (Gibco-BRL), 9800 Medical Center Drive, Rockville, MD 20850, USA.
Web site: www.lifetech.com/
Life Technologies Ltd, PO Box 35, Free Fountain Drive, Incsinnan Business Park, Paisley, PA4 9RF
Tel: 0800 269210 Fax: 0800 838380 Web site: www.lifetech.com
LKB See Amersham Pharmacia Biotech
Merck Sharp & Dohme Research Laboratories, Neuroscience Research Centre, Terlings Park, Harlow, Essex CM20 2QR
Microflex, PO Box 1865, San Francisco, CA 94083-1865, USA
Millipore Corporation, 80 Ashby Road, Bedford, MA 01730, USA
Tel: 001 800 645 5476 Fax: 001 800 645 5439 Web site: www.millipore.com
Millipore (UK) Ltd, The Boulevard, Blackmoor Lane, Watford, Hertfordshire, WD1 8YW
Tel: 01923 816375 Fax: 01923 818297 Web site: www.millipore.com/local/UK.htm
Molecular Dynamics, 928 East Arques Avenue, Sunnyvale, CA 94086, USA
Molecular Probes, Inc., 4849 Pitchford Avenue, PO Box 22010, Eugene OR 97402-9165, USA
MSD Sharp and Dohme GmbH, Lindenplatz 1, D-85540, Haar, Germany

Nalge Nunc (Europe) Limited, Foxwood Court, Rotherwas Industrial Estate, Hereford HR2 6JQ, UK.

Nalge Nunc International, 75 Panorama Creek Drive, Rochester, NY 14625, USA.
Web site: nunc.nalgenunc.com/

NEN, Du Pont NEN Research Products, 549 Albany Street, Boston, Massachusetts 02118, USA
Tel: 1-800-551-2121
United Kingdom
Tel: 44 0438 734865

New England Biolabs (UK) Ltd, Knowl Piece, Wilbury Way, Hitchin, Herts, SG4 0TY, UK; New England Biolabs Inc., 32 Tozer Rd, Beverly, MA 01915-5599, USA.
Web site: www.neb.com/

Nikon Corporation, Fuji Building, 2-3, 3-chome, Marunouchi, Chiyoda-ku, Tokyo 100, Japan
Tel: 00 813 3214 5311 Fax: 00 813 3201 5856
Web site: www.nikon.co.jp/main/index_e.htm

Nikon Inc, 1300 Walt Whitman Road, Melville, NY 11747-3064, USA
Tel: 001 516 547 4200 Fax: 001 516 547 0299 Web site: www.nikonusa.com

Nycomed Amersham plc, Amersham Place, Little Chalfont, Buckinghamshire, HP7 9NA
Tel: 01494 544000 Fax: 01494 542266 Web site: www.amersham.co.uk

Nycomed Amersham, 101 Carnegie Center, Princeton, NJ 08540, USA
Tel: 001 609 514 6000 Web site: www.amersham.co.uk

Park Scientific Ltd, 24 Low Farm Place, Moulton Park, Northampton NN3 1HY, UK.

PE Applied Biosystems, 850 Lincoln Centre Drive, Foster City, CA 94404, USA
PE Applied Biosystems, European Life Science Centre, Langen, Germany

Perkin Elmer Ltd, Post Office Lane, Beaconsfield, Buckinghamshire, HP9 1QA
Tel: 01494 676161 Web site: www.perkin-elmer.com

Pharmacia and Upjohn Ltd, Davy Avenue, Knowlhill, Milton Keynes, Buckinghamshire, MK5 8PH
Tel: 01908 661101 Fax: 01908 690091 Web site: www.eu.pnu.com

Pharmacia Biotech (Biochrom) Ltd, Unit 22, Cambridge Science Park, Milton Rd, Cambridge, Cambs, CB4 0FJ
Tel: 01223 423723 Fax: 01223 420164 Web site: www.biochrom.co.uk

Pierce Chemical Co., PO Box 117, 3747 N. Meridian Road, Rockford, IL 61105, USA

Pierce & Warriner (UK) Ltd, 44, Upper Northgate Street, Chester, Cheshire CH1 4EF, UK

Promega Corporation, 2800 Woods Hollow Road, Madison, WI 53711-5399, USA
Tel: 001 608 274 4330 Fax: 001 608 277 2516 Web site: www.promega.com

Promega UK Ltd, Delta House, Chilworth Research Centre, Southampton, SO16 7NS
Tel: 0800 378994 Fax: 0800 181037 Web site: www.promega.com

Qiagen UK Ltd, Boundary Court, Gatwick Road, Crawley, West Sussex, RH10 2AX
Tel: 01293 422911 Fax: 01293 422922 Web site: www.qiagen.com
Qiagen Inc, 28159 Avenue Stanford, Valencia, CA 91355, USA
Tel: 001 800 426 8157 Fax: 001 800 718 2056 Web site: www.qiagen.com

Research Genetics Inc., 2130 Memorial Pkwy SW, Huntsville, AL 35801, USA

Roche Diagnostics Ltd, Bell Lane, Lewes, East Sussex, BN7 1LG
Tel: 01273 484644 Fax: 01273 480266 Web site: www.roche.com
Roche Diagnostics Corporation, 9115 Hague Road, PO Box 50457, Indianapolis, IN 46256, USA
Tel: 001 317 845 2358 Fax: 001 317 576 2126 Web site: www.roche.com
Roche Diagnostics GmbH, Sandhoferstrasse 116, 68305 Mannheim, Germany
Tel: 0049 621 759 4747 Fax: 0049 621 759 4002 Web site: www.roche.com
Roche: F. Hoffmann-La Roche Ltd, CH-4070 Basel, Switzerland.
Web site: www.roche.com/
Sarstedt Ltd., 68 Boston Road, Leicester LE4 1AW
Sarstedt Inc., Route 2 St James Church Road, PO Box 468, Newton NC 28658, USA
Schleicher & Schuell: Schleicher & Schuell Inc, 10 Optical Ave, PO Box 2012, Keene, NH 03431, USA.
Shandon Scientific Ltd, 93-96 Chadwick Road, Astmoor, Runcorn, Cheshire, WA7 1PR
Tel: 01928 566611 Web site: www.shandon.com
Sigma-Aldrich Company Ltd, Fancy Road, Poole, Dorset BH12 4QH, UK
Sigma Chemical Company, PO Box 14508, St Louis, MO 63178, USA
Tel: 001 314 771 5765 Fax: 001 314 771 5757 Web site: www.sigma-aldrich.com
Sorvall: Kendro Laboratory Products Limited, International Centre, Boulton Road, Stevenage, Hertfordshire SG1 4QX, UK; Kendro Laboratory Products, 31 Pecks Lane, Newtown, CT 06470-2337, USA. Web site: www.sorvall.com/
Stratagene Europe, Gebouw California, Hogehilweg 15, 1101 CB Amsterdam Zuidoost, The Netherlands
Tel: 00 800 9100 9100 Web site: www.stratagene.com
Stratagene Inc, 11011 North Torrey Pines Road, La Jolla, CA 92037, USA
Tel: 001 858 535 5400 Web site: www.stratagene.com
Synteni (Incyte Pharmaceuticals), 6519 Dumbarton Circle, Fremont, CA 94555, USA
Tel-Test, Inc., 1511 County Road 129 Po Box 1421, Friendswood, TX 77546, USA
United States Biochemical, PO Box 22400, Cleveland, OH 44122, USA
Tel: 001 216 464 9277
Vysis Inc., 3100 Woodcreek Drive, Downers Grove, IL 60515, USA
Whatman International Ltd, Whatman House, St Leonard's Road, 20/20 Maidstone, Kent ME16 0LS, UK; Whatman Inc, 9 Bridewell Pl, PO Box 1197, Clifton, NJ 07014, USA.
Whittaker (Oxoid agar), Unipath Ltd, Wade Road, Basingstoke, Hampshire RG42 8PW, UK
Worthington Biochemical Corporation, Halls Mill Road, Freehold New Jersey 07728, USA
UK agents: Lorne labs, PO Box 6 Twyford RG10 9NL, UK

Index